CONSUMING SPACE

Consuming Space
Placing Consumption in Perspective

Edited by

MICHAEL K. GOODMAN
King's College London, UK

DAVID GOODMAN
University of California, Santa Cruz, USA

MICHAEL REDCLIFT
King's College London, UK

LONDON AND NEW YORK

First published 2010 by Ashgate Publishing

Published 2016 by Routledge
2 Park Square, Milton Park, Abingdon, Oxfordshire OX14 4RN
711 Third Avenue, New York, NY 10017, USA

First issued in paperback 2016

Routledge is an imprint of the Taylor & Francis Group, an informa business

British Library Cataloguing in Publication Data
Consuming space : placing consumption in perspective.
1. Consumption (Economics) 2. Social ecology.
I. Goodman, Michael K., 1969- II. Goodman, David, 1938-
III. Redclift, M. R.
306.3-dc22

Library of Congress Cataloging-in-Publication Data
Consuming space : placing consumption in perspective / edited by Michael K. Goodman, David Goodman and Michael Redclift.
p. cm.
Includes bibliographical references and index.
ISBN 978-0-7546-7229-6 (hardback) -- ISBN 978-0-7546-8911-9 (ebook) 1. Human geography. 2. Human territoriality. 3. Spatial behavior. 4. Consumption (Economics) I. Goodman, Michael K., 1969- II. Goodman, David. III. Redclift, M. R.
GF41.C5748 2010
304.2--dc22

2009037679

ISBN 13: 978-1-138-27945-2 (pbk)
ISBN 13: 978-0-7546-7229-6 (hbk)

Contents

List of Figures *vii*
List of Contributors *ix*
Preface *xi*
Acknowledgements *xiii*

INTRODUCTION – GROUNDING CONSUMING SPACE

1 Introduction: Situating Consumption, Space and Place 3
Michael K. Goodman, David Goodman and Michael Redclift

2 Multiple Spaces of Consumption: Some Historical Perspectives 41
Frank Trentmann

3 The Seduction of Space 57
David B. Clarke

PART I – THE CONSUMPTION OF SPACE AND PLACE

4 Frontier Spaces of Production and Consumption:
Surfaces, Appearances and Representations on the 'Mayan Riviera' 81
Michael Redclift

5 Recognition and Redistribution in the Renegotiation of Rural Space:
The Dynamics of Aesthetic and Ethical Critiques 97
John Wilkinson

PART II – CONSUMPTION IN SPACE AND PLACE

6 Ethical Campaigning and Buyer-Driven Commodity Chains:
Transforming Retailers' Purchasing Practices? 123
Alex Hughes, Neil Wrigley and Martin Buttle

7 The Cultural Economy of the Boutique Hotel:
The Case of the Schrager and W Hotels in New York 147
Donald McNeill and Kim McNamara

PART III – CONSUMPTION AS CONNECTION/ DISCONNECTION/RECONNECTION

8 Manufacturing Meaning along the Chicken Supply Chain: Consumer Anxiety and the Spaces of Production 163
Peter Jackson, Neil Ward and Polly Russell

9 Place and Space in Alternative Food Networks: Connecting Production and Consumption 189
David Goodman

PART IV – CONSUMPTION AS PRODUCTION AND PRODUCTION AS CONSUMPTION

10 Creating Palate Geographies: Chilean Wine and UK Consumption Spaces 215
Robert N. Gwynne

11 Consuming Burmese Teak: Anatomy of a Violent Luxury Resource 239
Raymond L. Bryant

12 Space for Change or Changing Spaces: Exploiting Virtual Spaces of Consumption 257
Angus Laing, Terry Newholm and Gill Hogg

Index *277*

List of Figures

3.1	The *vel* of alienation	64
8.1	Contemporary broiler shed	166
8.2	Domestic chicken production, c. 1925	166
8.3	Intensive chicken hatchery, 1970s	171
9.1	Maps of two London 'foodie neighbourhoods'	203
10.1	Hierarchical ranking and the price, quality, effort and risk classification for consumers of Chilean wine products in the UK market	221
10.2	Chilean wine: Value downstream to UK market	225
10.3	Chile's main wine regions	229
12.1	The 'balancing paradigm' in healthcare	260
12.2	Settled spaces	271
12.3	Contested spaces	272

List of Contributors

Raymond L. Bryant – Department of Geography, King's College London.

Martin Buttle – Impactt Limited, London.

David B. Clarke – School of the Environment and Society, Swansea University, Wales.

David Goodman – Environmental Studies Department (Professor Emeritus) and Department of Geography, King's College London (Visiting Professor).

Michael K. Goodman – Department of Geography, King's College London.

Robert N. Gwynne – School of Geography, Earth and Environmental Science, University of Birmingham.

Gill Hogg – School of Management and Languages, Herriot-Watt University, Edinburgh.

Alex Hughes – School of Geography, Politics and Sociology, Newcastle University.

Peter Jackson – Department of Geography, University of Sheffield.

Angus Laing – Department of Management, University of Glasgow.

Kim McNamara – Urban Research Centre, University of Western Sydney.

Donald McNeill – Urban Research Centre, University of Western Sydney.

Terry Newholm – Manchester Business School, University of Manchester.

Michael Redclift – Department of Geography, King's College London.

Polly Russell – The British Library, London.

Frank Trentmann – School of History, Classics and Archaeology, Birkbeck College, University of London.

Neil Ward – Centre for Rural Economy, University of Newcastle.

John Wilkinson – Department of Development, Agriculture and Society, Federal Rural University, Rio de Janeiro.

Neil Wrigley – School of Geography, University of Southampton.

Preface

Exploring and understanding the relational ties between and within space, place and consumption provides important insights into some of the most powerful forces and processes constructing the social and material worlds of contemporary societies. We 'produce' and 'consume' space just as we 'produce' and 'consume' nature in the development of economic relations. Space and place are made and remade, produced and re-produced through the iterative processes, iconographies, and materialities of consumption. According to these perspectives, 'space' has historically represented a challenge for capitalism and capitalism eventually filled it with the desiderata of late modernity's 'fetish' for fetishes: commercial imagery, brands and logos. Space has subsequently become occupied with images that we construct, or are constructed for us, to encourage the growth of the commodity form and commodity cultures. Yet, space is never a passive location—as the site for social activity—but, in the form of *social space*, it is the means through which economic and political systems establish hegemony and gain legitimacy.

In effect, the 'making' of space is part of the process through which societies transform nature and allocate consumption. This implies contradictions, which the market cannot easily solve, as social space is a unitary yet contingent concept embodying the *physical, the mental and the social.* In other words, space draws together different facets of the material and the cultural. Yet, simultaneously, the concepts of space exist within our heads as a function of mental processes. Thus, the *space of social practice* is occupied by sensory phenomena, including products of the imagination, such as projects and projections, symbols and utopias, which describe and contain consumption cultures.

On this reading, space is a highly complex concept that embodies and is embodied by cultural as well as physical properties and cannot be understood unless these interrelationships are recognized and disentangled. In addition, since the construction of space is an active, contested and transitive process, involving cultural meanings as well as territorial dimensions, it is best seen as a process linked to the development of societies themselves and as a thoroughly negotiated topographical project and outcome. Overall, then, the redrawing of space in physical terms and the construction of social space have given 'rise to a very specific dialectic' as Lefebvre would have it, especially with respect to consumption and its varied and entangled processes.

One of the central ideas in thinking about space and consumption together is that they often appear to have taken on a reality of their own within capitalist society. Space is regularly divested of its social nature and consumption can fall along a shifting spectrum, from spaces of intended political meaning, to ambivalence, to pure and essentialized spaces of pleasure. The relationships among consumption, space and place can indeed be a mystification since attention is often paid to *appearances*, rather than materiality, and both historical and contemporary analysis fails to capture the social forces that actually produce space. Thinking about consumption and space sometimes suffers from the *illusion of transparency*, which places the design of space in the foreground, and serves to hide the shadows behind the light, representing space as an 'innocent' domain. Hence ideas like those of 'discovery', 'settlement', and 'ethical consumption' can be reinterpreted, from this perspective, as ways of concealing as much as they reveal.

Space and place thus are contingent material and cultural categories, socially constructed and emergent, evolving with the dynamic interactions of production and consumption and their changing scalar dimensions. This constructed, emergent nature draws theoretical attention to the 'making' and 'remaking' of space and place as the social relations of production and consumption unfold. The analytical challenge is to understand both the social forces and power relations at the moment of 'making', as represented by the emergence of a coherent, contingently stable ordering of spatial and place-based structures and social practices, and the dynamic, contested processes of transition and reconfiguration. An understanding that the production and consumption of space and place over time have been geographically embedded in changing power relations and social struggle reveals what John Agnew describes as the 'historicity of spatiality'. Such historical embeddedness, and the corollary that space and place are active social processes always in formation, not preordained, 'purified' entities, is a prominent theme of a number of contributions to this volume.

Yet, one of the central questions the volume seeks to address is the following: how do cultures of consumption discover and rediscover *space*, and how do they construct and reconstruct *place* at different periods and in different ways? This question and the preceding set of ideas were put to the authors, either as participants of a seminar series in the Department of Geography, King's College London in 2006 or as subsequently invited contributors, as a challenge rather than a set of axioms to give maximum intellectual space in which to grapple with these ideas. The overarching project of the volume is to tease out the implications of conceptualizing consumption as a spatial, increasingly globally-scaled, yet intensely localized activity. Similarly, one of the aims here is the development of integrative approaches that articulate the relational and iterative processes involved in the production of space and place and their consumption. Thus, this volume brings together a varied, engaging and novel array of chapters to explore the spatiality and placed nature of consumption and its role in structuring contemporary capitalist political economies.

Acknowledgements

The editors would like to thank the ESRC/AHRC Cultures of Consumption Programme for its contribution to the original seminar series from which the majority of the volume's chapters emerged. In particular, we would like to acknowledge Frank Trentmann for his generosity and for agreeing to lead off the seminar programme; his excellent contribution definitely set the tone for the rest of the series. We would like to thank Val Rose, Katy Low and Aimée Feenan at Ashgate for their enthusiasm and patience with the book and Mike Raco for his comments on the introductory chapter. Finally, Mike Goodman would like to personally thank his co-editors—but especially the authors—for their hard work, perseverance and fortitude in seeing this project through to its end, especially since the production of the volume occurred in the midst of the birth of his second child, Wyatt, to whom we dedicate this volume.

INTRODUCTION
Grounding Consuming Space

Chapter 1
Introduction: Situating Consumption, Space and Place

Michael K. Goodman, David Goodman and Michael Redclift

> *Above all, the transformative force of consumption becomes evident as soon as one begins to take geography seriously.*
>
> (Clarke et al. 2003, 80)

Many social scientists and others have heeded Henri Lefebvre's (1991, 342) directive that 'the commodity needs its space'. Commodities—what Michael Watts (1999), in channelling Marx, has quite rightly called the 'DNA of capitalism'—have become and continue to be the focus of a great deal of research and scholarly engagement. From more popular accounts of things such as chewing gum (Redclift 2004), sushi (Corson 2007), 'luxury' items (Thomas 2007) and bananas (Koeppel 2008), to wide-ranging, critical engagements with (post)colonialism (Cook and Harrison 2003, 2007), empire (Domosh 2006), and globalization (Freidberg 2004a; Mansfield 2003) *through* specific 'tales' of goods, much of this work is about uncovering the cultural and political economies that make up the 'social lives' (Appadurai 1986) of particular commodities. Many scholars—and especially geographers—have sought to add to the scope and content of this work by exploring the 'spatial lives' of these commodities, clearly one aspect of Lefebvre's wide-open refrain; indeed, as he puts it, '[t]he commodity is a *thing*: it is *in space* and occupies a location' (Lefebvre 1991, 341; emphasis in original). Drawing variable inspiration from the anthropologically-inflected field of cultural materialism (e.g. Woodward 2007; Douglas and Isherwood 1996) to the muckraking *Jungle*-like exposés of authors such as Upton Sinclair and more contemporary investigative journalists, this spatialization of material culture has sought to investigate how spaces, places and materialities weave in and out of commodity cultures, circuits, networks and chains. Thus, it is not all that hard to argue that, fundamentally, commodities are *essentially* geographical: they inhabit, produce and embody space and spatial relationships in all their multitudinous ways and means across their travels and travails from production to consumption and beyond.

Echoing Lefebvre's statements and in parallel with his wider endeavours, many have started to argue that consumption, too, needs its space. In short, we are *where* we consume and investigating this aphorism is where this volume is specifically situated. Building on the existing writings on the geographies of consumption (e.g. Bell and Valentine 1997; Clarke 2003; Clarke et al. 2003, 79–131; Crang 2008; Jackson and Thrift 1995; Mansvelt 2005), the chapters in this volume argue that the inseparable and indelible relationships of consumption, space and place should perhaps not be as implicit—as they sometimes are in commodity network studies (e.g. Cook et al. 2004)—nor, even, as forgotten as they often can be in broader economic and other geographical 'stories'. Thus, seen from this angle, quite a bit more can and should be drawn out of David Harvey's (1990) famous passing comment that the grapes on our supermarket shelves are silent about the exploitative relations behind them. Indeed, this statement was arguably made from *inside* the spaces of consumption and *through* the figure of the consumer and is deserving of much greater empirical and theoretical engagement.

Rather than simply dismissing it as merely ephemeral eruptions of 'culture' on the wider landscapes of capitalism, understanding consumption and its geographies is *central* to understanding the powerful geographical imaginations and materialities of the contemporary 'society of consumers' (Bauman 2007). This is particularly true in the context of those working for more 'ethical' and 'alternative' socio-material futures *through* consumption (e.g. Barnett et al. 2005; Harrison et al. 2005) in our newfound role as 'citizen'/'political' consumers (Clarke et al. 2007; Mansvelt 2008). Here, ethical consumption might simply be a variation on the more general consumer-capitalist theme, whereby consumption, '… rather than being a freedom, a right or a liberty, … has become one's civic duty and a collective responsibility' (Doel 2004, 154).

Further, consumption must be and remain *equally central* to theorizations of space and place. In just one example, the practices and politics of consumption are clearly one of the fundamental, yet relatively unrecognized, processes that conjoins Soja's (1996) triptych of socio-spatial relations—i.e. spatial practice, representations of space, and spaces of representation—in the construction of the modern lived (urban) experience (Mansvelt 2005, 57). Indeed, as Jayne (2006, 157) has summarized, despite advances in spatial theory,

> [r]esearch has tended to overlook the practices and values of consumers … This failure to engage with the multiplicity of consumption cultures (and so much of their power and dynamism) is only now beginning to be rectified. Similarly, those seeking to link production with consumption, to identify consumptive subversions and the different meanings which different people assign to particular activities and practices (and how local[ities] and nation states mediate this), are beginning to show how such issues are historically and geographically constructed and negotiated.

In taking this a bit further, Clarke et al. (2003, 80; emphasis in original) make the no-bones-about-it claim that

> … consumption tends to *reconfigure* space and place, often disrupting, undermining and *displacing* consumption activities that were once thought of being related to specific places (think of "Italian" food, "exotic" fruit, or even the humble potato). The complexities of geography tend to undermine all simple

> all-or-nothing generalizations, not least when it comes to consumption. The geography of consumption frequently seems to pull in two different directions at once—setting up contrasts between spaces that are *spectacular* and *seductive*, on the one hand, and spaces that are *ordinary* and *mundane*, on the other; creating—paradoxically—*homogeneity* and *heterogeneity* at the same time; promoting both *mobility* and *fixity* without contradiction; and ensuring that *space* and *place* sit alongside each other in a way that challenges unreflexive assumptions about the way the world works.

But this too can be taken further, albeit by re-packaging some of what Clarke et al. describe above as an argument for the need to understand the inherent 'relationalities' among what they see as the characteristics, creations and effects inhabiting the geographies of consumption. Drawing on some of the wider theoretical trajectories and ontological interventions in geography (e.g. *Geografiska Annaler* 2004; Massey 2005; Murdoch 2006; Lee 2006; Whatmore 2002) and elsewhere (e.g. Goodman 2001), then, the focus shifts away from dualities and 'contrasts' to an exploration of not only the relationalities and relationships between and among consumption, space and place, but, for instance, how and in what ways the spectacular and the ordinary are often present and indeed, fundamentally *construct* and are *constructed by* consumption and its discrete practice(s). Here, eating is one of the most obvious illustrations of some of these relationalities: put simply, dining out in the spectacular spaces of a five-star, celebrity chef-run bistro in London (or New York or Las Vegas or Los Angeles or Paris or Hong Kong or …) is also about fuelling one's body with the most mundane of biological activities in the ingestion of calories in the form of (most-likely very expensive, yet tasty) foodstuff. Further, that local, organic chicken that briefly inhabits your plate 'space' and the act of consuming it contains, constructs and connects *both* spatial mobilities and fixities: the chicken was shuttled along a commodity chain from the sites of local production to your very table, while, at some point, it not only became 'sticky' through those socio-material processes that certified it as 'organic' but, more simply, it had to remain in one particular place in order to be eaten and further metabolized into the mobilities of your body. It is this conceptualization of consumption, space and place—that space and place *make* and are *made in* and *through* consumption—that is at the implicit and explicit theoretical heart of the chapters in this volume.

Yet, these relationalities are and remain anything but innocent; as Miller (1995, 10) has quipped, the consumer, in the not-so-surprising guise of the First World inhabitant, has become a kind of 'global dictator'. Thus, as Massey (2005, 101; emphasis in original) has put it, 'what is always at issue is the *content*, not the spatial form, of the *relations through which* space is constructed'. Neil Smith (2004, 28) echoes this point with aplomb: 'Space matters, of course, but even more so the processes and events that make nature and space'. Here, then, the relationalities among space, place and consumption, *in particular*, are littered with those that are and have been decidedly unequal and exploitative of people and nature; these are relations that are material in their very essence, power-full in their production and practice, political in their constitution and cultural in their deployment and engagement. Thus, Marcus Doel's (2004, 150–151; emphasis in original; see also Goss 2004, 2006) declaration—written as a sort of counter-weight to the perceived excesses of post-structuralism for which the topic of consumption is seen to hold a particularly unhealthy obsession—holds truth for us and for many of the authors here:

> We remain—as always—resolutely materialists. So, we are struck by the *force* of signs, by the *intensity* of image, and by the *affects* of language. … As fanatical materialists, *we are struck by everything*—nothing will be set aside from the play of force, nothing will be spirited away onto a higher plane or exorcised into a nether-world. One does not have to be a magician, market-maker or medium to know that onto-theology and diabolism act in our world. It is true that we take up signs, words, images, quantities, figures, maps, photographs, money, hypertext, gardening advice, lipstick traces, the exquisite corpse and so on and so forth—but we take them up as force: as strikes and counter-strikes; as blows and counter-blows.

It is how the forces of consumption 'strike' and are 'struck' by, in and through space and place that, in particular, weave their way in and out of the volume's chapters.

Building on this concern and the others mentioned above—and in order to contextualize the writings that follow—the remainder of this introduction is organized into four different discussions that work through some of the key relationalities of consumption, space and place and divide the volume up into discrete sections. These are as follows: *the consumption of space and place*, *consumption in space and place*, *consumption as connection/disconnection/reconnection,* and *consumption as production and production as consumption.*

The Consumption of Space and Place

> *The body serves both as point of departure and as destination. … A body so conceived, as produced and as the production of a space, is immediately subject to the determinants of that space: symmetries, interactions and reciprocal actions, axes and planes, centres and peripheries and concrete (spatio-temporal) oppositions.*
>
> (Lefebvre 1991, 194–195)

From the air we breathe to the food we eat, the metabolic necessities of life require us to be consumers in some shape or form. Space too is a requirement as it is the very consumption of spaces and places by our (non)mobile bodies that gives us somewhere to live and make our lives, act from and be acted upon. We make our body 'space' *literally* through practices of ingestion—what might be called 'eating' the world (Lupton 1996; Valentine 1999)—and the 'work' (e.g. exercise) we do (or don't) on our bodies. At the same time, we make our 'selves' *figuratively* through those very same culturally-specific foods, but also, for example, through the 'fashionable' clothes that adorn us to the hairstyles, jewellery and/or make-up that we choose (or don't). Bodies, much like commodities, are essentially geographical given that they occupy, move in and, indeed, *are* space and (a) place. Space and place mark and make our bodies at the very same time we *make* them by being *in* them and by *being* them, by consuming *in* them and by *consuming* them. Thus, Lefebvre (1991), as quoted above, only tells half the story of bodies and spaces; instead, here, corporeality implies the consumption of space and the consumption of space implies corporeality.

Yet, increasingly the corporeal body is becoming less important and indeed, untethered in some ways from consumption and the consumption of space and place. Virtual spaces and place have quickly become the norm for everything from shopping to aspects of one's social life to the living of portions of one's life in virtual realities and video games (Crang et al. 1999).[1] Here, in ways probably unthought-of by Lefebvre, absolute and representational spaces come to intermingle with absolute and representational materialities in the spaces of the virtual. Consuming (in) the virtual then, is not only about bits and bytes, electrons and electronics—the materialities of virtualism—but how the consumption *of* and *in* the virtual creates these spaces and offers up new representation, identities and ways of being but also how it is about consuming new, (dis)embodied spaces and places. From the 'ordinary' consumption of vegetables and the 'adoption' of distant sheep (Holloway 2002), to that of the Zapatista social movement in Mexico (Froehling 1999), to the virtualization of sustainable consumption (Hinton 2009) and of 'cool ways' to mitigate climate change (Boykoff et al. 2010), the mix of the 'real' and the virtual that has come to (re)define the everyday for much of the globe poses new challenges to the conceptualizations of the widening landscapes of both consumption and space (Doel and Clarke 1999).

A leading example of the way in which the consumption of place has influenced the way we think is provided by Lefebvre (1991). He has written about tourism and consumption in the following terms:

> This is the moment of departure—the moment of people's holidays, formerly a contingent but now a necessary moment. When this moment arrives, "people" demand a qualitative space. The qualities they seek have names: sun, snow, sea. Whether these are natural or simulated matters little. Neither spectacle nor mere signs are acceptable. What is wanted is materiality and naturalness as such, rediscovered in their (apparent or real) immediacy. Ancient names, and eternal—and allegedly natural—qualities. Thus the quality and the use of space retrieve their ascendancy—but only up to a point. In empirical terms, what this means is that neocapitalism and neo-imperialism share hegemony over a subordinated

1 See Carrier and Miller (1998) for alternative takes on virtualism and consumption.

> space split into two kinds of regions: regions exploited for the purpose of and by means of production (of consumer goods), and regions exploited for the purpose of and by means of the consumption of space. Tourism and leisure become major areas of investment and profitability, adding their weight to the construction sector, to property speculation, to generalized urbanization (not to mention the integration of capitalism of agriculture, food production, etc.). No sooner does the Mediterranean coast become a space offering leisure activities to industrial Europe than industry arrives there; but nostalgia for towns dedicated to leisure, spread out in the sunshine, continues to haunt the urbanite of the super-industrialized regions. Thus the contradictions become more acute—and the urbanites continue to clamour for a certain "quality of space". (Lefebvre 1991, 353)

It is clear that tourism today represents one of several 'spaces of consumption' in which the features that draw people to the area are increasingly contextualized, and at the same time hybridized, enabling the tourist consumer to experience them as part of a wider suite of experiences, of cuisine, costume, architecture, and music. In this sense 'space' has become invested with the cultural semiotics of 'place', it has acquired the elusive force of 'identity', so important to tourist destinations in the international tourist market. It also shows how in many respects tourism has developed a new vein of consumption, in rich and unexpected ways (Redclift 2006). The study of tourism has also forced us to confront the way we analyse place itself.

The discussion of place is closely linked to governing paradigms and systems of explanation. It thus possesses the potential to both signal something about location and the meaning which is attached to it. The dual conceptual role of place in consumption has been referred to as 'place confirmation', to underline the centrality of the idea of place both as location and the association of meanings with location (Manuel-Navarrete and Redclift 2010). Like gender and nature, the meaning of place may be negotiable but its importance in the canon of concepts suggests considerable room for further development.

In the absence of systematic quantitative methods, place acquired a largely positivist mantle before the 'ideological decades' of the 1970s and 1980s, and its apologists acquired a quantitative zeal. The 'cultural turn' and post-modernism revealed a new emphasis on the human face of 'place' and its social construction, in which rather than being buried by globalization it offered a new form of conceptual revival. For both Marxists and neoliberals, place has suggested the interface of global structures and localized pockets of resistance—a regrouping of social expression in a locus of space. Its derivatives have opened up a new lexicon—emplacement, displacement, sense of place—with which to slay the dragon of global, place-less modernity, all flows and essences. One of the routes into place-confirmation, then, is clearly through enlarging the way that the concept of place is employed.

Another, second point of entry is through recognizing the sociological processes which condition us to think about place: its naturalization. This naturalization is important not just in the more conservative, bounded sense of place as 'mosaic', the traditional way in which geographers viewed 'places', but also in the more relational way place is employed today: place and identity, place and memory, place and belonging. A sense of place clearly exists in memory (and is institutionalized in memorializing), and this sense of place appears and disappears as places are discovered, erased and rediscovered. A number of examples occur in the chapters in this volume, and might lead us to ask questions about what lies behind the erasure and discovery of place. What do these processes tell us about societies and their histories? To develop conceptually, the idea of place needs to be linked to alternative visions of spatial polity in which history is an essential element, rather than a later embellishment.

One possible way of understanding the highly diverse literature on place emerging in the last two decades is to look at the politics of place from an historical and evolutionary perspective. Throughout history, place construction has played practical, socio-cultural, and symbolic roles. At the foundations of place construction are the processes through which individuals and groups develop survival strategies, solve common problems, and make sense of their own existence. Place attachment, sense of place, affection, embeddedness, identification, and other concepts are appropriate for interpreting this fundamental dimension of place. However, as humans became more capable of controlling the environment, the construction of social and cultural meanings grew increasingly independent of physical settings. The social dimension of human experience even surpassed nature's importance in the shaping of place. For instance, the built environment served to substitute for the ecological context in some sacred places. At the same time, the colonization of vast territories by relatively small groups, in the cause of imperial expansion brought about the possibility of transposing meanings and cultural systems from one geographical setting to another, and facilitated cultural hybridization, as happened during the Roman Empire.

With science and modernity, place construction was increasingly perceived in terms of filling 'emptiness' with 'civilization'. The concept of space (as empty place), the production of maps, and the notion of private ownership of land were instrumental in the successful passage of colonialism. Concepts such as location, locale, or region were linked to the modern administration of place, which achieved its ultimate expression with the hegemony of the modern nation-state. In addition, colonization opened the doors for a diverse range of new power relations that would, in turn, lead to the construction of new places (of exclusion, domination, resistance, and so on).

In recent years, economic globalization is bringing the modern homogenization of place one step further by bypassing the constraints of national culture and state administration. Today, economic globalization is colonizing the 'empty space' spared by the modern state and constructing new places of consumption dominated by logics of extraction, and economic profit. It is also creating new places of resistance and struggle, as Arturo Escobar (2009) shows in his work in Colombia. However, this homogenization has never completely replaced the historic and alternative constructions of place which are grounded in personal attachment, sense of place, cosmologies, personal intimacy and familiarity. Rather, modern and globalized spaces are being superimposed on top of previous meanings. Furthermore, the process of individuation that started in the Early Modern period and developed under liberal democracies was further deepened with post-modernity. As a consequence, the modern homogenization of place is only apparent and superficial. It is a force constantly counteracted and reversed by people's impulse to find an existential meaning that the uniformity of mass consumption might never provide. It is in this context that the revival of academic interest in place construction is emerging.

The analysis of place requires the acknowledgement of ambiguities that are central to thinking in contemporary consumption studies. Places are collectively shared and contested. They do not necessarily mean the same thing to everybody. They are not 'owned' in the same way by everybody. This observation is also clearly true of the academic disciplines which have utilized place. In the world of academic discourse place is often part of a critique, and exists on an intellectual terrain. However, in the 'lived' world of experience place also has phenomenological import—it can be an affirmation of humanity, and in that sense critique alone does it a disservice. Acknowledging the hybridity of place provides another route into place-confirmation, distancing the concept from its more descriptive history, and opening up the possibility of place as a more heuristic device, a way of understanding society rather than a point from which to view it. It also provides a pointer for the analysis of the way that place is consumed.

Food, even more now than ever, is tied up with the consumption of space and particularly place. One food-scare or media exposé after another—components of what Susanne Freidberg (2004b) calls the 'ethical complex' of food—have caused what might be called the 'transparency revolution': consumers are asking for and/or food suppliers are providing greater knowledge about where food is coming from and how it is being produced. Here, both the literal and figurative consumption of space and place are counter-posed to the 'place-less' and 'face-less' commodities of the globalized food system, the latter quite often equated with environmental and human exploitation at numerous scales. Thus, coffee now exclaims it is from Nicaragua (or Columbia or Ethiopia or Indonesia or …) and traded 'fairly' (Bacon et al. 2008), milk now comes from 'happy' organic cows (e.g. DuPuis 2000) (quality) potato chips are now 'naked' (e.g. Illbery and Maye 2008) or specifically from small-scale, traditional farms and farmers (Goodman forthcoming) and vegetables are now often local to wherever it is you are shopping (e.g. Goodman and Goodman 2007; *Local Environment* 2008). Food is being sold through stories (Freidberg 2003) of not just its origins but the ins and outs of how it was made and by whom in order to further fill up the meanings attached to these commodities by 'placing' them, most often, in their ecological and social contexts. Thus, not only are the qualities and materialities of *terroir*—a defined, geographic place that produces distinct material characteristic in foods (Barham 2003)—being metabolized by consumers, but increasingly so too are the 'fairer' economic and labour processes as well as the images of producers and their farms plastered on food labels and (increasingly) web pages. Undeniably, less ecological and human exploitation, the latter troublingly less of a concern in organic production systems than it should be (Guthman 2004), leaves a better taste in one's mouth and a glowing imprint on one's (food) conscience.

This sort of 'place theory of value'—one that has begun to wrap up the use, exchange, labour and sign value of some of these meaningful foods into one neat 'spatialized' package—takes on and extends Cook and Crang's (1996) earlier cultural materialist work on the role of geographical knowledges in constructing our 'worlds on a plate'. Here, they discuss the 'displacement' and 'replacement' of various 'culinary cultures' through the three spatial processes of settings, biographies and origins embedded in foods and, now the many *other* commodities that work to 'unworry' their consumers about where it is that they come from.

Two points stand out here. First (food) commodity biographies and origins are becoming increasingly legislated and institutionalized through the processes of standardization, codification and verification. Standards like these set up their own political economies, not surprisingly, by dictating who can be in and who is left out of these networks from the spaces of production to consumption (Guthman 2007); inequalities of access follow quite closely in the wake of standardization at the same time it opens up spaces for some to enter into more 'defined' markets. In addition here, the ordinary 'messiness' of places and spaces are, through standards and codification, reduced and abstracted to logos, labels and texts, a process that also works to ossify the very places and people that end up on various food and other commodity narratives. Place becomes quantifiable and easily knowable, commodify-able and package-able for the consumer and the marketplace; as Prudham (2007, 414) puts it in the context of the commodification of nature more broadly, these processes 'render the messy materiality of life legible as discrete entities, individuated and abstracted from the complex social and ecological integuments'. These sorts of renderings have led in many cases, to the rather 'easy' buy-in by large multinational capitals into these markets; and as the continuing mainstreaming and institutionalizations of organic, fair trade and other niche food networks driven heavily by this corporatization of these sectors is showing, these processes don't seem to be slowing down.

Second, exploring the settings of culinary cultures—what Cook and Crang (1996, 142) define as '... the contexts in which they can and should be used'—might lead us to not only consider how the places and spaces of (food) consumption work, but that they can be riddled with inequalities in not only the levels of geographical knowledge that consumers may have, but by virtue of the economic access that many do not have to these meaningfully placed foods. Here consumption is not only culturally situated and contextualized in the places that we might find ourselves eating, *but also* economically situated, contextualized and produced. Thus, the consumption of food place and placed food is most often only available for those who can afford it and/or for those who can marshal the necessary knowledges to make it meaningful or value-laden. The consumption of place and space in food networks is as much about economic and cultural power as it is about the desire to eat, drink and be merry in novel and alternative ways.

While situated in the *Introduction* section of the volume, both Trentmann's and Clarke's chapters very much touch on the theme of the consumption of space and place. Trentmann introduces an historical dimension into the account of the relationalities of space and place and the ways they are consumed. Here, he not only argues for the need to get beyond the analytics of the consumerist 'present' in order to recognize its much deeper and more entangled historical trajectories, but that an historical approach is crucial to understanding how the practices of consumption have been complicit in the re-jigging of public and private spaces over time. To explore these histories, he deploys the three themes of the 'mental spaces' of the consumer, the 'remoralization' of trade practices and, by drawing on Lefebvre and de Certeau, the governance of everyday spaces through consumption. Overall, the historicization of the spatialities of consumption proposed in Trentmann's chapter provides fertile ground from which to build further analyses of the temporal dialectics of consumption and space.

Clarke's chapter endeavours to show how the spaces of consumption are much more than, as he puts it, the 'consummate spaces of capitalism'. In particular, he works to reinvigorate understandings of the commodity fetish and its spatialized pleasure- and desire-making abilities in consumption through the lenses of psychoanalysis. Fantasy and seduction, fostered by the fetish whose form is dedicated to their creation, are the engines that drive the production and consumption of space through the effervescent consumption-scapes we can't seem to look away from. In a rather justifiably fatalistic turn, instead of the fetish allowing us—through its critical analysis—to perhaps explore the social relations of production and consumption as a way out, we are 'absorbed and abolished' in the seductions of symptomatically overdetermined consumerist spaces that haunt us with their commodities.

Of the chapters specifically in *Part II* of the volume, Redclift examines the ways in which 'place' is constructed culturally as well as geographically through consumption, taking as his case the Mexican Caribbean. In the chapter, he shows how the meanings attached to 'place' and the way that it and its meanings are consumed have changed over time reflecting shifts in global consumption patterns and connections. Moving from the production-consumption of chicle/chewing gum to the consumption by tourists of a denominated space called 'The Mayan Riviera', this space now brings together both material and symbolic elements of place in the search to attract the global tourist. In a turn very much related to Trentmann's arguments, Redclift concludes that in order to understand how place and space are consumed they must be placed in an historical context sensitive to the spatialized dialectics of production and consumption.

Situating alternative food networks in the context of the global institutional architecture of the World Trade Organization and the imitative competitive strategies of powerful actors in the conventional food system, Wilkinson describes how these networks produce and consume alternative rural spaces. Social control of food provisioning and the capacity of alternative production-consumption networks to revitalize rural space are the stakes in these struggles to appropriate the consumerist-oriented symbolic capital vested in place. His analysis distinguishes networks pursuing ethical, universalistic critiques of conventional food provisioning, such as Fair Trade movements, from those adopting an aesthetic, particularistic critique predicated on place-based geographical specificities of inherited craft knowledges incorporated in farming, food processing and culinary practices. In particular, Wilkinson recounts the dialectic between markets and social movements as mainstream actors' attempts to replicate the symbolic values embedded in craft knowledges are countered by the efforts of alternative food networks to retain control of the economic rents generated in rural space by re-asserting the distinctive qualities of locality and geographical specificities.

Consumption in Space and Place

> *Will you survive the fire? The Shopocalypse!*
> *Can you feel the heat in this shopping list?*
> *The neighbors fade into the super mall.*
> *The oceans rise but I—I must buy it all.*
> *Shopocalypse, Shopocalypse ...*
>
> (First stanza from Reverend Billy's [2006, 57] Church of Stop Shopping's hymn *The Shopocalypse*)

As argued by many (e.g. Castree 2003; Massey 2005; Thrift 2003), space and place are not merely the stage on or containers in which we act out our social and material lives, but rather are actively negotiated, created and changed through all manner of relationships. And, as we and many of the authors in this volume argue, consumption is one of the key relationalities actively constructing and changing spaces and places which in turn recursively affect consumption practices. Yet, the acts of consumption are situated and contextualized in both time and space; they happen some*where* and some*when*. In Lefebvre's (1991, 341) words '[e]xchange with its circulatory systems and networks may occupy space worldwide, but consumption occurs only in this or that particular place'. Bell and Valentine (1997) work to engage with this analysis specifically by working through food consumption across and through the scales of the body, home, community, city, region, nation, and globe to suggest that consumption connects intimately across all of these spaces and places. Phil Crang (2008) takes a slightly less schematic approach to explore relations—through geographical knowledges and (dis)connections—between the creation, expression and practices of global and local consumption cultures. Mark Paterson (2006, 171; emphasis in original), in his excellent exploration of 'consumption and everyday life' engages with culture, consumption and space descriptively, yet even more diffusely:

> … culture is ordinary, about everyday life. We continually produce and reproduce our culture through activities, some of which are reflective and some unreflective, and this certainly applies to consumption. … [I]t's not just *what* we do but *where* we do it, so that the spatial contexts of cultural practices often help structure the activities occurring within them. The design and planning of spaces of consumption actually alter consumer behaviour, both intentionally and unintentionally. … [T]he *places* in which everyday life occurs are important. This point is crucial. The *spaces* which are given or imposed on us, the shopping malls and supermarkets designed from architectural blueprints, come to be *places* through the uses and activities of the people within them. Just like the use of commodities, the use of a space might significantly alter its intended meaning or purpose, and may change over time. [For example, s]hopping malls are famous for being the hangout for pensioners, walkers, mallrats and housewives at different times of the day, consuming to varying extents and using the given space in manifold ways.

Over time, the 'everyday' spaces and places of retailing, as keyed on by Paterson, have come under increasing scrutiny. This has been from several different perspectives, namely writings that engage with the 'external' economic spaces and places of (mainly) retail capital and its trans-nationalization through supermarkets, the 'internal' cultural spaces of places like department stores, malls, fast food joints and other chain retailers and, also, more integrative work that looks to cross this external/internal divide through a broadly conceived 'cultural political economy' (e.g. Hudson 2008; Jessop and Oosterlynck 2008) applied to retail settings. The story about external retail spaces has mostly been one about consolidation and globalization, from 'Do it Yourself' stores to supermarkets to 'discount' grocers (e.g. Wrigley and Lowe 2007) and how they work to embed themselves in the globalized spaces of shopping and consumption (Wrigley et al. 2005). In particular, in the UK, two stores (Tesco and Asda in 2005) accounted for 50 percent of the spending on food (Tallontire and Vorley 2005), with one in eight pounds in the economy travelling through a Tesco supermarket (Parsley 2008). Research and writing on the internal spaces of retail has focused on the 'palaces of consumption' in the spectacular places of the department store and the mall. Drawing a historical line through from Benjamin's Arcades Project, urban and suburban landscapes have been given over to the distinctive activities of shopping—what Paterson (2006, 174–185; emphasis in original) suggests now encompasses *looking* rather than *spending*, *desire* rather than *need* and the activities of leisure *and* pleasure specifically coded as *feminine*—through the spread of department store 'place' and the 'mall-ification' of much of the developed, consumerist world (see also Goss 1999). Here, malls produce and re-produce paradoxical spaces of both freedom, in the form of (still limited) choice, and control. Crang (2008, 388) expands here, describing the mall this way:

> There is the attempt to produce a fabricated space in which the individual consumer can be made to feel like consuming ... There is the emphasis on creating an internal, closed-off, privately owned but partially public environment, divorced from the harsh exterior world, not only climatically but also socially, through the operation of security systems that ensure the absence of anyone who might threaten this consumer paradise or disrupt the pleasure of the shopper. Thus, these privately owned and managed spaces rework the older public spaces of the street or market ...

He continues with a second important and related point, specifically that '... there is much more to going shopping than buying things ...'. Indeed,

> [s]hopping is in part about experiencing an urban space, seeing and being seen by other shoppers. It is a social activity. But they seek to manage this social experience. The street gets recast as a purified space of leisurely consumption, cleansed of nuisances that in habit the "real" streets, like "street people".

In short, Benjamin's strolling *flâneur*—in the more contemporary guise of the housewife and the teenaged 'mall-rat'—are as subtly and not-so-subtly controlled by these designed retail spaces as they are 'free' to produce and inhabit them through their own presences, identities and behaviours.

More integrative approaches—through the rise of 'new retail geographies' (Wrigley and Lowe 1996) and 'commercial cultures' (Jackson et al. 2000; Miller et al. 1998)—have urged us to take *both* 'economic and cultural geography seriously' (Lowe 2002) in analysing and conceptualizing retail and consumption space and place. Part of this project builds on Peter Jackson's theorizations (e.g. Jackson 1999, 2002; see also Crang 1997) of consumption and 'commodity cultures' where he attempts to muddy up and engage with the dualisms—production-consumption, culture-economy, material-discursive, local-global—that have tended to haunt research on consumption, shopping and retail. For Wrigley and Lowe (2002, 16–17), who echo both Jackson's and Phil Crang's work, analysing 'new' retail and consumption geographies, thus, involves the following sort of analytical framing:

> Our perspective on retailing and consumption, like Crang's (1997, 15) on economic geography in the era of the "cultural turn" more widely, rests on the view that "these spaces, places and practices are never purely economic, and nor are the surpluses they produce", but it is "the commitment to study these vital economic moments—production (in its broadest sense), circulation and consumption—and their regulation, by whatever means necessary, wherever they take place, and whatever materials are produced, circulated and consumed in them" that is essential.

One of the best examples of this sort of work is Michael Smith's (1996) critical engagement with the cultural economy of Starbucks. Here he produces a multi-form 'reading' of the spaces of the coffee retail giant that not only engages its connections to the colonial relations that (still) define coffee production and consumption, but how these relations are 'placed'—through marketing schemes and the 'Starbucks' experience' created by their retail settings—and made to create value for 'knowing' consumers and the company alike. Tasting the world through coffee denotes powerful historical, economic and geographical relations: Starbucks '… trades not only in the 300-year-old market for this tropical commodity but in an equally enduring if less tangible symbolic economy of images and representations that are the cultural correlates of Euroamerican domination' (515). Yet, simultaneously,

> the symbolic appeal of Starbucks coffee cannot be separated from the manner in which it is served, since service is so central to the Starbucks model … There is thus a performative element in Starbucks, an aestheticization of the commodity, as the "baristas" transform the formerly mundane acts of serving coffee into the theatrics of consumption. (506)

Here, the cool logics of post-colonial capitalism and cultural marketing are wedded to the middle-class, urbanite themes of 'coolness' and the European coffee 'encounter' through the sale of an over-priced and over-meaningful 'ordinary' cup of coffee. Indeed, much like the mall—and now the spaces of the home through TV shopping networks and Internet shopping—the spectacle-ized shopping experience for more ordinary, 'everyday' commodities, what Paterson (2006) calls 'McDisneyfication', seems to be the order of the day in Starbucks and many other retail settings.

Yet, lest one think that the increasing power of retail capital has gone quietly into the night, a host of different forms of resistances and different forms of alternatives have arisen in light of the overwhelming shadows of corporate retailer control over local and global lives and spaces. A multiplicity of strategies—what Littler (2005) calls 'beyond the boycott'—are in use here, many of which have drawn political inspiration from social critics such as Naomi Klein and her vastly influential *No Logo* (2000) and Eric Schlosser and *Fast Food Nation* (2001). For example, web-driven activist campaigns—two of the most popular being tescopoly.com and wakeupwalmart.com—have sought to successfully empower local movements interested in reclaiming and defending their community spaces from the retail giants of Tesco in the UK and Wal-Mart in the US. Other engagements include consumer pressure campaigns designed to get more environmentally and socially responsibly goods into stores; here, groups or individual consumers become 'guerrilla' shoppers who ask store staff for these more responsible goods and/or write letters to the retailers requesting these goods be stocked. This strategy, along with the threat of wide-scale picketing, caused Starbucks and several other supermarkets to begin sourcing fair trade coffees in the US (Goodman 2004). A more direct and performative strategy (beyond getting and choosing alternative goods) includes the work of the Reverend Billy of the Church of Stop Shopping (see Reverend Billy 2006). He and his flock perform street, shop, department store and mall '... activist-theatre events including anti-consumerist conversions, blessings on sidewalks and choreographed mobile phone actions' (Littler 2009, 80). As the good Reverend (2006, 3; see also www.revbilly.com) puts it,

> Do I have a witness? As the Smart Monks from here at the Slow Down Your Consumption School of Divinity have said "Stop! Stop shopping, Stop!" Now children, we are all Shopping Sinners, each of us is walking around in a swirl of gas and oil, plastics and foil. We should all hit our knees and weep and confess together. We are not evil people, but somehow we have allowed the Lords of Consumption to organize us into these mobs that buy and dispose, cry and reload. Yes, the Rapture of the Final Consumption, The Shopture, is underway.

Littler (2009, 81–82) conceptualizes their performances this way:

> Reverend Billy and his Church epitomize anti-consumerism in one of its most entertainingly camp forms. They exemplify the politics of "boycott culture" mixed prominently with a flamboyant advocation of consumer abstinence ... [I]t works as a "lever" or promotional tool to generate consideration of the effects of what consumers buy on them/ourselves, on the people who produce these goods and the environment; on the ties and alliances in question. Beyond this, its recommendations are either undefined as "open", depending on your point of view, although a variety of potential actions are pointed towards: lessened consumption, alternative forms of consumption (second-hand swap shops), and unionized activity.

The performance of Billy and the choir chanting refrains about the 'Shopocalypse' work to rough up the spaces of retail and act as an 'interventionary' force for good by getting consumers to engage with and think about their 'devotion' to shopping and retail spaces and places in a much more critical fashion.

A final but related strategy discussed here includes the development of growing 'alternative' retail-scapes throughout much of the consuming global North. From second-hand clothing and goods shops (Gregson and Crewe 2003), to fair trade worldshops (Goodman and Bryant 2009) to more mainstream alternatives such as the Body Shop (Littler 2009; Kaplan 1995) or Whole Foods (Johnston 2008), these shops offer not only different consuming experiences but also commodities of 'difference' in used goods and/or environmentally and socially friendly items on offer. The growing importance of 'alternative food networks' (Goodman and Goodman 2009) or AFNs has particularly spurred the development of alternative forms and spaces of retailing in the EU, UK and US; these include the growth of such places as farm shops, farmers' markets, regional food networks, food box schemes and local, small-scale retailers supplying 'local' foods. Much has been written about these emerging spaces and places in terms of their possibilities but also paradoxes, contradictions and problematics that inhabit these networks (Kneafsey et al. 2008; Maye et al. 2007; Guthman 2003, 2004, 2008a, 2008b).

Farmers' markets, in particular, as a sort of novel and/or revitalized alternative food production/consumption space have received much sustained popular and academic attention (Hinrichs 2000; Holloway and Kneafsey 2000; Kirwan 2006; Slocum 2007). Less (critical) attention, however, has focused on how supermarkets have so easily and thoroughly bought into AFNs, especially in the UK, by becoming some of the largest suppliers of organic, fair trade, free range and 'natural' foods. Thus, as they have moved into the mainstream spaces of the supermarket, AFNs have quickly become just another accumulation strategy, virtually devoid of their 'oppositional' or 'resistance-oriented' politics (Allen et al. 2003). One of the key analytical projects emerging from these observations, then, is how, in the spaces and places of consumption, the relationalities of the alternative and 'conventional' work to construct each other but also the very spaces and consumption practices they inhabit and make.

Utilizing a political economic lens to explore consumption in space and place, Hughes, Wrigley and Buttle examine how corporate retail buying practices undermine international codes of conduct on labour standards and the efforts of ethical trade movements and advocacy organizations to persuade UK-based retailers to view these practices as an integral part of their responsible sourcing policies. Drawing on Gary Gereffi's distinction between producer-driven and buyer-driven global commodity chains, Hughes et al. explore the buying power exploited by large transnational commercial capitals in UK food and clothing markets, which increasingly operate in the production and consumption spaces of the Global South in sourcing commodities and/or expanding their retail activities. Their analysis of corporate governance in buyer-driven UK supermarket and retail clothing chains and the 'brokering role' played by NGOs and ethical consultancy groups in the construction of more ethically-based purchasing practices reveals the fundamental spatiality of these interdependent yet asymmetrical consumer-producer relations. Here, even in the morally-inflected retail space economy, we only get what retail capitals are willing to pay for.

Moving in related directions, McNeill and McNamara analyse the contribution of aesthetic design to cultural production-consumption and surplus-value generation in the urban space economy of New York City in the 1980s. They document how and why hotel developers launched the new phenomenon of heavily differentiated 'boutique' hotels, a set of strategies which subsequently spread to global fashion centres in Europe. Avant-garde design of the lobby space as a dramatic, theatricalized 'scene' or public space, and the consequent blurring of public and private worlds, was at the heart of this cultural innovation designed to encourage consumption in the particularized produced spaces of the urban hotel. In the 'post-Fordist' era of flexible production and market segmentation, innovative hotel developers used lobby design aesthetics to create new nodes of cultural consumption as a strategy of product differentiation targeting specific demographic niches. Competitive imitation of the boutique format by major hotel groups later resulted in its corporatization and standardization. There are conceptual and empirical parallels here with the quality 'turn' in food consumption, the growth of alternative networks and supply chains and the corporatized appropriationist response by mainstream retailers as detailed in the chapters by Hughes et al. and Goodman.

Consumption as Connection/Disconnection/Reconnection

> *Laws are like sausages, it is better not to see them being made.*
>
> (Otto von Bismarck)

Bismarck and his famous declaration about our desire to have the making of sausage hidden could hardly be more wrong today; we not only like to know the places and people our commodities come from, but the trials and tribulations they underwent in their travel to get to us and our store shelves. In some ways, this is the recognition—and one used for political effect by activists, journalists and scholars alike—that the act of consumption is an act of *connection* both literally and figuratively. The purchase of the latest fashion connects one not only to the labour and material processes of the production, processing (e.g. dyeing), and manufacturing (e.g. sewing), often under some pretty poor conditions no matter where one is (Crewe 2004; Ross 2004), but also to the significance surrounding those clothes as the 'latest', 'trendy', and perhaps, 'cutting edge'. And, given what someone has called our 'obsessive branding disorder' (Conley 2008; see also Arvidsson 2006), if those clothes or goods are branded in a particular way, as, say, Nike or Louis Vitton (Thomas 2007), the meanings of these fetishized brands and our desired connections to them holds sway over pretty much everything else. Thus, even with this acknowledgement of consumption as fundamentally a form of connection, in practice according to Marx, the conditions and relations of production remain hidden behind commodity and brand fetishes when produced and consumed under the aegis of capitalism. Indeed, for many, this is the root of the current social and ecological crisis: we are unable to see, in short, we are 'disconnected', from the 'true' consequences of our consumption choices and patterns. Recovery often comes in the form, or at least tropes, of 'reconnection': consumers are able to 'see' and 'know' the impacts and results of their consumption, thus, armed with this information, they will make more progressive and 'better' choices (e.g. Hartwick 1998, 2000). In reconnected goods and their novel, more 'transparent' economies, space and place (once again) become known and knowable to the consumer.

Yet, supplying what Cook and Crang (1996) call commodity 'biographies' often does not operate to the same effects across the spaces of contemporary commodity cultures. This is particularly true for the labelling of goods, one of the most wide-spread forms of reconnection, which have quickly taken on different forms and different levels of 'depth'. At one level, there is Starbucks selling 'single-origin' coffees, i.e. 'Kenyan' or 'Costa Rican', or a leather coat announcing on its label that it is 'Italian-made'. Further, a more officially regulated instance of these biographies includes the development of 'Protected Designations of Origin'

(PDOs) and the similar *Appellation d'origine contrôlée* (AOC), each designed to delineate regional goods by geographical location and 'protect' the names under which they can market themselves. Very much related to the consumption of place and *terroir* discussed above, origins here become an important part of the biographies and cultural cachet of these goods. At another and perhaps 'deeper' level of (re)connection, labelled commodities can be more fully concerned with reporting on but also improving the social and ecological conditions under which particular things are produced and moved around. Here, these goods are often sold with a guarantee about their stories and the meanings they hold: most of these commodity biographies are standardized and audited by external bodies in order to boil down these stories into meaningful, transportable and protected logos relatively easily recognized by consumers. Yet, the 'lines' of (re)connection are quite diverse and change with the commodity under question; for example, fair trade is about consumers connecting to producers' livelihoods (Goodman 2004; cf. Lyon 2006, Varul 2008), Slow Food is about connecting to 'tradition' (Miele and Murdoch 2002) and organic goods are about connecting to the environment (Seyfang 2006).[2] Labelled goods such as these, particularly in the case of something like fair trade with the predominant images and stories of producers on and/or surrounding them, are about developing a more moral economy that can overcome spatial, social and economic distances (Barnett et al. 2005) and peel away the 'dazzling' fetishes that infect commodities. Consumption becomes, then, a form of space/time compression for the progressive forces of good.

Many of these issues have been aired and rehearsed in debates devoted to exploring how to describe and conceptualize the ways by which consumption, production and commodities are related and connected. Part of the task has been developing methods to engage with what Lefebvre (1991, 341; emphasis in original) points to in his thinking about commodities:

> Chains of commodities (networks of exchange) are constituted and articulated on a world scale: transportation networks, buying- and selling-networks (the circulation of money, transfers of capital). Linking commodities together in virtually infinite numbers, the commodity world brings in its wake certain attitudes towards space, certain actions upon space, even a certain concept of space. Indeed, all the commodity chains, circulatory systems and networks, connected on high by Gold, the god of exchange, do have a distinct homogeneity. Exchangeability … implies interchangeability. Yet each location, each link in a chain of commodities, is occupied by a *thing* whose particular traits become more marked once they become fixed, and the longer they remain fixed, at that site; a thing, moreover, composed of matter liable to spoil or soil, a thing having

2 Other (re)connecting goods might include sustainable wood (Klooster 2006), fair trade gold (Hilson 2008), non-conflict gems (Le Billion 2006), non-sweatshop clothing (Hale 2000) ethical flowers (Hughes 2001; Hale and Opondo 2005; Wright and Madrid 2007) and ethical 'veg' (Freidberg 2003).

> weight and depending upon the very forces that threaten it, a thing which can deteriorate if its owner (the merchant) does not protect it. The space of the commodity may thus be defined as homogeneity made up of specificities.

Thus, from commodity systems (Friedland 2001), to value chains (Ponte and Gibbon 2005; Gereffi 2001) and Global Production Networks (Coe et al. 2008) to more specifically culturally-inflected commodity networks (Hughes 2000), chains (Hughes and Reimer 2004), circuits (Cook et al. 2000) and cultures (Jackson 1999), to engagements with cultural materialist conventions (e.g. Murdoch and Miele 2004) and commodity qualities (e.g. Atkins 2010; Mansfield 2003), much ink has been spilt over the need to explicitly include the impacts of the cultural politics of consumption on the materialities of the world.[3] On the whole though, this 'geographical detective work' (Cook et al. 2007) has worked its own 'politics of (re)connection' by tracing and 'following the things' (Cook et al. 2004) that construct and produce contemporary political and material economies.

One growing and important form of (re)connection—albeit not always but becoming more predominantly labelled, especially with food (Allen and Hinrichs 2007; DuPuis et al. 2006)—includes those movements towards the 'relocalization' of economic and social relationships. Standing as a critique of the depersonalizing and abstracting forces of the globalized economy, examples include the 'alternative economic spaces' (Leyshon et al. 2003) of local and regional currencies (i.e. LETS programmes) (Lee et al. 2004; Williams et al. 2003), community-based, 'diverse' economies (Gibson-Graham 2006, 2008) and the so-called 'locavore' movement designed to 're-place' food through farmers' markets, food boxes and cooperatives, regional food networks and the '100 mile diet' (Smith and Mackinnon 2007; Feagan 2007; see also Morgan et al. 2006). In general, relocalization is about a spatialized restrictiveness designed to (re)embed economies in sets of social relations that might develop an 'ethics of care' of the community (Kneafsey et al. 2008), respond better to local concerns and tastes as well as build up local and regional production/consumption capacities, all with a defensive eye toward the parasitical nature of global markets (see Castree 2004; Escobar 2001).

Yet, scepticism abounds, especially in the context of the politics, processes and problematics of consumption as a form of and pathway to (re)embedded (re)connection; ambiguities and contradictions are rife here and they are worth the brief exploration of at least a few points. First, amongst her other critiques (Guthman 2002, 2003) Julie Guthman (2007) takes consumer-oriented labels to task for simply being about a neo-liberal value creation 'trick' through produced preciousness in the development of novel niche markets. Here, the labelization of sustainability is about the commodification of the 'right' kinds of space in nature and labour practices; sustainable development is simplified into simply another

3 For a more thorough discussion of these different approaches—and indeed a full-throated critique of more culturalist approaches to production/consumption networks—see Bernstein and Campling (2006a, b).

product we can buy off the shelves on our weekly shopping trip. In other words, labels, and indeed ethical consumption more broadly (Clarke et al. 2007), work to entrench the old adage of 'voting' with one's money whereby citizenship and consumerism are closely aligned (Trentmann 2007) in what should be thought of as some problematically classed ways. Second, in these networks of (re)connection, particular forms and types of privilege are cultivated and entrenched; namely, consumption is often constructed as *the* form of action required to foster wide-spread changes, with the provisioning of consumer and commodity knowledges *the* pathway to these progressive changes and 'knowledgeable', 'responsible' and 'ethical' consumers as *the* agents of these politicized actions (cf. Clarke et al. 2007). Here, commodity knowledges are very often particularly situated and contextualized; for example as scholars are beginning to point out (e.g. Guthman 2008a, b; cf. Slocum 2006), so-called 'alternative' foods and their consumption are culturally and materially coded as 'white, middle-class' and so have not penetrated as much into lower-income, African-American neighbourhoods in California as many activists would like. For Guthman (2008a, 443),

> … it may well be that the focus of activism should shift away from the particular qualities of food and towards the injustices that underlie disparities in food access. … My deeper concern is how whiteness perhaps crowds out the imaginings of other sorts of political projects that could indeed be more explicitly anti-racist.

Furthermore, knowledges on labels, in contrast to some of the arguments suggesting they 'defetishize' the commodities in question (Hudson and Hudson 2003), work to re-inscribe fetishes through the very 'transparent' knowledges provided on them. Thus, as some have argued, the re-working of these fetishes in goods such as fair trade advertisements and labels in the form of indigenous people and women, entrenches the old colonial relationships that fair trade was set up to economically subvert (Vural 2008; Lyon 2006). While far from settling these debates, what this suggests is a need to engage with the complexities of commodity fetishism and transparency—what Cook and Crang (1996) euphemistically term 'getting with the fetish'—especially given the success of 'alternative' foods like fair trade and its ability to 'seize the fetish', grow markets and promote development for these same indigenous farmers and women (see Goodman 2010).

Third, as Crang (2008, 392) states

> … the levels of knowledge required to understand the incredibly complex biographies of production are more amenable to academic study than they are to someone who has something else to do. Even the most committed ethical consumers therefore rely on shorthand rationales, judgements and knowledges …

What this points to is the suggestion that knowledge created and facilitated through commodities, while perhaps potentially overwhelming, is just one reason or motivating factor for more ethical consumption practices. Barnett and Land (2007), while on the one hand recognizing the importance of information and knowledge as motivational forces for shifting consumption patterns, say, suggest that there is much more going on. Conceptualized for them through new 'geographies of generosity' they state that

> ... precisely because the demonstration of a person's implication in labour exploitation in far away places or in environmental degradation only works by establishing the dependence of these consequences on myriad mediating actions, then strictly speaking the motivating force of the demonstration is fairly indeterminate. It might persuade a person that their actions contribute, in small ways, to the reproduction of those harms. It is just as likely for someone to conclude that their contribution is so highly mediated that not only are they not able to do much about it, but that this does not really count as being responsible in any reasonable sense at all. (1068)

Rather, '... at the very least, the reasons one might have for acting differently in light of causal knowledge are not likely to be reasons of knowledge alone. ... The fixation on chains of causality hides from view the degree to which responsible, caring action is motivated not in monological reflection on one's own obligations, but by encounters with others' (1069). Thus, for them,

> ... focusing on a modality of action such as generosity suggests, instead, a different programme, less exhortatory, more exploratory: one which looks at how opportunities to address normative demands in multiple registers are organised and transformed; at the ways in which dispositions to respond and to be receptive to others are worked up; and how opportunities for acting responsively on these dispositions are organised.

Some of what this points to is that even with these forms of consumer and commodity knowledges, progressive acts may either not follow nor be possible for a complex set of reasons.

Finally, relocalization movements have come under increasing scrutiny. Arguing against an overt 'geographical fetishism' (Castree 2004), DuPuis and Goodman (2005) suggest that the 'local' can take on the potentially ugly moves of an uncritical and divisive reactionary politics of spatial and social exclusivity. Similarly, as Hinrichs (2000) has pointed out, in local food movements and practices, there is often a conflation of social relations with spatial relations: for example, consuming local goods might allow us to be 'closer' and 'better' in terms of the spatial distances they have had to travel but be just as humanly, ecologically and socially exploitative as any other commodity. DuPuis and Goodman argue for a 'bringing back in' of the politics and processes of what gets constructed as 'local' through what they call a 'reflexive politics of localism'. Here, they caution

> … against the reification of the local found in normative and market-oriented perspectives and their naturalization as a bulwark against anomic global capitalism. … An inclusive and reflexive politics in place would understand local food systems not as local "resistance" against a global capitalist "logic" but as a mutually constitutive, imperfect, political process in which the local and the global make each other on an everyday basis. In this more "realist" open-ended story, actors are allowed to be reflexive about both their own norms and about the structural economic logics of production.

This need for the practice of reflexivity in the very politics of consumption—and not just in the politics of localism—is crucial and, indeed worthy of much greater reflection and research; this is especially true as we continue to hang many of our hopes for a better world on the likes of the 'conscious' ethical consumer.

In this part of the volume, Jackson, Ward and Russell are concerned with the formation and connectivities of food knowledges about the broiler chicken industry and how the different understandings found in the spaces of intensive commodity production, retail commerce and domestic consumption are negotiated and accommodated on the chicken's journey to the kitchen table. As they detail in the chapter, part of this negotiation involves retailer knowledge of consumers to overcome their anxieties and create 'trust' and consumer understandings of food quality as provenance and animal welfare. Jackson et al. use life histories to trace how food knowledges have evolved with the agro-industrial and retail transformation of these spaces since World War II in order to analyse the interplay of the material culture of the 'manufactured meanings' of chicken designed to foster (re)connections between the spaces and places of production-consumption.

Building on Jackson et al's chapter, the contested control of the knowledge economy of 'quality'—who defines how food is grown and how it is known—and how it is 'placed' to (re)connect consumers to their food is a central theme of Goodman's analysis of food production-consumption networks and their shifting retail geographies of the UK. Struggles over cultural understandings of quality hold the key to the constructions of value as UK supermarkets have been quick to realize as they build 'own label' lines of organic, local and fair trade products that can be argued to substitute convenience for social (re)connection. Such mainstreaming strategies erode the legitimacy and alterity of these knowledge claims and inhibit the growth of alternative spaces of consumption.

Consumption as Production and Production as Consumption

> *So, while economists have often taught us that consumption is simply the opposite of production and that demand should be coupled with supply, "consumption" actually retains a far richer set of meanings.*
>
> (Clarke et al. 2003, 1)

A final way to engage with the relationalities of space, place and consumption involves what might be considered the rather essentialized conceptualization of consumption *as also* production and production *as also* consumption. In short, in the 'moments' of *both* production and consumption are their so-called opposites—a position Clarke et al. (2003) urge us to consider, one we wish to highlight here and one that is highlighted in many of the chapters, albeit perhaps not specifically in this manner. For example, in the simplest of terms, consumption might be seen as the production of everything from pollution and garbage (e.g. Rogers 2005; Royte 2006) in the contemporary 'throw away' society, to the production of one's identity—and indeed the very (re)production of one's own existence—to the production of profits and particular forms of politics (Ekstrom and Brembeck 2004; Princen et al. 2002). Furthermore, in the context of the overall theme of the volume, consumption can be said to *produce* space and place, while production can be said to equally *consume* space and place. Clearly, we need to be careful in terms of how far we might take these arguments theoretically (e.g. Smith 1998), metaphorically and in the 'real' spaces of the everyday. Yet, at a minimum, conceptualizing production and consumption in this way allows us to open up these two processes and practices to fruitfully explore their dialectical and situated complexities, relationalities and spatialities. We do so here through three very brief explorations, focusing primarily on consumption as the production of identities and politics but also how the dialectics of consumption/production are materially related.

Taking this last, and seemingly somewhat obvious, point first, consumption is about the material production of space and place; anyone redecorating their house will recognize this fact rather quickly. The consumption of calories is about the literal (re)production of one's body 'space', a point argued earlier in this introduction. But more broadly and at a much wider scale, the consumption of goods—at least in the majority of the post-industrial North—is almost invariably the production of material wastes in the form of pollution and garbage. Indeed, it seems as if the contemporary over-determined, over-consumerist moment we live in,[4] much of it facilitated by media and an energetic marketing industry, is about the production of spectacularly 'negative', 'blighted' and/or 'toxic' spaces and places through, for example, deforestation, industrial farming and the filling of landfills. Yet, simultaneously, production is very much about the consumption of the materialities of space and place. Here then, in a general sense, is the dialectic at work through 'creative destruction': the production of hardwood furniture, strawberries and/or a mobile phone—all destined for consumption at some point in their travel—is at one and the same time the consumption and fundamental transformation of forests, soil nutrients and metals and minerals. Others, such as Redclift (1997) and Bakker and Bridge (2006; see also Bridge 2009) refer to the dialectics being described here as the 'metabolization' of nature: the consumption

4 It remains to be seen how the economic downturn will fundamentally alter consumption levels and behaviours throughout the global economy. One take on how this might affect more sustainable consumption is provided by Hinton and Goodman (2010).

of the material aspects of the environment in order to fuel the 'engines' of a productive and value-generating consumerist capitalism.[5] And, although some would perhaps not necessarily agree with the specific conceptualizations we have suggested here, further excellent work has specifically built on these discussions by exploring the 'nature of neo-liberal natures' (McCarthy and Prudham 2004) and the role and socio-economic stakes of privatization in facilitating this metabolization (*Antipode* 2007).

Altering the dialectics of consumption/production sits at the current centre of many activist movements, especially those concerned with fostering more sustainable consumption. More than just the instances of (re)connection of consumption and production as mentioned above, the seeming goal is to alter the social and ecological relations of production *through* consumption. In important ways, then, sustainable, green and ethical consumption are about the intention of fostering an alternative and more progressive production of material space and place. Carbon offsetting, one of the pillars of the new carbon economy, is the perfect example of these intentions (see Bumpus and Liverman 2008). Individual consumers can purchase (i.e. consume) carbon offsets in order to facilitate the 'soaking up' of their personal CO_2 emissions of, for instance, one's holiday air travel. These carbon offset purchases are more often than not now linked to development projects in parts of the Third World through environmental interventions such as the planting of trees that work to take the emitted CO_2 out of the atmosphere. And while many have been quick to rightly point out the complications and problematics of these sorts of offsetting schemes (Liverman 2009; Prudham 2009), not least in the further and deeper commodification of nature in the form of ecological processes, consumption of the 'negative' commodity of the offset produces a particular sort of (arboreal) nature in the places and spaces where the money in these offsetting networks touches down.

Fiddling with the consumption/production dialectic is also a part of those campaigns interested in less- or non-consumption and/or what some have called 'anti-consumption' (*Cultural Studies* 2008; see also Hinton and Goodman 2010). In short, while anti-consumption can be seen as the production of a form of politics, the point here is to materially consume less or nothing and so about materially *producing* less or *nothing* in terms of waste or consumer-oriented goods in the first instance. Non-consumption, then, in spirit if anything else, has the intention of the production of non-consumerist spaces and places, but also the spaces and places of non-production and thus, non-exploitation; in many of these campaigns, there is a rather deep questioning of the relationships of needs and desires and the desire to question the very tenets of consumerism and consumerist capitalism.

A second brief concern here is the ways and means by which consumption works to produce identities, the process of which Elliot (2004) calls 'making up people'. As Mansvelt (2005, 80) puts it '[c]onsumption is a medium through which

5 Clearly, political ecologists (e.g. Michael Watts) and eco-Marxists (e.g. Noel Castree) have been important contributors to these discussions.

people can create and signify their identities', with, '[t]he possibilities of casting off an old identity and adopting a new one [affording] yet another opportunity for consumption' (Clarke et al. 2003, 16). Drilling down a bit more deeply, Mansvelt works to question whether we consume *to become who we are* or do we consume according *to who we already are*. In both schemes, she says,

> … it is easy to posit the subject as an object of consumption, as one who (must) purchase identities to establish a coherent sense of self, or as consumers whose (classed) habitus is the source of all consumption. (83–84)

In some ways, drawing on Bourdieu, this is what Lury (1996; cited in Paterson 2006, 39) refers to as 'positional consumption' designed to signify class, taste and/or the particular sub-culture or 'neo-tribe' (Bennett 2003) one belongs to. Brands, branding processes and, indeed, the much wider marketing 'scape' is also incredibly important to consider here; Paterson (2006) refers to brands and logos as the 'poetics' of consumer capitalism but he also explores how they allow the production of politics through the rejection and critique of specifically branded goods such as Nike and Starbucks. The deployment of both poetics and politics have reached new heights in the development of the markets for sustainable and ethical goods; here Barnett et al. (2005) talk about the (re)branding of the self as an ethical consumer through the performances of purchasing and displaying one's consumption and commitment to fair trade coffee and the use of 'green' consumerism guides (Clarke et al. 2007).

Yet, as Mansvelt (2005, 84) continues:

> Both perspectives offer partial understandings of how consuming subjects may be constituted. Identities may be as much about belonging and sociality, practical knowledges and provisioning, as they are about representations, distinction and individuality … An emphasis on purchase of commodities for identity purposes (implicit in postmodern narratives) renders invisible the diverse ways in which subjectivities are affirmed and contested through consumption practices, rituals and discourses (Jackson and Thrift, 1995). Identities are also attached to bodies, making a consideration of bodies important to understanding relationships between consumption and processes of subject formation.

Consumption—while important at this contemporary moment—is thus just one of the ways that embodied identities can be produced and rendered fluid in space and time.

The final point to be made here, and one that works across much of the material discussed above, involves a consideration of how consumption can work to produce different forms of politics and politicized spaces and places. Drawing specifically on work by Michele Micheletti (e.g. Micheletti 2003; Micheletti and Stolle 2008) and more diffusely on that by others (e.g. Goodman and DuPuis 2002; Sassatelli 2006), Clarke et al. (2007; see also Clarke 2008 and Barnett et al., forthcoming) attempt to move beyond the 'consumer choice as politics' discussion to engage with consumer-driven politics more deeply. For them, 'doing politics in an ethical register' involves a kind of 'collectivization' of the economic and political forces of 'ethical consumption singularities':

> … consumer purchases are … worked into a form of politics by supporting the lobbying and media work of the NGOs and charity organizations economically connected to the markets for ethically-consumed goods. The politico-ethical subjectivities of consumers are materialized and actualized in acts of consumer choice which are then materialized in the wider political activities of organizations. (Goodman 2010, 106)

And, while this work has, at one level, suffered from a general lack of empirical engagement, geographical specificity and over-theorization, at another it has been useful in making the important claim that

> … we have failed to recognise the political character of such consumption—the way in which such consumption is *organised and mobilised by social movements and other organizations*, and the way it acts as a *medium through which commitments are registered, policy-makers lobbied, and claims made on legislators*. (Clarke 2008, 1877; emphasis in original)

Another important contribution in this area has been a reassessment of the civic dynamics of consumption and so 'broadening the political' (Trentmann 2007) in Kate Soper's (2004, 115) development of the notion of 'alternative hedonism' to describe the disaffection with consumerism and which provides 'one distinctive rationale for the shift to ethical and green consumption'. The premise of alternative hedonism is that consumers are increasingly disenchanted with the contradictions embedded in the consumerist imaginary as they come to realize that the production of their consumption—pollution, congestion, stress, overwork, health risks, global warming, for example—is compromising their quality of life and pre-empting other possible pleasures and satisfactions (Soper and Thomas 2006; Soper 2007, 2008).

Soper (2004, 116) is not so much concerned with disenchanted individuals who adopt greener and more ethical lifestyles for altruistic reasons but rather with those who are motivated by 'the more self-regarding forms of disenchantment' and how these can reinforce

> the project of sustainable development … hence … the relevance of fostering the dialectic of "alternative hedonism": of making explicit the desires implicit in current expressions of consumer anxiety, and of highlighting the alternative structure of pleasures and satisfactions towards which they gesture.

In drawing attention to a kind of 'inverted affluence', this work again reveals the complex subjectivities and performativities which inhabit consumption/production dialectics.

One of the important things that has received very little play in these discussions of consumption politics is the seeming double move now being ascribed to consumption and consumers. In short, consumption itself renders its own dialectic: on the one hand it is the cause of environmental and social problems that are, on the other hand, solvable through *more* consumption, albeit of the green and more sustainable kinds. This gets at Clarke et al.'s (2003, 1) statement of the ambivalences contained within consumption and, perhaps even within consumers themselves. For them,

> … one of the most interesting—and infuriating—things about consumer society is its ambivalence in almost every respect: socially, culturally, aesthetically, politically, economically and morally.

Ascribing further power and abilities to consumers—in their sustainable and ethical guises—and endowing them *specifically* with the 'politics of the possible' is a topic worthy of much further exploration.

As one of the chapters in this last section of the volume, Gwynne shows how taste, in the production and consumption of 'palate geographies', is not dead and indeed has become an accumulation strategy for UK retail capitals in the 'location', marketing and sale of Chilean vintages in the international commodity spaces of 'New World' wines. In this account of the production of a novel global value chain, the concept of palate geographies captures the key importance of geographical specificity, including British consumers' wine knowledges and the material skills and symbolic resources of Chilean wine producers in cultivating 'global' grape varieties and creating intellectual property in local place identity as *terroir*. These place specificities are exploited as a source of surplus value by oligopsonistic supermarkets in their role as the gatekeepers and producers of British palates and purses and as mediators of the knowledge flows that are so influential in the creation of these distinctive consumer cultures.

Bryant's incisive dissection of the export industry of Burmese teak speaks strongly to the common issue of expanding Western consumption and the making of globalized production networks, here characterized by the destruction of nature, extractive livelihoods and their cultural materialist foundations. The spatial supply relations between the luxury consumption of teak and violence in its worlds of production is the constant in this moral geography, whose contingency arises principally as the post-colonial state and elites succeed their imperial counterparts and the military as the consumption of this versatile hardwood gives way to corporate and personal consumption. Bryant unmasks the 'violent environments'—or, indeed, the 'taste' for violence and dispossession—produced by the commodification of nature in teak extraction and its contemporary militarization, a narrative of 'creative destruction' *par excellence* that has gained little political traction in the spaces of luxury consumption.

Shifting gears slightly for the final chapter in the book, Laing, Newholm and Hogg explore the ways in which the internet is producing a sort of 'paradoxically' empowered consumer with regard to the production and consumption of professional services. As they show through a series of focus group interviews based on legal, financial and healthcare service sectors, they explore how the so-called 'information revolution' of online knowledge provision has opened up what they call the 'spaces of opportunity' as well as the 'spaces of threat' and anxiety. In short, information gleaned from the internet has destabilized professional discourses as it can equally lead to more well-informed clients, but can also be 'dangerously misleading to the ill advised' complicating service providers' efforts. Here, they conclude that 'virtual' knowledge production and consumption produces the internet as a 'changing space' which has real world effects that, while potentially positive, are inherently complicated and complex, not the least in terms of how they work to entrench individuals as self-governing subjects.

To conclude this brief discussion of our particular take on the relationalities of consumption, space and place, we turn to Julie Mansvelt's (2005, 164) deft prose as a way to encapsulate the arguments and explorations put forth here:

> When consumption is conceived as a geography of embodiment, embeddedness, performance and travel, it does not restrict consumption to particular consumption sites and spaces, to the end of a commodity chain, to a circuit of culture, to practices of self-identification, to the symbolic and material appropriation of commodities in everyday life, or to following the biographies, histories and geographies of commodities. Rather knowledges about consumption can be seen as performative of these things, circulating and being invested with power in situated contexts and providing different insights on the world. Under such a schema, consumption does not come to mean everything and nothing, but takes its form and is actualized in the conditions of its reproduction. It is crucial to explore the (moral) spaces of visualization and embodiment and translation that result from the seeing, doing and becoming of consumption as situated social practice, and as part of the practice of geography. Doing so may assist in understanding and addressing the undesirable and uneven consequences of the power geometries that result.

References

Allen, P., FitzSimmons, M., Goodman, M. and Warner, K. (2003), 'Shifting Plates in the Agrifood Landscape: The Tectonics of Alternative Agrifood Initiatives in California', *Journal of Rural Studies* 19, 61–75.

Allen, P. and Hinrichs, C. (2007), 'Buying into "Buy Local": Engagements of United States Local Food Initiatives', in Maye, D., Holloway, L. and Kneafsey, M. (eds), *Constructing Alternative Food Geographies: Representation and Practice* (pp. 255–272) (Oxford: Elsevier).

Antipode. (2007), 'Special Issue on Privatization and Nature', 39.

Appadurai, A. (ed.) (1986), *The Social Life of Things* (Cambridge: Cambridge University Press).

Arvidsson, A. (2006), *Brands: Meaning and Value in Media Culture* (London: Routledge).

Atkins, P. (2010), *Liquid Materialities: A History of Milk, Science and the Law* (Aldershot: Ashgate).

Bacon, C., Mendez, V., Gliessman, S., Goodman, D. and Fox, J. (eds) (2008), *Confronting the Coffee Crisis: Fair Trade, Sustainable Livelihoods and Ecosystems in Mexico and Central America* (Cambridge, MA: MIT Press).

Bakker, K. and Bridge, G. (2006), 'Material Worlds? Resource Geographies and the "Matter of Nature"', *Progress in Human Geography* 30: 1, 5–27.

Barham, B. (2003), 'Translating Terroir: The Global Challenge of French AOC Labeling', *Journal of Rural Studies* 19, 127–138.

Barnett, C., Cloke, P., Clarke, N. and Malpass, A. (2005), 'Consuming Ethics: Articulating the Subjects and Spaces of Ethical Consumption', *Antipode* 37, 23–45.

Barnett, C., Cloke, P., Clarke, N. and Malpass, A. (forthcoming), *Globalizing Responsibility: The Political Rationalities of Ethical Consumption* (London: Wiley-Blackwell).

Barnett, C. and Land, D. (2007), 'Geographies of Generosity: Beyond the "Moral Turn"', *Geoforum* 38, 1065–1075.

Bauman, Z. (2007), *Consuming Life* (Cambridge: Polity).

Bell, D. and Valentine, G. (1997), *Consuming Geographies: We Are Where We Eat* (London: Routledge).

Bennett, A. (2003), 'Subcultures or Neo-Tribes?' in Clarke, D., Doel, M. and Housiaux, K. (eds), *The Consumption Reader* (pp. 152–156) (London: Routledge).

Bernstein, H. and Campling, L. (2006a), 'Review Essay: Commodity Studies and Commodity Fetishism I: "Trading Down?"' *Journal of Agrarian Change* 6: 2, 239–264.

Bernstein, H. and Campling, L. (2006b), 'Review Essay: Commodity Studies and Commodity Fetishism II: "Profits with Principles?"' *Journal of Agrarian Change* 6: 3, 414–447.

Boykoff, M., Curtis, I. and Goodman, M. (2010), 'The Cultural Politics of Climate Change: Interactions in the Spaces of the Everyday', in Boykoff, M. (ed.), *The Politics of Climate Change* (pp. 136–154) (London: Routledge).

Bridge, G. (2009), 'Material Worlds: Natural Resources, Resource Geography and the Material Economy', *Geography Compass* 3, 1–28.

Bumpus, A. and Liverman, D. (2008), 'Accumulation by Decarbonisation and the Governance of Carbon Offsets', *Economic Geography* 84: 2, 127–155.

Carrier, J. and Miller, D. (eds) (1998), *Virtualism: A New Political Economy* (Oxford: Berg).

Castree, N. (2003), 'Place: Connections and Boundaries in an Interdependent World', in Holloway, S., Rice, S. and Valentine, G. (eds), *Key Concepts in Geography* (pp. 165–186) (London: Sage).

Castree, N. (2004), 'Differential Geographies: Place, Indigenous Rights and "Local" Resources', *Political Geography* 23: 2, 133–167.

Clarke, D. (2003), *The Consumer Society and the Postmodern City* (London: Routledge).

Clarke, D., Doel, M. and Housiaux, K. (eds) (2003), *The Consumption Reader* (London: Routledge).

Clarke, N. (2008), 'From Ethical Consumerism to Political Consumption', *Geography Compass* 2: 6, 1870–1884.

Clarke, N., Barnett, C., Cloke, P. and Malpass, A. (2007), 'Globalizing the Consumer: Doing Politics in an Ethical Register', *Political Geography* 26, 231–249.

Coe, N., Dicken, P. and Hess, M. (2008), 'Introduction: Global Production Networks—Debates and Challenges', *Journal of Economic Geography* 8, 267–269.

Conley, L. (2008), *Obd—Obsessive Branding Disorder: The Illusion of Business and the Business of Illusion* (Jackson, TN: Public Affairs).

Cook et al. (2004), 'Follow the Thing: Papaya', *Antipode* 36, 642–664.

Cook, I. and Crang, P. (1996), 'The World on a Plate: Culinary Culture, Displacement, and Geographical Knowledges', *Journal of Material Culture* 1, 131–153.

Cook, I., Crang, P. and Thorpe, M. (2000), 'Constructing the Consumer: Category Management and Circuits of Knowledge in the UK Food Business', in Bryson, J., Daniels, P., Henry, N. and Pollard, J. (eds), *Knowledge, Space, Economy* (pp. 242–260) (London: Routledge).

Cook, I., Evans, J., Griffiths, H., Morris, R. and Wrathmell, S. (2007), '"It's More Than Just What It Is": Defetishizing Commodities, Changing Pedagogies, Situating Geographies', *Geoforum* 38, 1113–1126.

Cook, I. and Harrison, M. (2007), 'Follow the Thing: West Indian Hot Pepper Sauce', *Space and Culture* 10, 40–63.

Corson, T. (2007), *The Zen of Fish: The Story of Sushi, from Samurai to Supermarket* (New York: HarperCollins).

Crang, M., Crang, P. and May, J. (eds) (1999), *Virtual Geographies: Bodies, Space and Relations* (London: Routledge).

Crang, P. (1997), 'Cultural Turns and the (Re)Constitution of Economic Geography', in Lee, R. and Wills, J. (eds), *Geographies of Economies* (pp. 3–16) (London: Arnold).

Crang, P. (2008), 'Consumption and Its Geographies', in Daniels, P., Bradshaw, M., Shaw, D. and Sidaway, J. (eds), *An Introduction to Human Geography: Issues for the 21st Century* (pp. 376–398) (Harlow: Pearson).

Crewe, L. (2004), 'Unravelling Fashion's Commodity Chains', in Hughes, A. and Reimer, S. (eds), *Geographies of Commodity Chains* (pp. 195–214) (London: Routledge).

Cultural Studies. (2008), 'Cultural Studies and Anti-Consumerism: A Critical Encounter', 22: 5.

Doel, M. (2004), 'Poststructuralist Geogrpahies: The Essential Selection', in Cloke, P., Crang, P. and Goodwin, M. (eds), *Envisioning Human Geographies* (pp. 146–171) (London: Edward Arnold).

Doel, M. and Clarke, D. (1999), 'Virtual Worlds: Simulation, Suppletion, S(ed)uction and Simulacra', in Crang, M., Crang, P. and May, J. (eds), *Virtual Geographies: Bodies, Space and Relations* (pp. 261–283) (London: Routledge).

Domosh, M. (2006), *American Commodities in an Age of Empire* (London: Routledge).

Douglas, M. and Isherwood, B. (1996), *The World of Goods: Towards an Anthropology of Consumption*, 2nd Edition (London: Routledge).

DuPuis, E. M. (2000), 'Not in My Body: rBGH and the Rise of Organic Milk', *Agriculture and Human Values* 17: 3, 285–295.

DuPuis, E. M. and Goodman, D. (2005), 'Should We Go Home to Eat?: Towards a Reflexive Politics of Localism', *Journal of Rural Studies* 21: 3, 359–371.

DuPuis, E. M., Goodman, D. and Harrison, J. (2006), 'Just Values or Just Value?: Remaking the Local in Agro-Food Studies', in Marsden, T. and Murdoch, J. (eds), *Between the Local and the Global: Confronting Complexity in the Contemporary Agri-Food Sector* (pp. 241–268) (Oxford: Elsevier).

Ekstrom, K. and Brembeck, H. (eds) (2004), *Elusive Consumption* (Oxford: Berg).

Elliot, R. (2004), 'Making up People: Consumption as a Symbolic Vocabulary for the Construction of Identity', in Ekstrom, K. and Brembeck, H. (eds), *Elusive Consumption* (pp. 129–144) (Oxford: Berg).

Escobar, A. (2001), 'Culture Sits in Places: Reflections on Globalism and Subaltern Strategies of Localization', *Political Geography* 20, 139–174.

Escobar, A. (2009), *Territories of Difference: Place, Movements, Life, Redes* (Durham: Duke University).

Feagan, R. (2007), 'The Place of Food: Mapping out the 'Local' in Local Food Systems', *Progress in Human Geography* 31: 1, 23–42.

Freidberg, S. (2003), 'Not All Sweetness and Light: New Cultural Geographies of Food', *Social and Cultural Geography* 4: 1, 3–6.

Freidberg, S. (2004a), *French Beans and Food Scares: Culture and Commerce in an Anxious Age* (Oxford: OUP).

Freidberg, S. (2004b), 'The Ethical Complex of Corporate Food Power', *Environment and Planning D: Society and Space* 22, 513–531.

Friedland, W. (2001), 'Reprise on Commodity Systems Methodology', *International Journal of Sociology of Agriculture and Food* 9: 1, 82–103. Available at: http://www.otago.ac.nz/nzpg/csafe/ijsaf/archive/vol9/IJSAF-Vol9-No1-2001.pdf.

Froehling, O. (1999), 'Internauts and Guerrilleros: The Zapatista Rebellion in Chiapas, Mexico and Its Extension into Cyberspace', in Crang, M., Crang, P. and May, J. (eds), *Virtual Geographies: Bodies, Space and Relations* (pp. 164–177) (London: Routledge).

Geografiska Annaler. (2004), 'Special Issue on the Political Challenge of Relational Space', 86: 1.

Gereffi, G. (2001), 'Beyond the Producer-Driven/Buyer-Driven Dichotomy: The Evolution of Global Value Chains in the Internet Era', *IDS Bulletin* 32: 3, 30–40. Available at: http://www.soc.duke.edu~ggere/web/gereffi_ids_bulletin.pdf.

Gibson-Graham, J. K. (2006), *A Post-Capitalist Politics* (Minneapolis: University of Minnesota Press).

Gibson-Graham, J. K. (2008), 'Diverse Economies: Performative Practices for "Other Worlds"', *Progress in Human Geography* 32, 1–20.

Goodman, D. (2001), 'Ontology Matters: The Relational Materiality of Nature and Agro-Food Studies', *Sociologia Ruralis* 41, 182–200.

Goodman, D. and DuPuis, M. (2002), 'Knowing and Growing Food: Beyond the Production-Consumption Debate in the Sociology of Agriculture', *Sociologia Ruralis* 42: 1, 6–23.

Goodman, D. and Goodman, M. (2007), 'Localism, Livelihoods and the "Post-Organic": Changing Perspectives on Alternative Food Networks in the United States', in Maye, D., Holloway, L. and Kneafsey, M. (eds), *Constructing Alternative Food Geographies: Representation and Practice* (pp. 23–38) (Oxford: Elsevier).

Goodman, D. and Goodman, M. (2009), 'Alternative Food Networks', in Kitchin, R. and Thrift, N. (eds), *International Encyclopedia of Human Geography* (pp. 208–220) (Oxford: Elsevier).

Goodman, M. (2010), 'The Mirror of Consumption: Celebritization, Developmental Consumption and the Shifting Cultural Politics of Fair Trade', *Geoforum* 41, 104–116.

Goodman, M. (forthcoming), 'Towards Visceral Entanglements: Knowing and Growing the Economic Geographies of Food', in Lee, R., Leyshon, A., McDowell, L. and Sunley, P. (eds.), *The Sage Companion of Economic Geography* (London: Sage).

Goodman, M. and Bryant, R. (2009), 'The Ethics of Sustainable Consumption Governance: Exploring the Cultural Economies of "Alternative" Retailing', Environment, Politics and Development Working Paper Series,WP#15, Department of Geography, King's College London. Available at: http://www.kcl.ac.uk/content/1/c6/03/95/42/GoodmanBryantWP15.pdf.

Goss, J. (1999), 'Consumption', in Cloke, P., Crang, P. and Goodwin, M. (eds), *Introducing Human Geographies* (pp. 114–122) (London: Arnold).
Goss, J. (2004), 'Geographies of Consumption I', *Progress in Human Geography* 28: 3, 369–380.
Goss, J. (2006), 'Geographies of Consumption: The Work of Consumption', *Progress in Human Geography* 30: 2, 237–249.
Gregson, N. and Crewe, L. (2003), *Second-Hand Cultures* (Oxford: Berg).
Guthman, J. (2002), 'Commodified Meanings, Meaningful Commodities: Re-Thinking Production-Consumption Links through the Organic System of Provision', *Sociologia Ruralis* 42: 4, 295–311.
Guthman, J. (2003), 'Fast Food/Organic Food: Reflexive Tastes and the Making of "Yuppie Chow"', *Social and Cultural Geography* 4: 1, 45–58.
Guthman, J. (2004), *Agrarian Dreams? The Paradox of Organic Farming in California* (Berkeley: University of California Press).
Guthman, J. (2007), 'The Polyanyian Way?: Voluntary Food Labels and Neoliberal Governance', *Antipode* 39, 456–478.
Guthman, J. (2008a), 'Bringing Good Food to Others: Investigating the Subjects of Alternative Food Practice', *Cultural Geographies* 15, 431–447.
Guthman, J. (2008b), '"If They Only Knew": Color Blindness and Universalism in California Alternative Food Institutions', *Professional Geographer* 60: 3, 387–397.
Hale, A. (2000), 'What Hope for "Ethical" Trade in the Globalized Garment Industry?' *Antipode* 32: 4, 349–356.
Hale, A. and Opondo, M. (2005), 'Humanising the Cut Flower Chain: Confronting the Realities of Flower Production for Workers in Kenya', *Antipode* 37, 301–323.
Harrison, R., Newholm, T. and Shaw, D. (eds) (2005), *The Ethical Consumer* (London: Sage).
Hartwick, E. (1998), 'Geographies of Consumption: A Commodity Chain Approach', *Environment and Planning D: Society and Space* 16, 423–437.
Hartwick, E. (2000), 'Towards a Geographical Politics of Consumption', *Environment and Planning A* 32: 7, 1177–1192.
Harvey, D. (1990), 'Between Space and Time: Reflections on the Geographical Imagination', *Annals of the Association of American Geographers* 80, 418–434.
Hilson, G. (2008), '"Fair Trade Gold": Antecedents, Prospects and Challenges', *Geoforum* 39, 386–400.
Hinrichs, C. (2000), 'Embeddedness and Local Food Systems: Notes on Two Types of Direct Agricultural Markets', *Journal of Rural Studies* 16, 295–303.
Hinton, E. (2009), '"Changing the World One Lazy-Assed Mouse Click at a Time": Exploring the Virtual Spaces of Virtualism in UK Third-Sector Sustainable Consumption Advocacy', Environment, Politics and Development Working Paper Series, WP#16, Department of Geography, King's College

London. Available at: http://www.kcl.ac.uk/content/1/c6/03/95/42/Changing theworldWP16.pdf.

Hinton, E. and Goodman, M. (2010), 'Sustainable Consumption: Developments, Considerations and New Directions', in Woodgate, G. and Redclift, M. (eds), *International Handbook of Environmental Sociology*, 2nd Edition (pp. 245–261) (London: Edward Elgar).

Holloway, L. (2002), 'Virtual Vegetables and Adopted Sheep: Ethical Relation, Authenticity and Internet-Mediated Food Production Technologies', *Area* 34: 1, 70–81.

Holloway, L. and Kneafsey, M. (2000), 'Reading the Space of the Farmers' Market: A Case-Study from the United Kingdom', *Sociologia Ruralis* 40, 285–299.

Hudson, I. and Hudson, M. (2003), 'Removing the Veil?: Commodity Fetishism, Fair Trade, and the Environment', *Organization and Environment* 16: 4, 413–430.

Hudson, R. (2008), 'Cultural Political Economy Meets Global Production Networks: A Productive Meeting?' *Journal of Economic Geography* 8, 421–440.

Hughes, A. (2000), 'Retailers, Knowledges, and Changing Commodity Networks: The Case of the Cut Flower Trade', *Geoforum* 31: 2, 175–190.

Hughes, A. (2001), 'Global Commodity Networks, Ethical Trade and Governmentality: Organizing Business Responsibility in the Kenyan Cut Flower Industry', *Transactions of the Institute of British Geographers* 26, 390–406.

Hughes, A. and Reimer, S. (eds) (2004), *Geographies of Commodity Chains* (London: Routledge).

Ilbery, B. and Maye, D. (2008), 'Changing Geographies of Global Food Consumption and Production', in Daniels, P., Bradshaw, M., Shaw, D. and Sidaway, J. (eds), *An Introduction to Human Geography: Issues for the 21st Century*, 3rd Edition (pp. 159–179) (Harlow: Pearson).

Jackson, P. (1999), 'Commodity Cultures: The Traffic in Things', *Transactions of the Institute of British Geographers* 24, 95–108.

Jackson, P. (2002), 'Commercial Cultures: Transcending the Cultural and the Economic', *Progress in Human Geography* 26: 1, 3–18.

Jackson, P., Lowe, M., Miller, D. and Mort, F. (eds) (2000), *Commercial Cultures: Economies, Practices, Spaces* (Oxford: Berg).

Jackson, P. and Thrift, N. (1995), 'Geographies of Consumption', in Miller, D. (ed.), *Acknowledging Consumption* (pp. 204–237) (London: Routledge).

Jayne, M. (2006), *Cities and Consumption* (London: Routledge).

Jessop, B. and Oosterlynck, S. (2008), 'Cultural Political Economy: On Making the Cultural Turn without Falling into Soft Economic Sociology', *Geoforum* 39, 1155–1169.

Johnston, J. (2008), 'The Citizen-Consumer Hybrid: Ideological Tensions and the Case of Whole Foods Market', *Theory and Society* 37, 229–270.

Kaplan, C. (1995), '"A World without Boundaries": The Body Shop's Trans/National Geographies', *Social Text* 13, 45–66.

Kirwan, J. (2006), 'The Interpersonal World of Direct Marketing: Examining Quality at UK Farmers' Markets', *Journal of Rural Studies* 22, 301–312.

Klein, N. (2000), *No Logo* (New York: Picador).

Klooster, D. (2006), 'Environmental Certification of Forests in Mexico: The Political Ecology of a Nongovernmental Market Intervention', *Annals of the Association of American Geographers* 96: 3, 541–565.

Kneafsey, M., Holloway, L., Cox, R., Dowler, E., Venn, L. and Tuomainen, H. (2008), *Reconnecting Consumers, Producers and Food: Exploring Alternatives* (Oxford: Berg).

Koeppel, D. (2008), *Banana: The Fate of the Fruit That Changed the World* (New York: Hudson Street Press).

Le Billion, P. (2006), 'Fatal Transactions: Conflict Diamonds and the (Anti)Terrorist Consumer', *Antipode* 38, 778–801.

Lee, R. (2006), 'The Ordinary Economy: Tangled up in Values and Geography', *Transactions of the Institute of British Geographers* 31: 4, 413–432.

Lee, R., Leyshon, A., Aldridge, T., Tooke, J., Williams, C. and Thrift, N. (2004), 'Making Geographies and Histories?: Constructing Local Circuits of Value', *Environment and Planning D: Society and Space* 22, 595–617.

Lefebvre, H. (1991), *The Production of Space* (Oxford: Blackwell).

Leyshon, A., Lee, R. and Williams, C. (eds) (2003), *Alternative Economic Spaces* (London: Sage).

Littler, J. (2005), 'Beyond the Boycott: Anti-Consumerism, Cultural Change and the Limits of Reflexivity', *Cultural Studies* 19: 2, 227–252.

Littler, J. (2009), *Radical Consumption* (Maidenhead: Open University Press).

Liverman, D. (2009), 'Conventions of Climate Change: Constructions of Danger and the Dispossession of the Atmosphere', *Journal of Historical Geography* 35, 279–296.

Local Environment. (2008), 'Special Issue on Local Food Systems', 13: 3.

Lowe, M. (2002), 'Commentary—Taking Economic and Cultural Geographies Seriously', *Tijdschrift voor Economische en Sociale Geografie* 93: 1, 5–7.

Lupton, D. (1996), *Food, the Body and the Self* (London: Sage).

Lyon, S. (2006), 'Evaluating Fair Trade Consumption: Politics, Defetishization and Producer Participation', *International Journal of Consumer Studies* 30: 5, 452–464.

Mansfield, B. (2003), 'Spatializing Globalization: A "Geography of Quality" in the Seafood Industry', *Economic Geography* 79: 1, 1–16.

Mansvelt, J. (2005), *Geographies of Consumption* (London: Sage).

Mansvelt, J. (2008), 'Geographies of Consumption: Citizenship, Space and Practice', *Progress in Human Geography* 32: 1, 105–117.

Manuel-Navarrete, D. and Redclift, M. (2010), 'The Role of Place in the Margins of Space', in Woodgate, G. and Redclift, M. (eds), *International Handbook of Environmental Sociology*, 2nd Edition (London: Edward Elgar).

Massey, D. (2005), *For Space* (London: Sage).

Maye, D., Holloway, L. and Kneafsey, M. (eds) (2007), *Alternative Food Geographies: Representation and Practice* (Oxford: Elsevier).
McCarthy, J. and Prudham, S. (2004), 'Neoliberal Nature and the Nature of Neoliberalism', *Geoforum* 35, 275–283.
Micheletti, M. (2003), 'Shopping as Political Activity', *Axess* 9. Available at: http://www.axess.se/english/archive/2003/nr9/currentissue/theme_shopping.php.
Micheletti, M. and Stolle, D. (2008), 'Fashioning Social Justice through Political Consumerism, Capitalism and the Internet', *Cultural Studies* 22: 5, 749–769.
Miele, M. and Murdoch, J. (2002), 'The Practical Aesthetics of Traditional Cuisines: Slow Food in Tuscany', *Sociologia Ruralis* 42: 4, 312–328.
Miller, D. (1995), 'Consumption as the Vanguard of History: A Polemic by Way of an Introduction', in Miller, D. (ed.), *Acknowledging Consumption* (pp. 1–57) (London: Routledge).
Miller, D., Jackson, P., Thrift, N., Holbrook, B. and Rowlands, M. (1998), *Shopping, Place, and Identity* (London: Routledge).
Morgan, K., Marsden, T. and Murdoch, J. (2006), *Worlds of Food: Place, Power, and Provenance in the Food Chain* (Oxford: Oxford University Press).
Murdoch, J. (2006), *Post-Structuralist Geography* (London: Sage).
Murdoch, J. and Miele, M. (2004), 'Culinary Networks and Cultural Connections: A Conventions Perspective', in Hughes, A. and Reimer, S. (eds), *Geographies of Commodity Chains* (pp. 102–119) (London: Routledge).
Parsley, D. (2008), 'Tesco Is Moving on to the Estate Agents' Patch', *The Independent* 3 February.
Paterson, M. (2006), *Consumption and Everyday Life* (London: Routledge).
Ponte, S. and Gibbon, P. (2005), 'Quality Standards, Conventions and the Governance of Global Value Chains', *Economy and Society* 34: 1, 1–31.
Princen, T., Maniates, M. and Conca, K. (eds) (2002), *Confronting Consumption* (Cambridge, MA: MIT Press).
Prudham, S. (2007), 'The Fictions of Autonomous Invention: Accumulation by Dispossession, Commodification and Life Patents in Canada', *Antipode* 39, 406–429.
Prudham, S. (2009), 'Pimping Climate Change: Richard Branson, Global Warming, and the Performance of Green Capitalism', *Environment and Planning A* 41, 1594–1613.
Redclift, M. (1997), *Wasted: Counting the Costs of Global Consumption* (London: Earthscan).
Redclift, M. (2004), *Chewing Gum: The Fortunes of Taste* (London: Routledge).
Redclift, M. (2006), *Frontiers: Histories of Civil Society and Nature* (Cambridge, MA: MIT Press).
Reverend Billy. (2006), *What Would Jesus Buy?: Fabulous Prayers in the Face of the Shopocalypse* (New York: Public Affairs).
Rogers, H. (2005), *Gone Tomorrow: The Hidden Life of Garbage* (New York: The New Press).

Ross, R. (2004), *Slaves to Fashion: Poverty and Abuse in the New Sweatshops* (Ann Arbor: University of Michigan).

Royte, E. (2006), *Garbage Land: On the Secret Trail of Trash* (New York: Back Bay Books).

Sassatelli, R. (2006), 'Virtue, Responsibility and Consumer Choice: Framing Critical Consumerism', in Brewer, J. and Trentmann, F. (eds), *Consuming Cultures, Global Perspectives: Historical Trajectories, Transnational Exchanges* (pp. 219–250) (London: Berg).

Schlosser, E. (2001), *Fast Food Nation: The Dark Side of the All-American Meal* (Boston: Houghton Mifflin).

Seyfang, G. (2006), 'Ecological Citizenship and Sustainable Consumption: Examining Local Organic Food Networks', *Journal of Rural Studies* 22: 4, 383–395.

Slocum, R. (2006), 'Anti-Racist Practice and the Work of Community Food Organizations', *Antipode* 38, 327–349.

Slocum, R. (2007), 'Whiteness, Space and Alternative Food Practice', *Geoforum* 38, 520–533.

Smith, A. and Mackinnon, J. (2007), *Plenty: Eating Locally on the 100 Mile Diet* (New York: Three Rivers Press).

Smith, M. (1996), 'The Empire Filters Back: Consumption, Production, and the Politics of Starbucks Coffee', *Urban Geography* 17: 6, 502–524.

Smith, N. (1998), 'Nature at the Millennium: Production and Re-Enchantment', in Braun, B. and Castree, N. (eds), *Remaking Reality: Nature at the Millennium* (pp. 271–285) (London: Routledge).

Smith, N. (2004), 'Space and Substance in Geography', in Cloke, P., Crang, P. and Goodwin, M. (eds), *Envisioning Human Geographies* (pp. 11–29) (London: Edward Arnold).

Soja, E. (1996), *Thirdspace: Journeys to Los Angeles and Other Real-and-Imagined Places* (Cambridge: Blackwell).

Soper, K. (2004), 'Rethinking the "Good Life": The Consumer as Citizen', *Capitalism, Nature, and Socialism* 15, 111–115.

Soper, K. (2007), 'Rethinking the "Good Life": The Citizenship Dimension of Consumer Disaffection with Consumerism', *Journal of Consumer Culture* 7: 2, 205–229.

Soper, K. (2008), 'Alternative Hedonism, Cultural Theory and the Role of Aesthetic Revisioning', *Cultural Studies* 22: 5, 567–587.

Soper, K. and Thomas, L. (2006), '"Alternative Hedonism" and the Critique of "Consumerism"', ESRC/AHRC Cultures of Consumption Research Programme, Working Paper Series.

Tallontire, A. and Vorley, B. (2005), *Achieving Fairness in Trading between Supermarkets and Their Agrifood Supply Chains* (London: UK Food Group).

Thomas, D. (2007), *Deluxe: How Luxury Lost Its Lustre* (New York: Penguin).

Thrift, N. (2003), 'Space: The Fundamental Stuff of Geography', in Holloway, S., Rice, S. and Valentine, G. (eds), *Key Concepts in Geography* (pp. 95–108) (London: Sage).

Trentmann, F. (2007), 'Before "Fair Trade": Empire, Free Trade, and the Moral Economies of Food in the Modern World', *Environment and Planning D: Society and Space* 25, 1079–1102.

Valentine, G. (1999), 'A Corporeal Geography of Consumption', *Environment and Planning D: Society and Space* 17, 329–351.

Varul, M. (2008), 'Consuming the Campesino', *Cultural Studies* 22: 5, 654–679.

Watts, M. (1999), 'Commodities', in Cloke, P., Crang, P. and Goodwin, M. (eds), *Introducing Human Geographies* (pp. 305–315) (London: Arnold).

Whatmore, S. (2002), *Hybrid Geographies: Natures, Cultures, Spaces* (London: Sage).

Williams, C., Aldridge, T. and Tooke, J. (2003), 'Alternative Exchange Spaces', in Leyshon, A., Lee, R. and Williams, C. (eds), *Alternative Economic Spaces* (pp. 151–167) (London: Sage).

Woodward, I. (2007), *Understanding Material Culture* (London: Sage).

Wright, C. and Madrid, G. (2007), 'Contesting Ethical Trade in Colombia's Cut-Flower Industry: A Case of Cultural and Economic Injustice', *Cultural Sociology* 1: 2, 255–275.

Wrigley, N., Coe, N. and Currah, A. (2005), 'Globalizing Retail: Conceptualizing the Distribution-Based Transnational Corporation (TNC)', *Progress in Human Geography* 29: 4, 437–457.

Wrigley, N. and Lowe, M. (eds) (1996), *Retailing, Consumption, and Capital: Towards a New Retail Geography* (Essex: Longman).

Wrigley, N. and Lowe, M. (2002), *Reading Retail: A Geographical Perspective on Retailing and Consumption Spaces* (London: Arnold).

Wrigley, N. and Lowe, M. (2007), 'Introduction: Transnational Retail and the Global Economy', *Journal of Economic Geography* 7, 337–340.

Chapter 2

Multiple Spaces of Consumption: Some Historical Perspectives

Frank Trentmann

In the sizable historical literature on consumption, space occupies a peculiar position. A central category of inquiry and explanation in the Annales school on material life in the early modern period – most famously Fernand Braudel's *The Mediterranean* – space plays an ever more marginal role the closer historical writing moves to the twentieth century. Most modern histories of consumption have been written under the shadow of the nation-state, taking nationally defined subjects as their starting point, approached through distinct national historiographical traditions, and situated in clearly bounded national or metropolitan spaces.

This chapter explores some potential interdisciplinary interfaces between historical and geographical approaches. It offers a short, selective pathway through the literature on 'modern' consumption to foreground the role of space in recent historical writing. Space has especially featured in two variants. One has been to give attention to the interplay between urban space and modern sensibility in commercial consumer cultures. The other has been the popular crop of commodity biographies, a genre initially developed by historical anthropologists and geographers.

There are additional spatial perspectives which deserve greater historical recognition. I want to highlight three. First, is the mental space of the consumer itself. Most studies have taken the consumer for granted, but the identity, morality, and sensibility of the consumer had to be mapped out in the course of modern history before it could become a social category that could be taken for granted, circulated, embraced and resisted. This space was created in moral and political battles – some transnational (as in anti-slavery campaigns and battles over Free Trade), some at the intersection between private and civic space (as in battles over urban services and utilities, like water). The second dimension I want to highlight concerns the moral geographies of consumption. Geographers have recently turned their attention to consumption as a source for 'caring at a distance'. How the spatial gulf between consumers and producers was morally defined and bridged also has a history. The histories of imperial consumerism and of the 'world food problem' in the early and mid-twentieth century reveal earlier moral geographies and the changing mapping of production and consumption. By way of conclusion, I will return to the history of everyday practices as a useful field for thinking about the flow between private and public spaces. Henri Lefebvre's *La Production de l'espace* (1974) has been a major source of inspiration for new geographical approaches to space, including this volume. Importantly, Lefebvre's work on space arose out of a sustained engagement with everyday life and the relative autonomy of material practices in an age of mediated mass culture. It is worth relating back discussions of space to the politics of everyday life in consumer societies.

Space and Place

'Mountains come first', wrote Fernand Braudel (1972, 25) in the opening chapter of *The Mediterranean* in 1949, a book which put geography squarely at the centre of the discussion of material life in the sixteenth century. It is difficult to imagine a similarly direct and emphatic geographical point of entry for historical studies of consumption in the nineteenth and twentieth centuries. Most accounts have been framed in terms of social groups (retailers, the petite bourgeoisie, shoppers), commercial agencies (the small shop, department store, the supermarket, advertising), social processes of taste and distinction (luxury, status) or social identities and political movements (subcultures, consumer politics).[1] Of course, space (like time) is a feature inherent to all these subjects, but it has rarely been given the analytical weight and sustained attention it had in the tradition of the Annales school for earlier periods.

It might, of course, be suggested that this is no surprise, since historians, by training, deal with time not with space. But this would be a crude oversimplification. After all, the last decade has seen a 'spatial turn' among historians, as historians have rediscovered earlier traditions of geohistorical and geopolitical writing influential a century ago in earlier debates about the nature and future of globalization. So far, however, consumption and material culture has played at best a marginal role in the spatial reorientation of historians. The focus has mainly been on the spatial force of political and military power, the creation of maps and territorial boundaries for particular nations and for 'Europe', and, most recently, the imprint of national identity and racism on the German landscape (Osterhammel 1998; Maier 2000; Blackbourn 2006; Black 2000; Schenck 2002; Sahlins 1989). Notwithstanding much interesting work it has generated, it is noteworthy how much of the 'spatial turn' has remained tied to older historiographical questions of nation and statehood. What dominates are questions about the spatial configuration of the nation in eras of imperialism and globalization, and the different reach and character of sea-born empires versus territorial regimes. By contrast, the spatial contours and transnational dynamics of consumption have received surprisingly little attention. Most historical studies have looked at the spread of material goods and commercial cultures within well-defined national areas or individual cities treating questions

1 From the larger literature, see, e.g. de Grazia and Furlough (1996), Crossick and Jaumain (1999), Daunton and Hilton (2001), Berg and Eger (2003), and Strasser et al. (1998). See also http://www.consume.bbk.ac.uk/publications.html#bibliography.

of space as given. A more transnational focus exists for the flow of goods in the mid-Atlantic world in the eighteenth century and in studies examining the export of the American model of 'consumer society' in the mid-twentieth century (e.g. Styles and Vickery 2006; de Grazia 2005),[2] but these are only partial exceptions in what remains a heavily lopsided nationally-oriented literature. There has been little reception in historical writing of the shift in recent geographical approaches away from a bounded territorial definition to a more relational view of space and place, interconnected through flows and networks, as advanced in the work of Doreen Massey (1994, 2004).[3]

If we think of space and place not in terms of separate territorial containers – one abstract and the distant source of global dynamics, the other local and specific, and reacting to events unleashed afar – but as interpenetrating spheres that shape each other, consumption can be seen to play a critical role as channel and mediator between the two. For history, this role was first explored by anthropologists like Sidney Mintz (1985), Arjun Appadurai (1986) and Timothy Burke (1996). In his classic *Sweetness and Power*, Mintz explored amongst other things how an exotic commodity like sugar, produced by sweat and blood in a slave economy, became indigenized in British society, redressed as a white everyday luxury, and how it became tied to a whole range of national consumption rituals. In the process, sugar redefined imperial identity and the sense of 'home' and 'colony' (Rappaport 2006).

To view space and place as reciprocal and interconnected holds out a relational and pluralistic view of global consumption. In contrast to the social science, which had a sustained debate about 'MacDonaldization' and the role of the local in the global, historians' attention to multiple versions of consumer cultures has been limited. The dominant narrative has been that of Americanization, most recently in Victoria de Grazia's (2005) substantial *Irresistible Empire*. De Grazia puts political economy back into consumer culture, highlighting the role of American 'soft power' in spreading a particular vision of 'consumer society' as well as the contribution of social bodies like the Rotary Club and commercial market-oriented norms. At the same time, the book minimizes the significant diversity of consumer cultures within Europe that, far from being atavistic, had their own modern dynamism. Local places helped shape different versions of American and 'modern' consumer culture (Kroen 2006).

Where space has played a more significant role is in studies of consumer culture's contribution to modernity. Studies of shopping, and the department store in particular, have blended the physical with the more psychological reordering of human space. Perhaps the single most important intellectual influence in this regard was the coming together of gender studies with the rediscovery of Walter Benjamin's Arcade Project, originally undertaken in the 1930s. Benjamin's cultural criticism offered a multi-

2 For the importance of global flows for European consumption, see Berg (2005), Prakash et al. (2009) and Trentmann (2009).

3 For a rare historical reception of Massey's work, see Hall and Rose (2006, 26).

layered view of the combined physical and mental transformation of urban space in nineteenth century Paris. Commercial spaces, like the Paris arcades, here became connected to new ways of viewing and experiencing urban space, epitomized by the *flâneur* (Benjamin 2002). For later historians interested in retrieving the agency of women, Benjamin's sensibilities opened a window onto a more dynamic view of the shopping experience as a conduit between private and public spaces. Far from being just about the point of purchase or a numbing experience, Erika Rappaport's (2000) study of the commercial transformation of the West End in late nineteenth-century London showed how for women 'shopping for pleasure' went initially hand in hand with the opening of new public spaces like coffee shops, clubs, and toilets and washrooms.[4]

We can only note here that, ironically, this was precisely the moment when according to Jürgen Habermas' hugely influential account of the *Transformation of the Public Sphere* (1989; see also Habermas 1957; cf. Calhoun 1992), the public sphere contracted thanks to the commercial pressures of mass culture. Habermas' categorical error was to anchor consumption firmly in the market, making it appear part of a separate system with its own commercial logic that is always challenging and encroaching on the public sphere and private lifeworld (*Lebenswelt*), undermining their own critical properties in the process – a misconception that can be traced back to his Frankfurt school elders.

This oppositional model is ill-equipped to capture consumption as a process that flows between civil society and commercial life – it also has little to say about consumption as a practice that operates beyond the point of purchase in domestic and associational spaces and whose rhythms are not just a reflex of commercial signals or pressures. The long history of social movements and associational life that was fuelled by consumption, from consumer politics to cooperative societies, points to the synergies between consumption and citizenship (as well as to tensions). In some ways, then, the new urban and shopping experience of the department store could be read as an example of what Henri Lefebvre called the 'spatial practice' of a society, although as far as I am aware Lefebvre's work had little influence on gender histories of shopping, or for that matter other histories of consumption (Lefebvre 1991). The trip to the department store and window-shopping combined new daily routines, the perception of modern urban and cosmopolitan space, and new urban networks that linked private and public spaces.

The changing implications of commercial consumer culture for public space over time deserve emphasis. Consumption is not a one-way street: it can lead towards public opening at certain times just as much as it can end in the fragmentation, even closure of public space. Attempts by critics, especially those writing from within a communitarian tradition, to postulate some universal

4 See also Walkowitz (1992), Jackson et al. (2000) and Cook (2003). For earlier periods, see Ogborn (1998) and Stobart (1998). For the spatiality of black markets, see Zierenberg (2008). For the consumption of space by tourism, see Cross and Walton (2005), Walton (1983), Baranowski and Furlough (2001) and Anderson and Tabb (2002).

damaging effects of 'consumerism' on public life and civic politics are fraught with dangers and historically suspect. How the spatial dynamics of material cultures can be so quickly redirected deserves probably more attention than it has received (see Soper and Trentmann 2007).

A counterpoint to the way in which consumer culture has at times opened up public spaces, is the story of how it helped fence off and segment communities in American suburbia, a central thread in Liz Cohen's (2003) *A Consumer's Republic*. Drawing on developments in New Jersey in particular, Cohen shows how, in the aftermath of World War II, race and social policy (with the help of federal dollars for mortgage assistance) converged in a new racially and socially segregated spatial order of housing and material life. Town centres and the public life which it had supported became hollowed out. Segregated white middle-class suburbs came with new infrastructure networks, car use, and shopping malls, which, in many cases, asserted the rights of private and commercial spaces against freedom of expression and sociability associated with the public sphere. In London, the 'Swinging Sixties' provide a stark contrast. Here cheap metropolitan housing together with an established older network of wholesale traders and the 'rag trade' provided a fortuitous setting for fashionable small shops in the Soho district (see Gilbert 2006). How affluent consumer societies have proceeded towards different arrangements of commercial/public spaces or avoided mallification altogether deserves more comparative historical research.

The Mental Spaces of the Consumer

The interplay between physical space and imagined public space already hints at the significance of consumption (its practices, agencies, and objects) for the changing representation of space. The practice of shopping enabled a different way of viewing urban space; in the case of the pioneer department store magnate Gordon Selfridge, of envisaging cosmopolitan modernity more generally. Let us now turn more directly to the imagined spaces of consumption. I am especially interested in two dimensions: the mental space of the consumer, and the way in which consumption involves a moral mapping of the spatial relations between different actors.

The mental space of the consumer concerns the discovery, we might even say invention, of 'the consumer' as a distinctive social and discursive category – a shared space that can be occupied, appealed to, and that offers a common role with which people can identify themselves or be identified with. Of course, people have always consumed stuff, but this does not mean they have always thought of themselves (or others) as consumers with shared traits. Like the teenager, 'the people', or the working class, the consumer is a historical not a universal category. The universalist analytical hegemony of neo-classical economy has blinded us to this simple truth. People in the middle ages or early modern period do not speak of themselves or address others as 'consumers'. Most often the growing prominence of the consumer has been traced to advertising and marketing, including the marketing of choice aimed at the consumer of public services in recent public policy reforms. The social sciences, especially authors in the so-called 'governmentality school', have noted the expanding discursive space assigned to the 'consumer' in the neo-liberal 1990s. The consumer, according to Nicolas Rose, has been a quintessential creature of 'advanced liberalism' – a project in which government mutated into the discipline of individuals governing themselves. Individual choice was the (self-)disciplinary tool by which citizens became consumers. Rose (1999) traces this advanced liberal mentality back to the 1950s – some authors would go back further.[5] Of course, there is no denying that 'the consumer' is today targeted from a whole range of commercial, governmental and non-governmental organizations, including those seeking to mobilize the 'ethical consumer'. Still, to view the consumer simply as an outgrowth of twentieth-century advertising and a neo-liberal or advanced liberal commercialized mindset makes for a deeply presentist, truncated history of what is a far richer, more colourful, and ambivalent story.

In the first place, the voice of the consumer made itself literally heard in battles over the appropriate relationship between private and public spaces. Thus, in mid-Victorian Britain, it was middle-class propertied taxpayers who invoked their rights as consumers in conflicts with local monopoly providers in natural monopolies like water and gas. Some of this involved battles over the very definition of 'domestic' – was running water into a bath tub a 'domestic' use of water or was it an 'extra' for which water companies could charge additional higher rates? None of these battles had much to do with choice, nor with governmentality in a Foucauldian sense. They were over the appropriate system of provision, civic rights, and public accountability. Many consumer groups pressed for municipalization. Municipalization was imagined to create a civic contract between local government and responsible citizen-consumers and to put a stop to 'waste' and 'drought'. Civic sentiment had been equally prominent in transatlantic consumer movements that boycotted slave-produced sugar in the early nineteenth century, and later in the battles of consumer leagues who used 'white lists' and the power of their purse to press for better working conditions in Western cities (Trentmann 2006b; Sussman 2000; Chatriot, Chessel and Hilton 2004).

What is interesting for our purposes is how 'the consumer' for a long time was not some universal space that embodied general notions of choice or individual cost-benefit calculation. Rather it remained tied to specific material processes and social and political fields, especially battles over taxation and citizenship and battles over basic goods and entitlements. Once out of the jar, it proved difficult to put the genie back. The consumer took on a life of its own. Thus, in the popular Free Trade movement in Britain on the eve of the First World War, the appeal to the consumer began to reach beyond the 'cheap loaf' to branded consumer goods like Nestlé's Swiss milk and Colman's mustard (Trentmann 2008).

5 See also Miller and Rose (1997) and Joyce (2003).

It is noteworthy how diverse the career of the consumer has been (and remains) globally. Commercial societies with strong liberal and radical political cultures like the United States and Britain developed a more organic and powerful language of the consumer than did societies on the European continent or Japan and China with their own corporate and nationalist traditions, and a different sequence of social and political citizenship (Trentmann 2006c; Gerth 2003; Garon and Maclachlan 2006). All this is not to deny that in the last two decades neoclassical economics and neoliberalism has given a boost to a more individualist and instrumentalist version of the consumer. To what degree these discourses have managed to reorder relationships and identities on the ground is a matter of debate (e.g. Clarke et al. 2007). It is certainly wrong to see the birthplace of the consumer in this neo-liberal era. It amounts to a form of historical amnesia in the social sciences – a diagnosis that blinds us to the plural origins and histories of the consumer and to the contribution social movements and popular politics have played in its expansion.

Moral Spaces

What is the connection between this mental space of the consumer and the moral mapping of social relations between consumers and other actors? Clearly, the appeal to the consumer and consumer power is central in contemporary attempts to re-moralize and reconnect relations between consumers and producers, as in the FairTrade movement and 'alternative' food networks (Barnett et al. 2005; Malpass et al. 2007; Kneafsey et al. 2008). Once again, however, it is useful to recognize that these attempts to reorder and remoralize relations across space are not something novel, or formations sui generis to late modern capitalism or post-modern ethical consumers. They, too, have a history. It is a case where historical inquiry can complement the work currently done by moral geographers interested in the question of 'caring at a distance' – and complicate some of the story.

In enlightenment thought, most famously Montesquieu's *Esprit de lois* (1748), it was the merchant who played the pivotal role in connecting distant actors and in assisting the flow of human understanding between different cultures. A good hundred and fifty years later, 'the consumer' was grafted onto this tradition of the sweet and pacifying 'douceur' of commerce, most notably in the progressive 'new liberalism'. For progressives like J.A. Hobson (1909) in the early twentieth century, the consumer was a unifying human interest that operated in contrast to the divisive logic of the producer.

This moral upgrading of the consumer into a global chain of peace and harmony had implications for the spatial relationship between consumers and producers. Hobson was not blind to what many contemporaries saw as the danger of the music hall, sport and gambling on the character of private and public life. But ultimately he was convinced that people carried within them the potential of 'citizen-consumers' who would play an active role in civil society and adapt (he would have said 'raised') their consumption habits in ways that benefited the community as a whole. Distant producers were a different matter. Globalization had ruptured more immediate local ties that had made farmers morally responsive to the needs of their neighbouring consumers. The 'Dakota farmer, whose wheat will pass into an elevator in Chicago and after long travel will go to feed some unknown family in Glasgow or in Hamburg', Hobson wrote in 1909, 'can hardly be expected to have the same feeling for the social end which his tilling serves'. Free trade ensured that consumers would not lose out. Put differently, Hobson saw the dilemma of moral responsibility in the global marketplace differently from how they tend to appear in current campaigns for FairTrade where the burden of moral responsibility is placed on consumers and producers are easily portrayed as victims of global capitalism.

It is illuminating to place the current FairTrade vision in a longer history of the moral geographies of consumption, alongside earlier projects that have ranged from popular Free Trade to conservative imperialism (Trentmann 2007, 2008). The use of commodities as tangible symbols of larger social and political relations across space reaches back a long time. Attempts to remoralize trade have come in a variety of ideological guises. The adaptabilty of this genre of spatializing relations of power and reciprocity is impressive. From a historical perspective, FairTrade appears less as an innovator or a paradigm shift than as the latest chapter in the story of political consumerism. There are, for example, some striking continuities with the mass campaign conducted by British Conservative housewives in the inter-war years who in Empire shopping weeks urged hundreds of thousands of consumers to buy imperial produce to show their moral solidarity with producers in the colonies. Of course, the ideological substance and political direction of these campaigns were different from those of FairTrade today. Conservatives imagined imperial apples and raisins as a moral connection between English men and women with their white brethren in the colonies. Still, there also affinities in the techniques of communication and representation. Imperial shopping weeks, tasting fairs, cooking demonstrations, and stalls in local shops already established in the 1920s many of the same techniques familiar from FairTrade weeks more recently. Particular commodities, like coffee, were prepared and displayed in tasting sessions as visual points of entry into the world of distant producers, tools of exocitizing 'the other', still resonant in many ethical consumption campaigns today.[6]

One difference between FairTrade and earlier campaigns for imperial solidarity, concerns the imagined space of the consumer. While always couched in racial overtones, the moral economy of imperial consumerism imagined producers also always as consumers, and vice versa. For English people to buy their Empire apple was not only good for the Canadian farmer but also for English workers, since the Canadian farmer would then have the money to buy and consume English machinery, thus keeping order books full and workers employed. 'Eat an Empire apple a day and help to keep the dole away', as the Conservative women's magazine

6 See also Domosh (2006).

put it.[7] This was, of course, a popular mercantilist argument that entirely ignored the realities of tariffs in the Dominions and of the relatively small size of inter-imperial trade. By contrast, in FairTrade campaigns today, there is a much more pronounced spatial divide between Northern consumers and Southern producers. There is little sense that Southern producers are consumers too, or for that matter that most people who buy FairTrade coffee in the North are also producers whose material livelihood and employment are also shaped by exchange relations with the South. Seeking to bridge the spatial divide now comes with a more unilinear view of the flow of moral responsibility: Northern consumers are urged to show that they care for distant producers. Whether globally a moral economy can sustain itself with such unevenly divided roles is an open question.

The history of world food politics in the mid-twentieth century is a case that highlights what has been lost in such a more personalized and segmented spatial rendering of producer-consumer relations. The movement to end world hunger, which led to the establishment of the Food and Agriculture Organization (FAO) towards the end of the Second World War, looked towards an international coordination of food between surplus and deficit areas. It imagined a world of conjoined moral responsibility. Hunger, in this view, was not the defect in a particular region of the world but a global problem with shared responsibility. All societies, rich and poor, were consumers as well as producers. The problem was that excess capacities and deficit areas were unevenly distributed. The early founders of the FAO looked towards mechanisms of international coordination including a world food reserve and buffer arrangements that would ensure supply would be moved from excess to deficit areas.

The FAO and the social movements like the cooperatives that supported such a new economic world order emphasized that all societies were part of the world food problem. At the same time, the focus was on international coordination not on the individual choices of consumers. In fact, individual ethics or charitable motivation, such as elderly ladies saving food for the hungry poor in India or Africa, were now criticized as old-fashioned, even counterproductive, out of touch with the more macro-economic and institutional mechanisms of governance required to rebalance world food. Propaganda films visualized the transfer of food across space like a global board game in the form of bushels of wheat and heads of cattle being moved as if by a magic hand from excess areas in one part of the world to fill deficit areas elsewhere. This more mechanistic representation of economic space and dependence appropriated the mercantilist mapping of commodities and their movement used by proponents of imperial development in the previous generation. There were many points of contact in ideas and personnel between the project of a world food plan and the earlier focus on the empire as an arena for coordinating trade (Trentmann 2006a; see also Vernon 2007).

7 Home and Politics (1925, 2).

Public and Private Spaces

The discussion in the previous sections has progressed outwards, from the commercial space of the shopper to the moral mapping of commercial relations between consumers and producers across vast global spaces. By way of conclusion, it may be interesting to reverse the direction and, instead, look at the spaces of everyday life in which the quotidian and ordinary practices of consumption take place, and to consider their relation to public life and politics. Lefebvre's work on *The Production of Space* (1974), it is worth recalling, grew out of a longer interest in the sociology of everyday life. Both for Lefebvre, especially in his second volume of *The Critique of Everyday Life* (1961), and for Michel de Certeau, in the first volume on *The Practice of Everyday Life* (1984), the overarching goal was to capture the time-space relations of everyday practices and assert their relative autonomy, agency, and potential source of resistance vis-à-vis an increasingly mediated commercial consuming culture epitomized by adverts and TV. They emphasized how, far from being just passive slaves, people through their everyday practices, 'tactiques' and 'ruses' created their own world with their distinct rhythms and spaces of living. The interest in practices and in space was always symbiotic. '[S]pace is a practiced place', in Certeau's words (1984, 117; see also Sheringham 2006). Walking and reading were privileged case studies precisely because they promised to reveal how practices, mentally and physically, turned places like a map or a text into spaces. The city street becomes a space as the walker appropriates it through the movement and senses of the body.

This focus on the dynamics of a practice in relation to space continues to have a lot of untapped potential for studies of consumption. Let us take urban networks and the rearrangement of private and public spaces as an example. In the 'developed world' consumer life-styles are intimately tied to networks of gas, water, and electricity that have their roots in the nineteenth century. Indeed, running water and ready supply of electricity at the flick of a switch are often deemed to be the very signs of modernity and civilization. What recent sociologists have called 'ordinary consumption', such as cooking, eating, washing, gardening or personal hygiene all depend on these networks (Gronow and Warde 2001) .

The question I want to raise here is how we should think about the influence of these networks on the relationship between public and private spaces. One view that has been applied by a group of recent historians is that of governmentality. As infrastructural conduits of governmentality, networks created new spaces for citizens to govern and discipline themselves. Networks connected homes and lives in an integrated map of order and public health. But even more importantly, in this view, they facilitate the privatization of discipline and rule as individuals move behind the closed doors of their private homes to apply to themselves the new technologies of hygiene associated with the internal bathroom and toilet. The rise of new urban technologies, new forms of self and self-discipline, and the closing off of private spaces from a public sphere are all part of the same momentum of advancing governmentality (Joyce 2003).[8]

What disappears in this approach is attention to the practices of consuming themselves, their rhythms and contradictions, as well as to the dynamics they feed back into public life. In addition to their construction and imagined order, we need to know about the uses of these spaces made possible by integrated urban networks. Here a focus on everyday practices as a field of conflict, agency and negotiation is useful, rather than viewing new technologies of consumption purely as a disciplinary device. Unlike for Foucault, 'the goal is not to make clearer how the violence of order is transmuted into a disciplinary technology, but', Certeau (1984, xiv, f) argued, 'rather to bring to light the clandestine forms taken by the dispersed, tactical, and make-shift creativity of groups or individuals already caught in the nets of "discipline"'.

In his last work on rhythms in the modern city, Lefebvre (1992, 40–42; cf. Shove et al., 2009) emphasized the illusory presence of liberty in modern cities. Humans, like horses, were products of 'dressage'. An 'automatism of repetitions' took over as 'space and time thus laid out make room for humans, for education and initiative: for liberty'. But this was never more than just a 'little room'. Dressage, according to Lefebvre, '*determines the majority of rhythms*'. In the street, people can turn right or left, but their walk, the rhythm of their walking, their movements [*gestes*] do not change for all that. Of course, it is debatable whether rhythms are purely disciplinatory and might not also do many other things at the same time, such as shaping personality, coordinating practices, and giving people a chance to devote their energies to creative and emancipatory ends. Still, in other work, Lefebvre, too, was keen to emphasize that the quotidian 'micro' was not simply determined by the 'macro'. Rather, interpersonal relations and monetary relations overlap, but without the latter necessarily dominating the latter.

In the specific context of metropolitan cities like late Victorian London, such a perspective brings to light forms of practiced space that run counter to the thesis of governmentality, order, and rule.[9] Networks were porous and vulnerable, giving rise to repeated droughts, breakdowns, and battles over water. Far from flowing smoothly and aiding liberal self-rule and restraint, new technologies and practices (like the bath) triggered conflicts between consumers and providers about what was appropriate domestic use, and what a legitimate price. Should civilized, respectable people be able to run a bath and pay for it as part of their normal 'domestic' water use, or was this a practice to be charged at an 'extra' rate? In cities like London and Sheffield, middle-class ratepayers came out in consumer leagues challenging new networks and demanding new modes of governance and accountability. Rather than being just privatizing, the bath and new bathing practices thus opened new political channels between private and public spaces.

---

8 See also Crook (2007), Rose (1999) and Burchill et al. (1991).

9 The following draws on Taylor and Trentmann (forthcoming) and Trentmann and Taylor (2006).

The late Victorian battles over water illustrate many of the types of 'antidiscipline' and 'ruses' so important to French critiques of everyday life. Instead of just acting out the technological script of new networks, for example, working class consumers would leave their taps open in times of drought. For water companies this was 'waste', pure and simple. For workers it was a perfectly rational and adaptive behaviour to catch water during the limited but often undeclared times when companies switched it on, enabling them to go to work or do their domestic chores in the interim. Ordinary forms of consumption, like the use of water, illustrate more generally the tension Lefebvre articulated for the city and urban practices: a contradiction between the collective mode of administration and an individual mode of reappropriation (de Certeau 1984, 96).

At the same time, we must guard against an overly romantic picture of 'antidiscipline', where the diversity of everyday practices is always at the cusp of becoming genuine resistance to dominant power and liberating itself from the dominant norms and practices of consumer culture. Yes, everyday practices spill over into consumer mobilization and reform politics – most dramatically leading to demands for public take-overs. But, at the same time, the increasingly resource-intensive demands of consumers (ruses or no ruses) also contributed to the growth of systems of provision with all their human and environmental costs and dilemmas (Taylor et al., 2009; Taylor and Trentmann, forthcoming). 'Water famines' did not come to an end with public takeover, as municipal reformers wished. Nor did 'waste'. In inter-war Britain, for example, citizens repeatedly ignored warnings to lower their water use, or filled up tubs and buckets in advance, thus calling the drought into existence the authorities had hoped to avoid.

Conclusion

A greater appreciation of space can offer fruitful new perspectives and questions for histories of consumption. In addition to established inquiries into the spaces of shopping and into the cross-spatial biography of a commodity, this chapter has highlighted three interconnected levels of inquiry: the mental space of the consumer, the imagined moral connection between consuming and producing relations, and the spillover of private practices into public politics. A generation ago, in the context of debates about 'consumer society' and their 'hidden persuaders', it was fashionable to look to everyday life as a field of practised spaces that retained its relative autonomy from the deluge and dictates of commercial culture and formed a potential for human resistance and emancipation. Today, the emphasis on use and how people construct space through their practices remains an essential element for an understanding of consumption. At the same time, the relationship between everyday life and the public life beyond is not just oppositional. By transforming private spaces, consumption has helped change public spaces, and vice versa.

References

Anderson, S. and Tabb, B. (eds) (2002), *Water, Leisure and Culture: European Historical Perspectives* (Oxford: Berg).

Appadurai, A. (ed.) (1986), *The Social Life of Things* (Cambridge: Cambridge University Press).

Baranowski, S. and Furlough, E. (eds) (2001), *Being Elsewhere: Tourism, Consumer Culture and Identity in Modern Europe and North America* (Ann Arbor: University of Michigan Press).

Barnett, C., Cloke, P., Clarke, N. and Malpass, A. (2005), 'Consuming Ethics: Articulating the Subjects and Spaces of Ethical Consumption', *Antipode* 37, 23–45.

Benjamin, W. (2002), *The Arcades Project* (New York: Belknap Press).

Berg, M. (2005), *Luxury and Pleasure in Eighteenth-century Britain* (Oxford: Oxford University Press).

Berg, M. and Eger, E. (eds) (2003), *Luxury in the Eighteenth Century: Debates, Desires and Delectable Goods* (Basingstoke: Palgrave Macmillan).

Black, J. (2000), *Maps and History: Constructing Images of the Past* (New Haven: Yale University Press).

Blackbourn, D. (2006), *The Conquest of Nature: Water, Landscape and the Making of Modern Germany* (London: Jonathan Cape).

Braudel, F. (1972), *The Mediterranean and the Mediterranean World in the Age of Philip II* (New York: Harper and Row).

Burchill, G., Gordon, C. and Miller, P. (eds) (1991), *The Foucault Effect: Studies in Governmentality* (London: Harvester Wheatsheaf).

Burke, T. (1996), *Lifebuoy Men, Lux Women: Commodification, Consumption, and Cleanliness in Modern Zimbabwe* (Durham, NC: Duke University Press)

Calhoun, C. (ed.) (1992), *Habermas and the Public Sphere* (Cambridge: MIT).

Certeau, M., de (1984 [1974]), *The Practice of Everyday Life* (Berkeley: University of California Press).

Chatriot, A., Chessel, M. and Hilton, M. (eds) (2004), *Au Nom du Consommateur: Consommation et politique ẹn Europe et aux États-Unis au XX Siècle* (Paris: Éditions La Découverte).

Clarke, J., Newman, J., Smith, N., Vidler, E. and Westmarland, L. (2007), *Creating Citizen-Consumers: Changing Publics and Changing Public Services* (London; Sage).

Cohen, L. (2003), *A Consumers' Republic: The Politics of Mass Consumption in Postwar America* (New York: Alfred A. Knopf).

Cook, D. (2003), 'Spatial Biographies of Children's Consumption: Market Places and Spaces of Childhood in the 1930s and Beyond', *Journal of Consumer Culture* 3: 2, 147–69.

Crook, T. (2007), 'Power, Privacy and Pleasure: Liberalism and the Modern Cubicle', *Cultural Studies* 21: 4/5, 549–569.

Cross, G. and Walton, J. (2005), *The Playful Crowd: Pleasure Places in the Twentieth Century* (New York: Columbia University Press).

Crossick, G. and Jaumain, S. (eds) (1999), *Cathedrals of Consumption: The European Department Store, 1850–1939* (Aldershot: Ashgate).

Daunton M. and Hilton, M. (eds) (2001), *The Politics of Consumption: Material Culture and Citizenship in Europe and America* (Oxford: Berg).

Domosh, M. (2006), *American Commodities in an Age of Empire* (London: Routledge).

Garon, S. and Maclachlan, P. (eds) (2006), *The Ambivalent Consumer: Questioning Consumption in East Asia and the West* (Ithaca, NY: Cornell University Press).

Gerth, K. (2003), *China Made: Consumer Culture and the Creation of the Nation* (Cambridge: Harvard University Asia Centre).

Gilbert, D. (ed.) (2006), *The London Journal* 31: 1.

Grazia, V., de (2005), *Irresistible Empire: America's Advance through 20th-Century Europe* (Cambridge, MA: Belknapp Press).

Grazia, V., de and Furlough, E. (eds) (1996), *The Sex of Things: Gender and Consumption in Historical Perspective* (Berkeley and London: University of California Press).

Gronow, J. and Warde, A. (eds) (2001), *Ordinary Consumption* (London: Routledge).

Habermas, J. (1957), 'Konsumkritik – Eigens zum Konsumieren', *Zeitschrift für Kultur und Politik* 12: 9, 641–645.

Habermas, J. (1989), *The Structural Transformation of the Public Sphere* (Cambridge: MIT).

Hall, C. and Rose, S. (eds) (2006) *At Home with the Empire* (Cambridge: Cambridge University Press).

Hobson, J. (1909), *The Industrial System: An Inquiry into Earned and Unearned Income* (London: Walter Scott).

Home and Politics (1925).

Jackson, P., Lowe, M., Miller, D. and Mort, F. (eds) (2000), *Commercial Cultures: Economies, Practices, Spaces* (Oxford: Berg).

Joyce, P. (2003), *The Rule of Freedom* (London: Cambridge University Press).

Kneafsey, M., Holloway, L., Cox, R., Dowler, E., Venn, L. and Tuomainen, H. (2008), *Reconnecting Consumers, Producers and Food: Exploring Alternatives* (Oxford: Berg).

Kroen, S. (2006), 'Negotiations with the American Way: The Consumer and the Social Contract in Post-War', in Brewer, J. and Trentmann, F. (eds), *Consuming Cultures, Global Perspectives* (pp. 231–277) (London: Berg).

Lefebvre, H. (1981), *The Critique of Everyday Life: From Modernity to Modernism, Volume 3* (London: Verso).

Lefebvre, H. (1991 [1974]), *The Production of Space* (Oxford: Blackwell).

Lefebvre, H. (1992), *Rhythmanalysis: Space, Time and Everyday Life* (Paris: Verso).

Maier, C. (2000), 'Consigning the Twentieth Century to History: Alternative Narratives for the Modern Era', *American Historical Review* 105: 3, 807–831.

Malpass, M., Barnett, C., Clarke, N. and P. Cloke, P. (2007), 'Problematizing Choice: Responsible Consumers and Sceptical Citizens', in Bevir, M. and Trentmann, F. (eds), *Governance, Citizens, and Consumers: Agency and Resistance in Contemporary Politics* (pp. 231–256) (Basingstoke: Palgrave Macmillan).

Massey, D. (1994), *Space, Place and Gender* (Cambridge: Polity).

Massey, D. (2004), 'Geographies of Responsibility', *Geografiska Annaler* 86, 5–18.

Miller, P. and Rose, N. (1997), 'Mobilizing the Consumer: Assembling the Subject of Consumption', *Theory, Culture and Society* 14: 1, 1–36.

Mintz, S. (1985), *Sweetness and Power: The Place of Sugar in Modern History* (New York: Penguin).

Montesquieu (1989 [1748]), *The Spirit of the Laws* (Cambridge: Cambridge University Press).

Ogborn, M. (ed.). (1998), *Spaces of Modernity: London Geographies, 1680–1780* (New York: Guilford Press)

Osterhammel, J. (1998), 'Die Wiederkehr des Raumes: Geopolitik, Geohistorie und historische Geographie', *Neue Politische Literatur* 43, 374–397.

Prakash, O., Riello, G., Roy, T. and Sugihara, K. (eds) (2008), *How India Clothed the World: The World of South Asian Textiles, 1500–1850* (Leiden: Brill).

Rappaport, E. (2000), *Shopping for Pleasure: Women and the Making of London's West End* (Princeton: Princeton University Press).

Rappaport, E. (2006), 'Packaging China: Foreign Articles and Dangerous Tastes in the Mid-Victorian Tea Party', in Trentmann, F. (ed.), *The Making of the Consumer: Knowledge, Power and Identity in the Modern World* (pp. 125–156) (Oxford: Berg).

Rose, N. (1999), *Powers of Freedom: Reframing Political Thought* (Cambridge: Cambridge University Press).

Sahlins, P. (1989), *Boundaries: The Making of France and Spain in the Pyrenees* (Berkeley: California University Press).

Schenk, F. (2002), 'Mental Maps: Die Konstruktion von geographischen Räumen in Europa seit der Aufklärung', *Geschichte und Gesellschaft* 28, 493–514.

Sheringham, M. (2006), *Everyday Life: Theories and Practices from Surrealism to the Present* (Oxford: Oxford University Press).

Shove, E., Trentmann, F. and Wilk, R. (eds) (2009), *Time, Consumption, and Everyday Life* (Oxford: Berg).

Soper, K. and Trentmann, F. (eds) (2007), *Citizenship and Consumption* (Basingstoke: Palgrave Macmillan).

Stobart, J. (1998), 'Shopping Streets as Social Space: Leisure, Consumerism and Improvement in an Eighteenth-century County Town', *Urban History*, 25: 1, 3–21.

Strasser, S., McGovern, C. and Judt, M. (eds) (1998), *Getting and Spending: European and American Consumer Societies in the Twentieth Century* (Cambridge: Cambridge University Press).

Styles, J. and Vickery, A. (eds) (2006), *Gender, Taste and Material Culture in Britain and North America, 1700–1830* (New Haven: Yale University Press).

Sussman, C. (2000), *Consuming Anxieties: Consumer Protest, Gender and British Slavery, 1713–1833* (Stanford: Stanford University Press).

Taylor, V. and Trentmann, F. (forthcoming), 'Liquid Politics: Water and the Politics of Everyday Life in the Modern City', *Past and Present.*

Taylor, V., Chappells, H., Medd, W. and Trentmann, F. (2009). 'Drought is Normal: The Socio-Technical Evolution of Drought and Water Demand in England and Wales, 1893–2006', *Journal of Historical Geography* 35: 3.

Trentmann, F. (2006a), 'Coping with Shortage: The Problem of Food Security and Global Visions of Coordination, c. 1890s–1950', in Trentmann, F. and F. Just (eds), *Food and Conflict in Europe in the Age of the Two World Wars* (pp. 13–48) (Basingstoke: Palgrave Macmillan).

Trentmann, F. (ed.) (2006b), *The Making of the Consumer: Knowledge, Power and Identity in the Modern World* (Oxford and New York: Berg).

Trentmann, F. (2006c), 'The Modern Genealogy of the Consumer: Meanings, Knowledge, and Identities', in Brewer, J. and Trentmann, F. (eds), *Consuming Cultures, Global Perspectives: Historical Trajectories, Transnational Exchanges* (pp. 19–69) (Oxford and New York: Berg).

Trentmann, F. (2007), 'Before "Fair Trade": Empire, Free Trade, and the Moral Economies of Food in the Modern World', *Environment and Planning D: Society and Space* 25, 1079–1102.

Trentmann, F. (2008), *Free Trade Nation: Commerce, Consumption and Civil Society in Modern Britain* (Oxford: Oxford University Press).

Trentmann, F. (2009), 'Crossing Divides: Consumption and Globalization in History', *Journal of Consumer Culture* 9: 2, 187–220.

Trentmann, F. and Taylor, V. (2006), 'From Users to Consumers', in F. Trentmann, *Making of the Consumer* (pp. 53–79) (Oxford: Berg).

Vernon, J. (2007), *Hunger: A Modern History* (Cambridge: Harvard University Press).

Walkowitz, J. (1992), *City of Dreadful Delight: Narratives of Sexual Danger in Late-Victorian London* (Chicago: University of Chicago Press).

Walton, J. (1983), *The English Seaside Resort: A Social History, 1750–1914* (Leicester: Leicester University Press).

Zierenberg, M. (2008), *Stadt der Schieber: Der Berliner Schwarzmarkt 1939–50* (Goettingen: Vandenhoeck & Ruprecht).

Chapter 3
The Seduction of Space

David B. Clarke

The person objectifies himself in production; the thing subjectifies itself in consumption.[1]

Marx (1973, 89)

The commodity needs its space too.

Lefebvre (1991, 342)

Spaces of consumption are consummate spaces of capitalism. But they are much more than this. There is, nonetheless, a widespread tendency to appropriate or – which amounts to the same thing – to disavow this excess: to portray everything as an aspect of commodity fetishism, and hence in train to the logic of capital or, in keeping with the structure of disavowal [*Verleugnung*] that defines fetishism, to fail to see that there is anything to see (alternatively, to insist on seeing that there is nothing to see). The purpose of this chapter is to cultivate a disposition that allows the irreducible excess of spaces of consumption to be taken for what it is. Initially, the focus is on Marx's formulation of commodity fetishism; particularly on the way in which recent attempts to deny its veracity reject any notion of false consciousness in favour of full consciousness on the part of the consumer. The first section aims to disclose the way in which this particular framing of the debate serves to scotomize the operation of the unconscious, with a view to exposing the dangers of so doing. Whilst the commensurability of Marx's notion of commodity fetishism and Freud's understanding of fetishism is notoriously treacherous territory,[2] Marx's account is undoubtedly far closer to acknowledging unconscious effects than many recent theorizations of consumption. Consequently Marx, whom Lacan credited with the discovery of the symptom, remains far 'more subversive than the majority of his contemporary critics who discard the dialectics of commodity fetishism as outdated' (Žižek 1994, 310). In line with the tenor of this argument and extending it considerably, the second section focuses on the alignment of the pleasure principle and the reality principle accomplished by consumerism. For who would have thought, to borrow Bauman's (2002, 187) words, 'that pleasure could be miraculously transmogrified into the mainstay of reality?' The fantasy

1 The English-language translation of the *Grundrisse* has 'in the person'. The present amendment is in line with *Marx–Engels Werke XIII* (Dietz Verlag, Berlin, 1961).

2 A territory opened up by Reich (1972 [1929]) and Fromm (1970 [1932]), and subsequently explored by the Frankfurt School (Marcuse 1966 [1955]; Adorno and Horkheimer, 1979 [1946]), Althusser (1996 [1964]), Jameson (1981), and Žižek (1989).

of abundance and choice articulated by spaces of consumption has played an increasingly vital role in perpetuating the very system that the unruly forces of desire were once unambiguously seen as threatening to undermine. In effect, the consumer society has put 'the thief in charge of the treasure chest. Instead of fighting vexing and recalcitrant but presumably invincible irrational human wishes, it has made them into faithful and reliable (hired) guards of rational order' (Bauman 2002, 187). Yet the consequences of this situation are infinitely more complex than many have allowed. Accordingly, the final section considers the way in which the reality principle and the pleasure principle have been deflected by their due alignment. This situation might best be captured in terms of the deregulation of the reality principle and the consequent emergence of an unprincipled reality.

Modernity was characterized by the instigation of the reality principle; by the immense effort to eradicate the power of the illusory and to lend the world the force of reality. That effort took place under the sign of production. As Baudrillard (1987a, 21) reminds us, however, 'The original sense of "production" is not in fact that of material manufacture; rather, it means to render visible, to cause to appear and be made to appear: *pro-ducere*'. This attempt to force reality to show itself ultimately rebounded in overexposure and overproduction: in a reality glut that would work to undermine the status of the real itself. Indeed, all such attempts to impose an unopposed logic undo themselves in this way. 'There is a kind of reversible fatality for systems', writes Baudrillard (1993a, 91), 'because the more they go towards universality, towards their total limits, there is a kind of reversal that they themselves produce, and that destroys their own objective'. Thus, as everything becomes equally important, everything becomes equally unimportant. When everything is privileged, nothing is. And when everything is afforded the status of reality, the status of reality loses its meaning. For the durability, solidity and reliability that we routinely attribute to reality ultimately depend upon the contrast marked with the flighty, friable, untrustworthy nature of everything deemed unworthy of that name. It is as a consequence of the liminal reversibility induced by the operationalization of the real that consumption has assumed the significance it has. Indeed, from this perspective, it is ultimately unsurprising that the pleasure principle should have come to be harnessed to the reality principle; that the vicissitudes of untrammelled desire could become 'the material for the lasting and solid, tremor-proof foundations of routine' (Bauman 2002, 187). However, to the extent that it is the pleasure principle that is enlisted in the service of the reality principle, the erasure of the distinction between them takes place from this unilateral direction – against which the seductive force of the pleasure principle is destined to exact its revenge. As Baudrillard (1990, 22) has suggested, it is not consumption but seduction – '*se-ducere*: to take aside, to divert from one's path' – that is opposed to production (opposed in an agonistic rather than a dialectical sense). And it is the divertive force of seduction that underlies the liminal reversibility of a system based on the limitless expansion of the real. '*Seduction* is that which is everywhere and always opposed to *production*; seduction withdraws something from the visible order and so runs counter to production, whose project is to set everything up in clear view' (Baudrillard 1987a, 21). It is in line with this characterization that, far from being an outmoded frame of reference, fetishism has come to saturate social space in its entirety: 'If it were possible, in the past, to speak of the fetishism of the commodity, of money, of the simulacrum and the spectacle, that was still a limited fetishism (related to sign-value). There stretches beyond this for us today the world of radical fetishism, linked to the de-signification and limitless operation of the real' (Baudrillard 2005, 72). Attuning itself to the logic of seduction, the present chapter marks this movement from the limited fetishism of the commodity to 'the total promiscuity of things' (Baudrillard, 1993a, 60); from the scene of consumption to its obscenity; from the production of space to the overexposure of pornogeography. It is, however, to the Marxian thesis that we first turn.

The Haunt of the Commodity

> *If it is any point requiring reflection ... we shall examine it to better purpose in the dark.*
>
> Poe (1985, 185)

If the dream worlds of mass consumption were once seen as the primary locus of commodity fetishism, such conceptions have long since fallen out of favour. Marx (1983, 77) defined commodity fetishism in precise terms, selecting a concept from the 'mist-enveloped regions of the religious world' to suggest that the same animism takes hold of the commodity as was held to apply to totemic objects within the 'primitive' religious purview. 'In that world the productions of the human brain appear as independent beings endowed with life, and entering into relation both with one another and the human race' (Marx 1983, 77). The same is true, Marx proposes, of the supposedly rational, disenchanted world of capitalism, where commodities also take on a life of their own, independently of the human labour that went into their production. Just as 'the light from an object is perceived by us not as the subjective excitation of our optic nerve, but as the objective form of something outside the eye itself' (Marx 1983, 77), the commodity, likewise, appears as something external. Whereas in the case of vision, however, an actual transfer of light from the object to the eye is involved, 'the existence of the things *quâ* commodities, and the value-relation between the products of labour which stamps them as commodities, have absolutely no connexion with their physical properties and with the material relations arising therefrom' (Marx 1983, 77). This is because, uniquely in the case of *commodity* production, 'each producer's labour does not show itself except in the act of exchange'.

> In other words, the labour of the individual asserts itself as a part of the labour of society, only by means of the relations which the act of exchange establishes directly between the products, and indirectly, through them, between the

> producers. To the latter, therefore, the relations connecting the labour of one individual with that of the rest appear, not as direct social relations between individuals at work, but as what they really are, material relations between persons and social relations between things. (Marx 1983, 77–8)

Hence the stress that Marx laid on the fetishism of commodities, strategically placing the topic at the end of the first chapter of the first volume of *Capital*. For the 'metaphysical subtleties and theological niceties' surrounding the commodity are a crucial and unavoidable consequence of a system based on value (Marx 1983, 76). 'Value is', says Marx (1983, 79n), 'a relation between persons expressed as a relation between things'. Thus, as Geras (1971, 76) notes, 'For the capitalist, the worker exists only as labour-power, for the worker, the capitalist only as capital. For the consumer, the producer is commodities, and for the producer the consumer is money'. For the consumer, milk comes from shops, not cows; for the capitalist, products worth their salt walk off the supermarket shelves unaided. As the perverse reflection of the relation between persons expressed as a relation between things, commodities really do have commerce amongst themselves. Despite the dilution of the English translation (Rose 1978), when Marx (1983, 77) specifies 'a definite social relation between men, that assumes, in their eyes, the fantastic form [*die phantasmagorische Form*] of a relation between things', he deliberately invokes the phantasms conjured up in the rarefied darkness of the nineteenth-century visual attraction, the phantasmagoria. For phantasms really do stalk the marketplace (*agora*). Their visibility, however, depends upon a particular distribution of the sensible. For Rancière (2004a, 34), it was the repartition of perception accomplished within the sphere of literature – and specifically in relation to the modern novel – that first revealed 'the phantasmagoric dimension of the true'. Hence Rancière's (2004b, 21) claim that 'Marx's commodity stems from the Balzacian shop'.

At the basis of Rancière's comment is a recognition of the discovery of the symptom, not in Freud or Marx, but in literature as such: literature conceived 'as a historical mode of visibility of writing' that establishes a particular framing of 'the relation between the sayable and the visible' (Rancière 2004b, 12). This modern regime of writing, dating from the start of the nineteenth century, effectively established a new sense of meaning: 'Meaning was no longer a relationship between one will and another. It turned out to be a relationship between signs and other signs. The words of literature had to display and decipher the signs and symptoms written in a 'mute writing' on the body of things and in the fabric of language' (Rancière 2004b, 17). This is best illustrated by example:

> This new regime and new "politics" of literature is at the core of the so-called realistic novel. Its principle was not reproducing facts as they are, as critics claimed. It was displaying the so-called world of prosaic activities as a huge poem – a huge fabric of signs and traces, of obscure signs that had to be displayed, unfolded and deciphered. The best example and commentary of this

> can be found in Balzac's *The Wild Ass's Skin.* At the beginning of the novel, the hero, Raphael, enters the showrooms of an antique shop. And there, Balzac writes, "this ocean of furnishings, inventions, fashions, works of art and relics made up for him an endless poem". The shop was indeed a mixture of worlds and ages ... The mixture of the curiosity shop made all objects and images equal. Further, it made each object a poetic element, a sensitive form that is a fabric of signs. (Rancière 2004b, 18–19)

It is as a consequence of this particular way of seeing that the possibility of reading the commodity *qua* symptom first arose. Hence Rancière's (2004b, 10) insistence upon the imbrication of the political and the aesthetic: 'Politics is first of all a way of framing, among sensory data, a specific sphere of experience. It is a partition of the sensible, of the visible and the sayable, which allows (or does not allow) some specific data to appear; which allows or does not allow some specific subjects to designate them and speak about them. It is a specific intertwining of ways of being, ways of doing and ways of speaking'. This is especially evident in the case of Marx. Indeed, for Rancière (2004a, 34), 'The Marxian theory of fetishism is the most striking testimony to this fact: commodities must be torn out of their trivial appearances, made into phantasmagoric objects in order to be interpreted as the expression of society's contradictions'. Marx's insight, therefore, inherits this possibility of a symptomatic reading from literature: 'the Balzacian paradigm of the shop as a poem had to exist first, to allow for the analysis of the commodity as a phantasmagoria, a thing that seems obvious at first glance but actually proves to be a fabric of hieroglyphs and a puzzle of theological quibbles' (Rancière 2004b, 21). At first glance, we see nothing untoward, for as Agamben (2000, 75) astutely observes, commodity fetishism is 'a secret that capitalism always tried to hide by exposing it in full view'. Yet if, as Rancière (2004a, 34) suggests, 'the ordinary becomes beautiful as a trace of the true' only if 'it is torn from its obviousness in order to become a hieroglyph, a mythological or phantasmagoric figure', the 'mysteriousness' of this process is vital, insofar as it problematizes what would otherwise remain unproblematic. In Derrida's (1994, 155) words, 'if the "metaphysical character" of the commodity, if the "enigmatic character" of the product of labour *as commodity* is born of "the social form" of labour, one must still analyse what is mysterious or secret about this process'.

The secret of commodity fetishism, says Derrida (1994, 155), 'has to do with a "quid pro quo"'. The sense of the Latin phrase, lost in the English translation, does not merely allude to a 'tit for tat' exchange but, rather, to the 'mistaken substitution of one thing for another, in particular in the theatre when characters reply to each other out of some misunderstanding' (as the translator of *Specters of Marx* points out: Derrida 1994, 193n). In delineating the properties of commodity fetishism, Marx (1983, 63n) took inspiration from that class of relations that Hegel termed reflex categories: 'For instance, one man is king only because other men stand in the relation of subjects to him. They, on the contrary, imagine that they are subjects because he is king'. The same reflexivity is apparent in commodity fetishism.

> There is a mirror, and the commodity form is also this mirror, but since all of a sudden it no longer plays its role, since it does not reflect back the expected image, those who are looking for themselves can no longer find themselves in it. Men no longer recognize in it the *social* character of their *own* labor. It is as if they were becoming ghosts in their turn. The "proper" feature of specters, like vampires, is that they are deprived of a specular image, of the true, right specular image (but who is not so deprived?). How do you recognize a ghost? By the fact that it does not recognize itself in a mirror. Now that is what happens with the *commerce* of the commodities *among themselves*. These ghosts that are commodities transform human producers into ghosts. And this whole theatrical process (visual, theoretical, but also optical, *optician*) sets off the effect of a mysterious mirror: if the latter does not return the right reflection, if, then, it phantomalizes, this is first of all because it naturalizes. The "mysteriousness" of the commodity-form as presumed reflection of the social form is the incredible manner in which this mirror sends back the image (*zurückspiegelt*) when one thinks it is reflecting for men the image of the "social characteristics of men's own labor": such an "image" objectivizes by naturalizing. Thereby, this is its truth, it shows by hiding, it reflects these "objective" (*gegenständliche*) characteristics as inscribed right on the product of labor, as the "socio-natural properties of these things" (*als gesellschaftliche Natureigenschaften dieser Dinge*). Therefore, and here the commerce among commodities does not wait, the returned (deformed, objectified, naturalized) image becomes that of a social relation among commodities. (Derrida 1994, 155–6)

Commodity fetishism, then, is a kind of conversion disorder. For Žižek (1994, 310), the 'social relations between things' amount to a symptom, 'the point of emergence of the truth about social relations'. The way in which 'relations of domination and servitude are *repressed*' with the bourgeois institution of the free labourer, 'whose interpersonal relations are discharged of all fetishism', ensures that 'the repressed truth – that of the persistence of domination and servitude – emerges in a symptom which subverts the ideological appearance of equality, freedom, and so on ... [H]ere we have a precise definition of the hysterical symptom, of the 'hysteria of conversion' proper to capitalism'. As with all symptoms, this is a symptom we have come to enjoy. The upshot of this argument is that spaces of consumption are themselves symptomatic spaces: articulations of a particular set of social relations between things and material relations between persons. Yet this is far from the kind of formulation that has found favour in recent work on spaces of consumption, which typically seeks to have done with commodity fetishism (for example, Miller et al. 1998). Such a response bears closer scrutiny, for where there are no spectres there is no call for exorcism – and exorcisms have a habit of resurrecting what they seek to expel.

For Rorty (1992, 40), 'the theory of the fetishism of commodities ... only looked convincing as long as one thought that Marxism offered a feasible proposal for an alternative material structure of society'. It is undoubtedly the case that a loss of conviction in the prospect of the kind of unalienated existence held out by Marxism explains the weight of recent attempts to discard commodity fetishism. The response to this argument is, however, typically reactive: if alienation is not a passing phase, to be transcended at some point in the future, the alternative is held to be a fully self-conscious, constitutive subjectivity; a form of transcendental subjectivity. Such a conception maintains an exaggerated respect for the rationality of the individual. This is evident from the way in which any alternative conception is alleged to usher in an unpalatable notion of false consciousness that is incompatible with the impossibility of an unalienated future. Yet it is simply not the case that these polarized positions – either full consciousness or false consciousness – are the only alternatives on offer. Indeed, this particular framing merely reproduces the terms of a longstanding idle debate – by beginning with the unquestioned premise of a pre-given individual who may or may not be duped, and coming down on one side of the argument or the other. Alongside reductive renditions of the commodity fetishism thesis (and a self-congratulatory tone on the part of those allegedly capable of crediting the consumer with sufficient intelligence to avoid being duped by the market), innumerable recent contributions assert that the consumer alone is best placed to pronounce the final word on consumption, which is typically accompanied by a kind of justificatory eudemonism ('if it feels good, it is good'). Yet, barring the vagaries of fashion, there is nothing to recommend between the polarized positions of false consciousness and full consciousness. Psychoanalytic thought, in contrast, acts transversally to this misleading opposition by refusing to begin with a pre-given subject. For psychoanalysis, the subject is constituted as much as it is constitutive, and is never fully present to itself. There is always the unconscious. The next section expounds this conception in detail, but it is worth drawing an initial contrast with a representative account of the recent tendency to assert a fully conscious subjectivity in refutation of false consciousness. Thus, for example, Miller (1998, 128) characterizes shopping as a practice creating 'relationships of love between subjects rather than some kind of materialistic dead end which takes devotion away from its proper subject – other persons'. Alongside the crude caricature of commodity fetishism as devotion to a materialistic cult of consumption, the ethnographically derived 'argument that objects are the means for creating the relationships of love between subjects' (Miller 1998, 128) presumes that being is directly amenable to meaning. Miller's formulation may be usefully set against Lacan's (1991, 147) insistence that to love is 'to give what one does not have', to which we will return in due course. Underlying Lacan's contention is an understanding that there is no possible adequation between being and meaning and, hence, no possibility of a non-alienated existence. This is captured in Lacan's (1977, 211) '*vel* of alienation',[3] which posits an apparent choice between being and meaning – a 'choice' analogous to that proffered by the highwayman's '*Your money or your life!* If I choose the money, I lose both' (see Figure 3.1).

3 The Latin *vel* is the logician's inclusive 'or', in contradistinction to the exclusive 'or' (Latin *aut*).

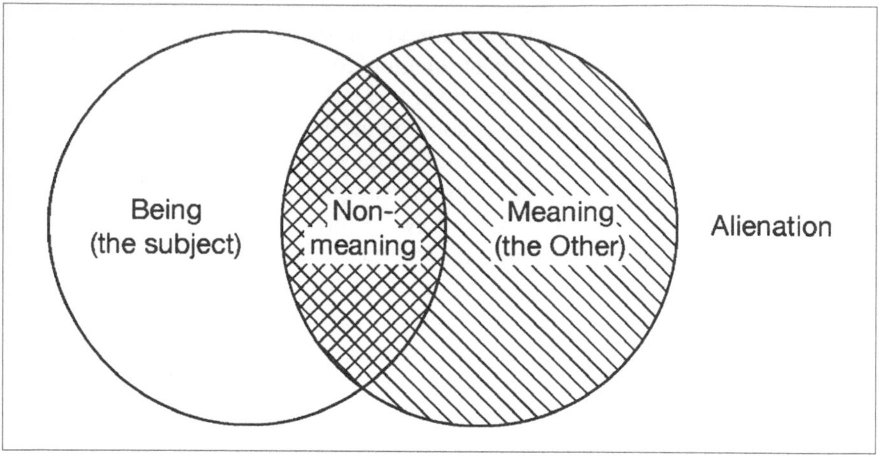

Figure 3.1 The *vel* of alienation (after Lacan [1977, 211])

By analogy, 'If we choose being, the subject disappears, it eludes us, it falls into non-meaning. If we choose meaning, the meaning survives only deprived of that part of non-meaning that is, strictly speaking, that which constitutes in the realization of the subject, the unconscious' (Lacan 1977, 211–2). This formulation escapes the terms of the opposition outlined above, refusing the possibility of both a fully constitutive subjectivity and a non-alienated subjectivity-to-come – and thus allows for a fuller understanding of the fetishism that attaches to spaces of consumption. This is pursued in the next section.

Dream Worlds

> *She had come in with clear eyes ... but they had gradually been seared by the spectacle.*
>
> Zola (2001, 262)

The thesis of commodity fetishism is sometimes dismissed on the basis that it permits the impression that consumption is a secondary issue, a distraction to be explained away in order that the real business of class and relations of production may take centre stage. Although essentially a misinterpretation, as the previous section serves to demonstrate, this sentiment nonetheless reflects the tension felt by the nineteenth-century capitalist between the worker as embodiment of labour-power and as consumer. As Harootunian (2004, 31) notes, Marx bore witness to 'the great illusion of every capitalist who wished only that other workers, and not his own, confront him as consumers'. Today, however, the very notion of the worker as consumer has been recast. In a fully fledged consumer society 'people are consumers first – and workers only a very distant second' (Bauman 1997, 24). Far from being an empirical statement – and infinitely far from being an apologia for the triumph of the market – this characterization alludes to the way in which the pleasure principle has been drafted into the service of the reality principle. In effect, 'Capitalism discovered that the morbid urge for distraction, that major scourge of profit-making from the exploitation of productive labour, may become the largest and perhaps an inexhaustible source of profit once it becomes the turn of the consumers, rather than the producers, to be exploited' (Bauman 2002, 187). That the pursuit of pleasure has come to serve as the principal mode of social reproduction is articulated, perhaps more clearly than anywhere else, in the spaces that are given over to allowing the impulsive desire that was once seen as in need of curbing to run free. Objections are occasionally levelled against this association of consumption with pleasure – which reputedly results in an overemphasis on spectacular 'cathedrals of consumption' such as megamalls – primarily on the basis that the greater part of consumption activity takes the form of routine, mundane, everyday chores; undertaken, for the most part, in equally quotidian surroundings. Yet consumption has consistently presented itself as a source of pleasure; that pleasure deriving as much from the sense of freedom offered within the sphere of consumption as from the more tangible elements of the consumption experience or the objects of consumption themselves. As for the mundane, routine aspects of consumption, one may say that it is pleasure-seeking itself that has become routine. Yet none of this is to say that the pleasures of consumption should be taken at face value. Far from it, in fact, for the freedom and autonomy offered by the sphere of consumption initially arose as a means of compensating for the heteronomy in the sphere of production ushered in by industrial capitalism. The sphere of consumption has never lost the imprint of its origins. Insofar as it remains perennially unlikely, as Bauman (1983, 39) avers, that 'the emancipatory urge, originated and perpetually re-fuelled by the heteronomy of productive activity, will ever be quenched by a success, however spectacular, of its surrogate form', consumerism possesses an inherently self-perpetuating character. That consumerism promises what it ultimately cannot deliver means, paradoxically, that dashed hopes and unfulfilled expectations strengthen rather than weaken its grip. Thus, although consumption presents itself as a system dedicated to pleasure, as 'a social logic, the system of consumption is established on the basis of the denial of pleasure' (Baudrillard 1998, 78). A consumer primed to seek pleasure yet destined never to find satisfaction would be perfectly attuned to the systemic requirements of capitalism. And although the nature of the consumer is far from determined by any such systemic requirements, the extent to which the nature of the subject expounded by psychoanalysis accords with this characterization is remarkable. This fundamental consonance guides the exploration of the unconscious mechanisms at work within the sphere of consumption that forms the remainder of this section.

What Freud termed the pleasure principle, and suggested might equally be labelled the unpleasure principle, refers to the way in which the *unconscious* is primed to seek pleasure and avoid unpleasure. If this is the overriding concern of the unconscious, which knows nothing of reality, the ego, Freud (*SE* XII, 219) held, amounts to a means of militating against the frustrated expectations that would arise from confusing the hallucinatory with the real:

> It was only the non-occurrence of the expected satisfaction, and the disappointment experienced, that led to the abandonment of this attempt at satisfaction by means of hallucination. Instead of it, the psychical apparatus had to decide to form a conception of the real circumstances in the external world and to endeavour to make a real alteration in them. A new principle of mental functioning was thus introduced; what was present in the mind was no longer what was agreeable but what was real, even if it happened to be disagreeable.

In other words, the 'reality principle does not … involve the suppression of the pleasure principle' (Laplanche and Pontalis 1973, 380) but, rather, its modification. It aims at increasing satisfaction by ensuring *real* gratification, even if this is deferred: a benefit thus accrues from the arrangement set in sway by the ego, though at a cost. It is the necessity of this trade-off, however, that the consumer society would efface. Whilst consumerism is evidently geared to the reality principle (it concerns real, not hallucinated, commodities; real, not hallucinated, pleasures), it is equally concerned with harnessing the pleasure principle. Indeed, consumerism's proclivity for pushing opportunities for instant gratification to the fore is one of its paramount features. For given that satisfaction is a typically short-lived state of affairs, instant gratification is particularly well attuned to intensifying the recurrent desire for satisfaction on which the market depends. As Bauman (2007, 98) notes, 'For a … society which proclaims consumer *satisfaction* to be its sole motive and paramount purpose, a *satisfied* consumer is neither motive nor purpose – but the most terrifying menace'. A consumer society requires consumers constantly in search of satisfaction if it is to reproduce itself as consumer society. And whilst the hope of satisfaction may always be kept alive by its deferral (by the promise of future fulfilment), instant gratification is a far more effective means of ensuring the expanded reproduction of the system. Hence Wilde's quip that 'A cigarette is the perfect type of the perfect pleasure. It is exquisite, and it leaves one unsatisfied' (cited in Bowlby 1993, 7). Whilst some sources of satisfaction are, of course, less socially acceptable than they once were, the consumer society as a whole has become ever more committed to ensuring that the time-span of satisfaction is minimized; to accelerating the turnover time of short-lived satisfactions.

The appeal consumerism makes to the pleasure principle renders unsurprising the fact that spaces of consumption have frequently been figured in terms of the imagery of dreams and reveries (Williams 1982). There is, however, little mileage in recounting superficial resonances between spaces of consumption and the state of dreaming, which are inevitable compromised by early-twentieth-century commercial applications of pseudo-Freudian ideas by figures such as Ernest Dichter (Horowitz 1998). Moreover, whilst Freud regarded dreams as enacting the disguised fulfilment of an unconscious (repressed) wish, the open promise of wish-fulfilment staged by consumerism is a largely conscious affair. A clue to the significance of the relation between the unconscious and consumerism is, however, already suggested by Marx's invocation of the phantasmagoria – which, as Hetherington (2005, 191) reminds us, is named after Phantasos, the Greek god 'responsible for sending dreams to people of inanimate objects, or things'. As with all human activity, over and above its conscious meanings, consumerism is shot through with unconscious effects. The discovery that the unconscious accords to entirely its own logic is thus vital to understanding the workings of consumerism. For Freud, this unconscious logic is most vividly revealed by dreams – 'the royal road to a knowledge of the unconscious activities of the mind' (*SE* v, 608). The fundamental importance of the unconscious is that it concerns drives, which are not under the sway of the rational ego but which are not simply a matter of biological instinct either.[4] Drives are, says Freud (*SE* xiv, 121), 'on the frontier between the mental and the somatic', and necessarily involve mental images or ideas. As an example of the latter, Easthope (1999, 6) offers the experience of the instinctual suckling infant for whom '[a]n idea or image of the nipple (along with associations of fulfilment) becomes remembered, a signifier which can become pleasurable in its own right'. The extent to which unconscious drives influence the subject is clearly a pressing matter, and dreams – along with jokes, parapraxes, and the like – are the prime means of grasping the way in which the unconscious works.

Freud specifically cautioned against the temptation 'to find the essence of dreams in their latent content' (*SE* v, 506–7n). The recovery of the latent content of a dream (typically a repressed wish) from its manifest content (as dreamt, recalled or recounted by the dreamer) does not grant some form of direct access to the unconscious, for this is to overlook 'the distinction between the latent dream-thoughts and the dream-work' (ibid.). It is the mental activity or 'dream-work' responsible for transforming a dream's latent content into its manifest content that serves to disguise the wish that the dream satisfies, and this process of disguisement offers an insight into the logic of the unconscious. For Freud (ibid.), the process reveals that 'At bottom dreams are nothing other than a particular *form* of thinking, made possible by the conditions of the state of sleep'. As a particular form of thinking – albeit a highly particular one – dreaming necessarily involves ideas. Recall again that a drive 'cannot be represented otherwise than by an idea [*Vorstellung*]' – for if a drive 'did not attach itself to an idea or manifest itself as an affective state, we could know nothing about it', says Freud (*SE* xiv, 177). The fact that dreaming responds to psychic stimuli that are *repressed*, entails that

4 Unhelpfully, the English *Standard Edition* indiscriminately uses 'instinct' to translate two German terms clearly differentiated by Freud: *Trieb* ('drive') and *Instinkt* ('instinct'). Instincts are biological: I do not know about instinctual responses such as my pupils dilating in the dark. Drives, however, are cultural: they involve 'the psychical representative [*Repräsentant*] of … stimuli originating from within the organism and reaching the mind, as a measure of the demand made upon the mind for work in consequence of its connection with the body' (*SE* xiv, 121–2).

the 'ideas' involved in dreams are thought in terms other than those available to consciousness. Things happen in dreams that cannot happen in words: the ideas involved are in excess of the strictures of conscious discourse. This point underlies Freud's suggestion that the ideas dealt with in dreams attach to visual signifiers, to 'thing-presentations' (*Sachvorstellung*), rather than 'word-presentations' (*Wortvorstellung*): 'the conscious presentation comprises the presentation of the thing plus the presentation of the word belonging to it, while the unconscious presentation is the presentation of the thing alone' (*SE* XIV, 201). What this somewhat tenuously formulated distinction attempts to capture is the fact that dreams involve '*the transformation of a thought into an experience*' (*SE* XV, 129).

Without, at this stage, delving any further into the process, this already provides a basic indication of the way in which consumerism accords to the pleasure principle. For however much consumption may be thought of as concerning the satisfaction that may be derived from things, as Baudrillard (1998, 193) suggests, 'the only objective reality of consumption is the *idea* of consumption' – the dream or fantasy that things might offer fulfilment. This initial insight suggests a series of further resonances between consumerism and the various features of the unconscious as adumbrated by Freud. For example, it was Freud's contention that the law of non-contradiction is entirely absent from the unconscious. Dreams contain conflicting elements which nonetheless register the same plea, piling up contradictory notions that nonetheless point to the same underlying desire.

> The whole idea … remind[s] one vividly of the defence put forward by the man who was charged by one of his neighbors with having given him back a borrowed kettle in a damaged condition. The defendant asserted, first, that he had given it back undamaged; secondly that the kettle had a hole in it when he borrowed it; and, thirdly, that he never borrowed a kettle from his neighbor at all. So much the better if only a single one of these three lines of defense were to be accepted as valid the man would have to be acquitted. (*SE* V, 119–120)

Precisely the same logic permeates consumerism. At its heart lies a fundamental contradiction between the promise of satisfaction and the persistence of unfulfilment, the proffering of pleasure and its withdrawal. This is ramified into a million promissory advertising messages, simultaneously offering satisfaction via indulgence ('Naughty… but nice') and self-restraint ('Be good to yourself').[5] The underlying message, however, is *to enjoy* – though, true to form, this is expressed in the contradictory guise of an injunction: 'Enjoy!'[6] Similarly, Baudrillard (1987b, 44) notes that, 'In advertising, there are no negatives; even when irony's involved, everything is positive and affirmative. In present-day society, as in the images that bind it together, the negative is absent'. The currency of consumerism, like the

5 'Naughty… but nice': a slogan, reputedly written by Salman Rushdie, used to advertise cream-cakes. 'Be good to yourself:' the slogan of a healthy-choice food range.

6 'Imagine, putting pleasure in the imperative like that', writes Calle (1999, 246).

register of dreams, is essentially *figural*, a matter of 'thing-presentations'. Since an image of nothing is nonetheless present *qua* image, the unconscious operates without negation, absence or denial (*Verneinung*). The distinction between 'is' and 'is not' belongs to consciousness. It is entirely absent from the unconscious. In response to his patient's statement, 'You ask who this person in the dream can be. It's *not* my mother', Freud remarked, 'We emend this to: "So it *is* his mother"' (*SE* XIX, 235). Conscious negation is similarly absent – or, better, absented – from the world of consumerism, which is presented directly to consciousness as a wholly affirmative affair.

Whilst such features highlight the appeal consumerism makes to the pleasure principle, consumerism's ability to draw the pleasure principle and the reality principle into alignment ultimately depends on its ability to mesh with the structure of desire. It is this point of contact – and the necessity of fantasy that is induced by desire – that is, finally, crucial to consumerism. Here, Freud's conception of fantasy as unconscious wish-fulfilment is surpassed by Lacan's formulation. If, for Freud, fantasy is a means of satisfying an unconscious wish – something the subject is lacking is granted in fantasy – for Lacan, fantasy stages desire: it is responsible for inducing the very lack that makes the fantasy desirable. This retroactive structure replicates the way in which the subject loses something on entering meaning schematized in Lacan's *vel* of alienation: the power of coherent expression promises to make good the selfsame lack that is introduced by the subject's accession into language. Despite misrecognizing itself as capable of finding meaning directly present to consciousness, the subject necessarily emerges in the field of the Other (the pre-existing reserve of signifiers from which all meaning derives). However, insofar as the signifier that would be capable of expressing the subject in its plenitude is lacking, a constitutive lack is introduced into being. In Lacan's (1989, 86) words, 'I identify myself in language, but only by losing myself in it like an object'. It is this lack that 'constitutes in the subject the eternalization of his desire' (Lacan 1989, 104).[7] Desire is insatiable. It is an expression of the impossibility of refinding, within meaning, that part of being that has been surrendered upon entering language. The subject is, nonetheless, compelled to attempt to make good this lack. Consequently, desire attaches to a sequence of objects that hold out the impossible hope of repairing the lack in the subject. 'Desire is the metonymy of the want-to-be [*manque-â-être*]', says Lacan (1989, 259), since each object stands in as a representative of what Lacan designates algebraically as *objet petit a*. Since *objet petit a* does not exist – it is 'the excess of *jouissance* which has no "use value"' (Evans 1996, 125) – the sequence is interminable. To put it another way, *objet petit a* exists only in fantasy: the subject is compelled to fantasize that *objet petit a* does, in fact, exist. This is why Lacan (1977, ix) claims that 'the only conceivable idea of the object is that of the object as cause of desire, of that which is lacking'. Whilst *objet petit a* is figured as the

7 The formulation is not intended to be restricted to the masculine gender: 'subject' is masculine in French (*le sujet*), whence 'his desire'.

lost object capable of repairing the subject's lack, it is only once it is found that is it afforded, in fantasy, such a status: 'The object is by nature a refound object. That it was lost is a consequence of that – but after the fact. It is thus refound without our knowing, except through the refinding, that it was ever lost' (Lacan 1992, 118). Its status arises as 'a retrospective effect of the subject's fantasy outside which the object does not exist as a real cause' (Easthope 2002, 150). Accordingly, the object the subject seeks provokes desire but makes it impossible to fulfil. This is, of course, essentially compatible with the structure of consumerism. Rather than an instance of false consciousness, consumerism taps into the structure of desire that defines the modern subject.[8] This is reinforced if it is recognized that vision itself functions as a drive, thus occasioning desire (*Schaulust*, scopophilia).

It has become commonplace to note that consumerism is constantly engaged in an attempt to catch the eye of the consumer. That the visual sense should have been privileged is significant, for the characteristic property of the scopic drive is that, in fostering the impossible search for the lost object, it is dependent on maintaining a distance from the object. This lends it particular significance with respect to fetishism, which involves not the desire to see but the desire not to see. For Freud, fetishism is a response to the castration anxiety induced by the infantile understanding of sexual difference. A fetish object effectively masks the absence of the maternal phallus, yet nonetheless marks that absence by its presence. According to Metz (1985, 88), 'Freud considered fetishism the prototype of the cleavage of belief: "I know very well, *but* …"'. In other words, fetishism involves the kind of willing suspension of disbelief that Freud expressed in terms of disavowal (*Verleugnung*): a certain pleasure is attendant on maintaining a belief that the subject is nonetheless aware is fictive.[9] This will be seen to have far-reaching implications, to which we return in the final section. However, although Freud's anatomical formulation underlies accounts of the fetishization of women as the object of male scopophilia, in which consumerism is undoubtedly complicit (Bowlby 1985), it hardly tallies with the equally strong connections between women and shopping (Radner 1995). It renders implausible, moreover, any attempt to superimpose Freud's understanding onto Marx's account of commodity fetishism (Billig 1999). Most significantly of all, it also curtails a fuller understanding of the scopic drive, which is generally – and not only in relation to fetishization as an aspect of visual mastery – caught up in the subject's desire to imagine itself complete. As Lacan

8 Lacan's formulation is far from a universal model and refers specifically to the centrality afforded to the ego since 'the middle of the sixteenth, beginning of the seventeenth centuries' (Lacan 1988, 7). The modern ego is characterized by fantasies of autonomy and mastery, and a self-deceptive narcissism.

9 Fetishism represents, for Freud, a perversion in the technical sense of that term, which designates deviation with respect to the aim of sexuality – though, importantly, Freud (*SE* VII, 171) regarded 'the disposition to perversions' as forming 'part of what passes as the normal constitution'. Sexuality proceeds by detour, deferral, diversion, and displacement – and so too does consumption.

(1977, 72) notes, as a subject, 'I see only from one point, but in my existence I am looked at from all sides'. The subject is necessarily situated within the visual field of the Other, yet the vantage point of the Other remains permanently unavailable to the subject. Insofar as I desire to see myself through the eyes of the Other, the gaze of the Other itself functions as *objet petit a*, provoking desire whilst making it impossible to fulfil. In consequence, the subject must attempt to negotiate the castrating gaze of the Other in order to render whatever lies within the visual field of the subject available for possession. If this desire is revealed in the displays of abundance and profusion characterizing spaces of consumption, it is notable that those who are enjoined to look are also the object of the gaze of a variety of representatives of the Other, from store detectives to a proliferation of prosthetic eyes. From the off, the profligacy of display gave rise to temptation and multiplied the opportunities for indulging it (Abelson 1989).

Whilst Freud's anatomical formulation restricted castration anxiety to boys (Freud detecting a predominance of fetishism amongst men), Lacan's (1989, 282) reformulation holds that castration anxiety is 'established without regard to the anatomical difference of the sexes'. For Lacan, castration anxiety derives from the disruption of the preoedipal bond, wherein the child identifies itself with the phallus it imagines to be the (symbolic, not biological) mother's object of desire. Castration amounts to the removal of this possibility by the (symbolic) father: the child can no longer identify with the imaginary phallus if it becomes apparent that the mother's desire goes beyond the child. The signifier that the child imagines would identify it is, of necessity, an imaginary signifier: the discovery, on the part of the subject, of the absence of the signifier that would adequately express it amounts to the discovery of a lack in the Other. It is this absence that Lacan equates to castration, to the subject's renunciation of the attempt to be the phallus, and this paves the way to Oedipalization. 'Castration means that *jouissance* must be refused, so that it can be reached on the inverted ladder (*l'échelle renversée*) of the Law of desire' (Lacan 1989, 324). Yet the Law of desire regulates by way of fantasy. Whilst the disavowal of the lack in the Other that defines fetishism entails the subject attempting to evade the gaze of the Other, selecting a fetish object to stand in for the imaginary phallus, the normalization of sexual difference attempted by Oedipalization entails that the lack in the subject induced by the lack in the Other is sutured in the imaginary. The subject maintains the pretence that the symbolic phallus, as 'the signifier of the desire of the Other' (Lacan 1989, 290), may be embodied in another subject: that another can possess what the subject desires and that the subject can possess what the other desires. Unable to evade the gaze of the Other, the subject engages in the fantasy of being looked at in the way the subject sees himself, as if through the eyes of the Other. Yet only within fantasy is there another who embodies *objet petit a*, the lost object capable of repairing the subject's constitutive lack; of replicating the dyadic bond experienced before the intervention of the Other. Only in fantasy is *jouissance* obtainable. Outside of fantasy, as Lacan (1982, 143) insists, 'there is no sexual relation' (*il n'y a pas de rapport sexuel*). In the real, the subject does not meet the demands of the other, nor are the other's demands met by the subject. Whilst fetishism amounts to an attempt to evade the impossibility of a sexual relation, the fantasy that a sexual relation is possible is played out in love. Love is as Lacan (1991, 147) suggests, giving 'what one does not have'. When Miller (1998) proposes that consumption is caught up in relations of love, he is not far wide of the mark – except that Miller's formulation simply reproduces, rather than analyses, the fantasy of love that is constantly replayed in the references to love that saturate contemporary culture. If there is one thing that consumerism does, it is to constantly reinforce the fantasy that the subject can have it all. Love is, undoubtedly, caught up in the desultory promise that nothing need stand between us and our fulfilment. But, as the final section proceeds to demonstrate, it is the perverse structure of desire encapsulated in fetishism that has the most fundamental bearing on the way in which spaces of consumption promote the fantasy that the subject can evade the real of its being by buying into the promise that satisfaction can be guaranteed. On this account, one should not forget that 'the failure [*défaillance*] of the real is the basis for the reality principle' (Baudrillard 2008), and that the harnessing of the pleasure principle to the reality principle is the continuation by other means of the same forced realization of the world.

Through the Looking-glass

> *Irreality no longer belongs to the dream or the phantasm, to a beyond or a hidden interiority, but to the hallucinatory resemblance of the real to itself.*
>
> Baudrillard (1993b, 72)

The very notion of fetishism, as Baudrillard (1981, 90) notes, has a habit of turning against those who use it: 'Instead of functioning as a metalanguage for the magical thinking of others, it ... surreptitiously exposes their own magical thinking'. Any analytical usage of the term today thus depends on disentangling it from a variety of theoretical frameworks in order to expose its inner logic. The most obvious danger lies in the fact that metaphorical invocations of the term – understood in the classical sense as the endowment of objects with a supernatural force 'in which the subject projects himself and is alienated' (ibid., 91) – retain the traces of its rationalist, humanist origins. The earliest, Christian uses of the term involved the 'condemnation of primitive cults by a religion that claimed to be abstract and spiritual' (ibid., 88). Consequently, wherever the term is called upon, it tends to produce 'a fetishization of the conscious subject or of a human essence' insofar as it 'presupposes the existence, somewhere, of a non-alienated consciousness of an object in some 'true', objective state' (ibid., 89). The Marxian account falters on precisely these grounds. It is one thing to say that 'Production ... not only creates an object for the subject, but also a subject for the object' (Marx 1973, 92) – that the human species creates itself in creating its world – and to acknowledge that this relation has become a source of alienation. It is quite another to propose that this need not be the case; that a non-alienated existence is possible. As Baudrillard (1998) proposes, Marxism merely reproduces the central myth that has dominated Western culture since the Middle Ages: the myth of a pact with the Devil. Enormous wealth accrues from selling one's soul. Having sold one's soul, however, one loses sight of oneself, and one's image no longer appears in the mirror.

> The mirror image here symbolically represents the meaning of our acts. These build up around us a world that is *in our image*. The transparency of our relation to the world is expressed rather well by the individual's unimpaired relation to his image in the mirror: the faithfulness of his reflection bears witness, to some degree, to a real reciprocity between the world and ourselves. Symbolically, then, if that image should be missing, it is a sign that the world is becoming opaque, that our acts are getting out of control and, at that point, we have no perspective on ourselves. Without that guarantee, no identity is possible any longer: I become another to myself; I am *alienated*. (Baudrillard 1998, 188)

It is on this basis that Marxism has held on to the diagnostic potential of the notion of fetishism. Only psychoanalysis has escaped the fate of the majority of invocations of fetishism 'by returning fetishism to its context within a perverse *structure* that perhaps underlies all desire' (Baudrillard 1981, 90). This is not simply because, since Lacan (1989, 6), psychoanalysis has recognized that there is no essence of the subject; that the self is necessarily an effect of its reflection in the Other, and that the imaginary unity conveyed by the image in the mirror relates to 'the *méconnaissances* that constitute the ego'. It is also because of the particular way in which psychoanalysis has conceived of fetishism, which has a broader bearing than on its clinical definition alone.

The psychoanalytic definition involves a crucial recognition of the nature of fetishism; one that has been lost in the majority of uses of the term. For what matters to the operation of fetishism is not the substance of the fetish object but its form: 'the perverse psychological structure of the fetishist is organized, in the fetish object, around a mark, around the abstraction of a mark that negates, bars and exorcises the difference of the sexes' (Baudrillard 1981, 92). Although, from its original condemnatory usage, the term fetishism rapidly acquired the sense of an object possessed of supernatural powers, and hence capable of dispensing happiness, health, prestige, security, etc. – a sense that is still operative, albeit critically, in the Marxian discourse – the etymology of the term fetishism originally 'signified exactly the opposite: a *fabrication*, an artefact, a labour of appearances and signs' (ibid., 91). It is this sense 'that is at the origin of the status of the fetish object, and thus also plays a part in the fascination it exercises' (ibid.). Thus, in a manner homologous to the fetish object as understood by psychoanalysis, the object of consumption hinges upon

> the (ambivalent) fascination for a form (the logic of the commodity or system of exchange values), a state of absorption, for better or worse, in the restrictive

> logic of a system of abstraction. Something like a desire, a perverse desire, the desire of the code is brought to light here: it is a desire that is related to the systematic nature of signs, drawn towards it, precisely though what this system-like nature negates and bars, by exorcising the contradictions spawned by the process of real labour. (Ibid., 92)

This recognition is vital if fetishism is to escape its humanist implications, for its fundamental importance lies in its seductive potential, which relates purely to its form. 'Thus, fetishism is actually attached to the sign-object, the object eviscerated of its substance and history, and reduced to the state of marking a difference, epitomizing a whole system of differences' (ibid., 93). It is this characteristic that is responsible for the seductive power exerted by consumerism.

According to Freud (*SE* XIV, 87), 'The charm of a child lies to a great extent in his narcissism, his self-contentment and inaccessibility, just as does the charm of certain animals which seem not to concern themselves about us, such as ... large beasts of prey'. Seductiveness, in other words, is directly proportionate to inaccessibility and indifference – such that what fascinates us most is 'that which radically excludes us in the name of its internal logic of perfection' (ibid., 96). Fetishism, therefore, takes on its full force at the point at which, undercut by its own overexposure, the artifice of the real comes into play – not in terms of any divergence from reality but in terms of a diversion of reality – insofar as it permits the fascination elicited by the perfection of a system defined by its own systematicity to take hold. In its radical form, this is what fetishism is – and it attains its consummate form in spaces of consumption:

> There is no longer any mirror or looking-glass in the modern order in which the human being would be confronted with his image for better or worse; there is only the *shop-window* – the site of consumption, in which the individual no longer produces his own reflection, but is absorbed in the contemplation of multiple signs/objects, is absorbed into the order of signifiers of social status, etc. He is not reflected in that order, but absorbed and abolished. (Baudrillard 1998, 192)

Acknowledgements

I am grateful to the editors, to Timon Beyes, and to Marcus Doel for their perceptive comments.

References

Abelson, E. S. (1989), *When Ladies Go A-thieving: Middle-class Shoplifters in the Victorian Department Store* (Oxford: Oxford University Press).

Adorno, T. and Horkheimer, M. (1979 [1946]), *Dialectic of Enlightenment*, trans. Cumming, J. (London: Verso).

Agamben, G. (2000), *Means without End: Notes on Politics*, trans. Binetti, V. and Casarino, C. (Minnesota: The University of Minnesota Press).

Althusser, L. (1996 [1964]), 'Freud and Lacan' in Corpet, O. and Matheron, F. (eds), trans. J. Mehlman, *Writings on Psychoanalysis: Freud and Lacan* (pp. 7–32) (New York: Columbia University Press).

Baudrillard, J. (1981), *For a Critique of the Political Economy of the Sign*, trans. Levin, C. (St. Louis: Telos).

Baudrillard, J. (1987a), *Forget Foucault*, trans. Dufresne, N. (New York: Semiotext(e)).

Baudrillard, J. (1987b), 'Softly, Softly', trans. Imrie, M., *New Statesman* 113, 6 March, 44.

Baudrillard, J. (1990), *Seduction*, trans. Singer, B. (London: Macmillan).

Baudrillard, J. (1993a), *Baudrillard Live: Selected Interviews* (ed.), Gane, M. (London: Routledge).

Baudrillard, J. (1993b), *Symbolic Exchange and Death*, trans. Hamilton Grant, I. (London: Sage).

Baudrillard, J. (1998), *The Consumer Society: Myths and Structures*, trans. Turner, C. (London: Sage).

Baudrillard, J. (2005), *The Intelligence of Evil or the Lucidity Pact*, trans. Turner, C. (Oxford: Berg).

Baudrillard, J. (2008), 'The Vanishing Point of Communication' in Clarke, D. B., Doel, M. A., Merrin, W. and Smith, R. G. (eds), *Jean Baudrillard: Fatal Theories* (pp. 15–23) (London: Routledge).

Bauman, Z. (1983), 'Industrialism, Consumerism and Power', *Theory, Culture and Society* 1: 3, 32–43.

Bauman, Z. (1997), 'The Haunted House', *New Internationalist* April, 24–25.

Bauman, Z. (2002), *Society Under Siege* (Cambridge: Polity).

Bauman, Z. (2007), *Consuming Life* (Cambridge: Polity).

Billig, M. (1999), 'Commodity Fetishism and Repression: Reflections on Marx, Freud and the Psychology of Consumer Capitalism', *Theory & Psychology* 9: 3, 313–329.

Bowlby, R. (1985), *Just Looking: Consumer Culture in Dreisser, Gissing and Zola* (London: Methuen).

Bowlby, R. (1993), *Shopping with Freud* (London: Routledge).

Calle, S. (1999), *Double Game* (London: Violette Editions).

Derrida, J. (1994), *Spectres of Marx: The State of the Debt, the Work of Mourning, and the New International*, trans. Kamuf, P. (London: Routledge).

Easthope, A. (1999), *The Unconscious* (London: Routledge).

Easthope, A. (2002), *Privileging Difference* (Basingstoke: Palgrave).

Evans, D. (1996), *An Introductory Dictionary of Lacanian Psychoanalysis* (London: Routledge).

Freud, S. (1953–74), *The Standard Edition of the Complete Psychological Works of Sigmund Freud*, ed. and trans. Strachey, J., 24 Volumes (London: Hogarth Press and the Institute of Psycho-analysis).

Fromm, E. (1970 [1932]), 'The Method and Function of an Analytical Social Psychology: Notes on Psychoanalysis and Historical Materialism' in *The Crisis of Psychoanalysis. Essays on Freud, Marx, and Social Psychology* (pp. 135–162) (New York: Holt, Rinehart and Winston).

Geras, N. (1971), 'Essence and Appearance: Aspects of Fetishism in Marx's *Capital*', *New Left Review* 65, 69–85.

Harootunian, H. (2004), 'Karatani's Marxian Parallax', *Radical Philosophy* 127, 29–37.

Hetherington, K. (2005), 'Memories of Capitalism: Cities, Phantasmagoria and Arcades', *International Journal of Urban and Regional Research* 29, 187–200.

Horowitz, D. (1998), 'The Émigré as Celebrant of American Consumer Culture: George Katona and Ernest Dichter' in Strasser, S., McGovern, C. and Judt, M. (eds), *Getting and Spending: European and American Consumer Societies in the Twentieth Century* (pp. 149–166) (Cambridge: Cambridge University Press).

Jameson, F. (1981), *The Political Unconscious* (London: Methuen).

Lacan, J. (1977), *The Four Fundamental Concepts of Psycho-analysis* (ed.), Miller, J.-A., trans. Sheridan, A. (Harmondsworth: Penguin).

Lacan, J. (1982), *Feminine Sexuality* (eds), Mitchell, J. and Rose, J., trans. Rose, J. (New York: Norton).

Lacan, J. (1988), *The Seminar of Jacques Lacan. Book II. The Ego in Freud's Theory and in the Technique of Psychoanalysis 1954–1955* (ed.), Miller, J.-A., trans. Tomaselli, S. (Cambridge: Cambridge University Press).

Lacan, J. (1989), *Écrits: A Selection*, trans. Sheridan, A. (London: Tavistock/Routledge).

Lacan, J. (1991), *Le Séminaire. Livre VIII. Le transfert, 1960–61* (ed.), Miller, J.-A. (Paris: Seuil).

Lacan, J. (1992), *The Seminar of Jacques Lacan. Book 7. The Ethics of Psychoanalysis, 1959–1960* (ed.), Miller, J.-A., trans. Porter, D. (London: Tavistock/Routledge).

Laplanche, J. and Pontalis, J.-B. (1973), *The Language of Psycho-analysis*, trans. Nicholson-Smith, D. (London: Hogarth Press and the Institute of Psycho-analysis).

Lefebvre, H. (1991), *The Production of Space*, trans. Nicholson-Smith, D. (Oxford: Blackwell).

Marcuse, H. (1966 [1955]), *Eros and Civilization: A Philosophical Inquiry into Freud* (Boston: Beacon Press).

Marx, K. (1973), *Grundrisse: Foundations of the Critique of Political Economy*, trans. Nicolaus, M. (Harmondsworth: Penguin).

Marx, K. (1983), *Capital: A Critique of Political Economy Volume 1*, trans. Moore, S. and Aveling, E. (London: Lawrence and Wishart).
Metz, C. (1985), 'Photography and Fetish', 34: *October*, 81–90.
Miller, D. (1998), *A Theory of Shopping* (Cambridge: Polity).
Miller, D., Jackson, P., Thrift, N. J., Holbrook, B. and Rowlands, M. (1998), *Shopping, Place and Identity* (London: Routledge).
Poe, E. A. (1985), 'The Purloined Letter' in *The Complete Tales and Poems of Edgar Allan Poe* (pp. 185–197) (Edison: Castle Books).
Radner, H. (1995), *Shopping Around: Feminine Culture and the Pursuit of Pleasure* (London: Routledge).
Rancière, J. (2004a), *The Politics of Aesthetics: The Distribution of the Sensible*, trans. Rockhill, G. (London: Continuum).
Rancière, J. (2004b), 'The Politics of Literature', *SubStance* 33: 1, 10–24.
Reich, W. (1972 [1929]), 'Dialectical Materialism and Psychoanalysis' in Baxandall, L. (ed.), *Sex-Pol: Essays, 1929–1934* (pp. 1–74) (New York: Random House).
Rorty, R. (1992), 'We Anti-representationalists', *Radical Philosophy* 60, 40–42.
Rose, G. (1978), *The Melancholy Science: An Introduction to the Thought of Theodor W. Adorno* (New York: Columbia University Press).
Williams, R. H. (1982), *Dream Worlds: Mass Consumption in Late Nineteenth-century France* (Berkeley: University of California Press).
Žižek, S. (1989), *The Sublime Object of Ideology* (London: Verso).
Žižek, S. (1994), 'How Did Marx Invent the Symptom?' in Žižek, S. (ed.), *Mapping Ideology* (pp. 296–331) (London: Verso).
Zola, É. (2001), *Au Bonheur des Dames (The Ladies' Delight)*, trans. Buss, R. (Harmondsworth: Penguin).

PART I
The Consumption of Space and Place

Chapter 4

Frontier Spaces of Production and Consumption: Surfaces, Appearances and Representations on the 'Mayan Riviera'

Michael Redclift

Introduction

On 12 September 2001 the massed taxi ranks at Cancun airport lay abandoned. After the appalling events of 9/11 nobody was arriving in Mexico from the United States, and the absence of tourists was to remain a problem for the tourist industry in Cancun and the 'Mayan Riviera' to the south for several more months. This is an area constantly exposed to hurricanes and tropical storms, and one familiar with the need to adapt rapidly to them. September 11th was a different order of 'disaster', and it took fully two years for this area of mass tourism to fully adapt to the consequences. However, by March 2004 tourism was 'booming' again, and in March 2004 the Mexican Government reported a record monthly income of one billion US dollars from foreign tourists. The Yucatan Peninsula, and particularly the Mexican Caribbean, remained the main tourist destination for international tourists to Mexico, four-fifths of whom are American. The events subsequent to 9/11 also raise, in a particularly vivid way, some of the peculiarities of thinking about 'space' in the context of consumption in general, and tourism in particular.

One example is the burgeoning effects of cruise ships and new cruise ports. In the wake of 9/11 there was an accelerated trend away from room occupancy in 'all-inclusive' hotels, and towards more passengers on cruise ships. There was a significant increase in cruise passengers: almost two million cruise passengers came ashore in Mexico in 2004, an increase of two hundred thousand over the previous year. Cruise ships represent a particular kind of 'space', within the lexicon adopted in this volume: secure and hermetically sealed, and allowing only limited contact with 'real' Mexicans. Cruise ships are also constantly, if not continuously, on the move; in this respect at least they represent the apotheosis of mobile 'space', and of 'place' as humanly occupied space (Lefebvre 1991). With their in-built security, international cuisine, and 24 hour access to entertainment and pleasure, the cruise ship provides a balm against the unpredictable horrors of alien cultures, as well as those of September 11th. At the same time cruise ships provide an example of a wider phenomenon which this chapter explores: the relationship between *physical space* and its *cultural assimilation.*

In this chapter I want to explore space and its relationship to consumption through a narrative account of one geographical space (the Mexican Caribbean coast) that demonstrates some of the ways in which distance and time can be compressed, producing 'layered' histories that tell us as much about we the 'historians', who construct our spaces, as about the spaces themselves. The chapter also examines some of the *hybridizations*, through which nature and society meet and refashion space. Following Lefebvre (1991) we seek to explore 'what lies beneath the surfaces of appearance'.

As economic relations develop, we both produce and reproduce nature, frequently converting it into an object of consumption. In the process we reconfigure both space and place, through refashioning them both culturally and geographically. Lefebvre makes the point that 'space' has always represented a way in which capitalism can represent itself, providing the images and (more lastingly) the imaginaries of modernity and post-modernity in today's icons, brands and logos. This process is a cultural one endemic to capitalism, since the production of commodities and the processes of commoditization are always accompanied by images, narratives and texts. Space, then, can be seen as an active location, rather than a passive one, a site for social and political activity and the means through which economic and political systems re-establish their hegemony and their wider legitimacy.

Recent research, particularly in geography and history, have benefited from a more reflective view of space, and an active search for its properties and significance over time. Space is no longer a 'given' in intellectual history, the blank parchment on which human purposes are written. Some writers even argue that space should be seen principally as enactment or performance: constructions of the human imagination, as well as materiality. In the view of Nicholas Blomley, for example, 'space (is present) in both property's discursive and material enactments' (Blomley 2003). Space like property, is active, not static (and), 'spaces of violence must be recognised as social achievements, rather than as social facts' (Blomley 2003, 126). Space thus assumes a role previously denied it, and performs a transitive role in the making of historical events.

This 'active', transitive conceptualization of space carries implications for the way in which we view resource peripheries, particularly within the context of 'globalization', a process that is increasingly seen as pre-dating modernity, rather than as an outcome of it. Geographical frontiers are ascribed, figuratively, temporally and spatially, in ways that serve to influence succeeding events. Their 'discovery' and 'invention' are acknowledged as part of powerful myths, which are worked and re-worked by human agents, serving to create environmental histories as important as the material worlds that they describe.

Elsewhere I have suggested that the re-working of space in cultural terms consists of separate but linked processes: the analogue, digital and virtual descriptions of space (Redclift 2006). Each of these provides a different construction of space and in the Mexican Caribbean is associated with distinctive 'pioneer' generations of settlers. In charting the resource histories of places, and the histories of the visitors and tourists who have 'discovered' them, we are engaged in continually re-working a narrative. The social processes through which we come to identify space over time resemble a series of 'successions' or 'layers' (Jones 2003; Martins 2000; Salvatore 1996).

The creation of existential spaces, as part of the fabric of environmental history, is seen clearly in the accounts of the Caribbean coast of Mexico, today's state of Quintana Roo. Over time we see: first, a 'wilderness', discovered by archaeologists, second, a 'natural resource' frontier of *chicle* extraction for the manufacture of chewing gum, third, an 'abandoned space' identified and exploited by early tourist entrepreneurs, and fourth, a 'tropical paradise' promising escape to international tourists, and ultimately turning nature into a commodity, as theme park, leisure complex and cruise liner.

There is also a darker side to this space of consumption and production, represented by the legacy of the Caste War, one of the great indigenous revolutions of the nineteenth century, which brought the 'rebel' Maya into conflict with their white masters. This conflict still resonates in the region to this day and constitutes another part of the narrative.

Space as Wilderness

The tourist 'pioneers' of the mid twentieth century were beating a track that had been followed by earlier pioneers, the most famous of whom were John Stephens and Frederick Catherwood, the 'giants' of Mayan archaeology in the mid nineteenth century. Stephens and Catherwood had already explored the major Mayan sites of northern Yucatan, such as Chichen Itza and Uxmal, and arrived in Valladolid at the end of March 1841. They made enquiries about getting to the Caribbean coast, no mean feat at the time since there were no roads. 'It is almost impossible to conceive what difficulty we had in learning anything definite concerning the road we ought to take', Stephens reported to his diary (Stephens 1988).

The coastal location that they aimed for was the settlement of Tankah, where a pirate named Molas had sought to evade the authorities in Merida, where he had been convicted of smuggling. Since there was no road they had to journey to the northern (Gulf) coast and take a 'canoa' down to the Caribbean, past today's Cancun and Isla Mujeres, to the Mayan fortress of Tulum. The journey took them two weeks, and was accomplished despite every privation known to explorers of the time: no wind, no protection against the sun, so much provisioning that there was no space for the human occupants, and little idea of where they were headed. Stephens says their objective was '… in following the track of the Spaniards along this coast, to discover vestiges or remains of the great (Mayan) buildings of lime and stone (that had been reported) …'.

They sailed first past Isla Mujeres, or 'Mugeres' as Stephens described it, an island notorious as the resort of Lafitte, another pirate who (rather like Molas) was well regarded by the Mayan fishing communities of the coast, and 'paid them well for all he took from them…'. Next was Cancun, or Kancune, as Stephens described it, which left a very poor impression on the travellers. It was nothing but 'a barren strip of land, with sand hills, where the water was so salt we could barely drink it …' Whenever they landed, usually in search of water, they were pursued by hordes of 'moschetoes', that made life difficult, and would continue to have done so 130 years later, if the Mexican Government had not intervened and sprayed them into oblivion.

They went on to land on Cozumel, at the only inhabited spot, the ranch of San Miguel, where they record that 'our act of taking possession was unusually exciting'. Here they stopped to feast on turtle and fresh water, strolled along the shoreline picking up shells, and went to sleep in their hammocks, 'as piratical a group as ever scuttled a ship at sea'.

The island of Cozumel had been 'discovered' several times before; once 'by accident' it is said, when Juan de Grijalva caught sight of it in March 1518. He had set sail from Cuba. Unlike Grijalva, three centuries earlier, John Stephens knew where he was in 1841 and noted for the benefit of the 'Modern Traveller' that they alone had proprietorship of 'this desolate island'.

As we shall see Cozumel reappears a century later, in the 1950s, as the location of some of the first successful tourist enterprises on the Mexican Caribbean coast. The intervening century, however, saw the area developed for quite different purposes: as part of the enclave economies associated with hardwoods and *chicle*, the resin from which chewing gum was made.

Space as a Natural Resource Frontier: Chicle and Chewing Gum

The 'boom' in *chicle* production, to meet North American consumer demand, began during the first two decades of the twentieth century, and reached its peak in the early 1940s. *Chicle*, the raw material from which chewing gum was derived, came from the Yucatan Peninsula and Central America, where the *chicozapote* tree grew in the high, tropical forests. The demand for *chicle* from the United States, served eventually to transform the landscape and ecology of the east and south of the Yucatan peninsula of Mexico, and paved the way for new land uses on the tropical frontier. It led to harvesting and production practices which are of contemporary importance, especially for protected tropical forest areas, in which forest products represent a growing market activity.

Most consumers in the twentieth century were doubtless oblivious of its origins, but nevertheless, by stimulating these distant commercial links chewing gum illustrates the way in which 'nature' is actively produced as both material artefact and discursive construct (Bridge and Jonas 2002).

Research has recently emphasized the way in which consumer markets, especially for products of extractive industries, are linked in complex ways with environmental and other policies (Simonian 1995; Bridge 2001; Redclift 2001). The areas from which raw materials are sourced have been described as 'the marginal spaces in, and through which, broader processes of socio-spatial order are worked out' (Bridge 2001, 2149). Indeed, it is suggested that today these spaces are rendered even more marginal by the prospect of plenty: 'already rendered distant, shadowy spaces by the value of the commodity chains, these commodity supply zones are pushed further out of sight by the emergence of a post-scarcity discourse that celebrates material abundance' (Bridge 2001, 2153). In the case of chewing gum its close association with the values of the twentieth century: leisure, independence and private indulgences, seem almost to be precursors of the 'post-scarcity' and 'post-material' age.

The impact of the enormous surge in consumption during the 1930s and 1940s, and the later depression in sales, when synthetics derived from hydrocarbons replaced the natural gum base, was felt particularly acutely in the east of the Yucatan Peninsula. Here, early production had been associated, like many extractive forest products, with transient labour working under onerous conditions and in an unregulated fashion, like so many 'informal sector' activities today.

Most of the first commercial *chicleros* (tappers) were natives of Veracruz on the Gulf coast, and they often arrived in the Yucatan Peninsula by boat after dangerous sea crossings. They worked under contract to men who provided the equipment for tapping gum, and lived for six months of the year (the wet season from June to December) in camps located deep in the mature tropical forest. Working in groups of about a dozen men in each camp, they tapped the milky white resin from the *chicozapote* trees within range of their camp. Using rope and a machete, they climbed these trees, cutting zigzags in the bark and collecting the tasteless resin in cups underneath. This was then boiled in vats until it had congealed, and could be transported in 'bricks' on mule-back. The contractors were allocated areas of forest for tapping, or entered it illegally, for there were few workable laws in what was very much a frontier area.

The principal zone of production was a stronghold of rebel Maya chieftains, veterans of the Caste War between whites and Mayan followers of the 'Talking Cross'. Their leader in the south of the peninsula, until 1931, was the notorious 'General' May, who had developed close relations with American gum manufacturers, such as Wrigley's, and whose revenues from *chicle* helped to fund armed opposition (Ramos Diaz 1999; Reed 2001). However, the containment, and suppression, of the rebel Maya, and the enlarged role of the Mexican state, especially under President Cardenas in the 1940s, brought the harvesting of *chicle* within the compass of organized cooperatives, and increasing measures of state regulation. In 1942 nearly four million kilos of *chicle* from Yucatan was sold to four large American-owned companies: Beechnut, Wrigley's, American Chicle Co. and Clark Bros. The commercial, and strategic, importance of these sources, at their height, can be gauged from the fact that, in June 1943 representatives of *chicle* cooperatives travelled to the United States to 'discuss and defend the price of *chicle*, one of the most appreciated wartime materials in the United States' (Encyclopaedia 1998, 101).

During the 1940s and 1950s the Mexican Government sought to control both the production and the export of gum, through the Agricultural Ministry and the Banco de Comercio Exterior. *Chicleros* were encouraged to organize themselves into marketing cooperatives and greater controls were exercised over their production by the Federal Government determined to 'settle' the forest frontier of Quintana Roo (and, by the late 1960s, to pave the way for mass tourism on the Caribbean coast south of Cancun). Most of the trees from which the resin was tapped, grew on land held by *ejidos* (peasant communities) or on federal lands making them a common property resource. Access to the forests, which was once governed by tradition and personal influence, became officially regulated. Production of *chicle* was increasingly managed through establishing production quotas and targets, and using more competitive tendering.

This period of state regulation, however, did nothing to reverse the fortunes of the industry. By the 1970s a forest industry that was potentially sustainable ecologically, and capable of providing livelihoods for poor families without causing wide-scale forest destruction, was in sharp decline, and secondary to the demands of global tourism (Primack et al. 1998).

Chewing gum sourced from *chicle* replaced products that were also native to the indigenous cultures of the Americas, notably spruce gum. However, because *chicle* was sourced from the Yucatan Peninsula, several thousand miles from its main market in urban America, its origins were almost invisible to those who consumed it. It appeared, like other manufactured commodities, to have come into being to meet a need of consumers, rather than a livelihood for producers. Few commodities were more material; but because of the distance (culturally as well as geographically) that separated consumers from producers, and the form taken by its commercial transformation into 'product', chewing gum was also invisible.

It was invisible, but not abandoned. The boom in *chicle* production eventually gave way to other forms of production and consumption, notably in the development of international tourism on an altogether more ambitious scale.

'Abandoned Spaces' and the Early Tourist Entrepreneurs

The coast of today's state of Quintana Roo had never been fully 'abandoned' by the Spanish, although the distance from Merida, made it difficult for them to govern the area effectively. Before the Conquest this part of the coast had been among the most densely settled areas of the Mayan world, a fact that was commented on by the Spanish ships which first observed the Mayan city of Tulum, in the sixteenth century. However, after the Conquest the population was decimated by war, epidemics introduced from Europe, and the gradual movement of much of the population towards the interior of the Yucatan Peninsula. After the Caste War, in 1851, the whole coastal zone was converted into a refuge for those Mayans fleeing bondage on the henequen plantations to the north (and, in the case of the island of Cozumel, people fleeing the 'rebel' Maya). It was then left to English hardwood traders and buccaneers, and settled by indigenous fishing communities (Andrews and Jones 2001).

One of the 'abandoned spaces' of the Mexican Caribbean was eventually renamed as the coastal resort of Playa Del Carmen, today one of the most rapidly urbanizing coastal settlements in Latin America. Playa was not 'discovered' until the summer of 1966, according to one account in a tourist magazine:

> Playa was discovered by a sixteen year old boy, in the summer of 1966. A momentous event, which changed forever the face of history for this small fishing village ... In 1966 Fernando Barbachano Herrero, born of a family of pioneers, arrived there and found it inhabited by about eighty people, with a single pier made of local (chico) zapote wood. Fernando befriended the local landowner, Roman Xian Lopez, and spent the next two years trying to talk him into relinquishing some of his land ... (*Playa Magazine*, August 1999, 7)

Two years later, in 1968, Fernando Barbachano bought 27 hectares of this land adjacent to the beach for just over $13,000 (US), or six cents a square meter. In 2003 it was worth about $400 (US) a square metre, an increase of over 6,000 per cent.

Today this piece of real estate constitutes less than 10 per cent of Playa's prime tourist development. As Playa developed, piers were built for the increasing number of tourist craft, and game fishers, hotels and bars were constructed fronting the 'virgin' beach, and clubs were opened a short way from the shoreline. The first hotel to be constructed was Hotel Molcas, in the 1970s, next to the little ferry terminal to Cozumel. Today the town possesses shopping malls, selling designer clothes and global brands. International gourmet restaurants compete for the lucrative tourist business; over twenty million tourists visit Mexico today. The beaches draw migrants from all over Mexico, particularly the poorer states such as Chiapas, and the town's hinterland contains squatter settlements as large as any in urban Latin America. These areas have names which sometimes suggest wider political struggles such as 'Donaldo Colosio', a 'squatter' area named after a prominent politician in the PRI (Party of the Institutional Revolution) who was murdered in 1994 in Tijuana by a crime syndicate.

Tourist 'pioneers' had taken an interest in the Mexican Caribbean coast even before Fernando Barbachano stumbled upon the resort potential of Playa Del Carmen. In the longer view tourist expansion on the coast of Quintana Roo can be compared with the trade in dyewood three hundred years ago, or of mahogany and chicle, the raw material for chewing gum, during the last century. All three were milestones in the development of the region, and linked it with global markets and consumers. Each possessed their own 'pioneers', like Fernando Barbachano, who 'discovered' a land of rich natural resources, apparently unworked by human hand. To some extent, however, these timber and gum pioneers not only paved the way for tourism; they re-entered the story at a later date as pioneers of tourism themselves. It is worth recalling that the account of Playa's 'discovery' in the passage above refers to a 'single pier made of local *zapote* wood …'. Chicozapote was, of course, the tree from which *chicle* (chewing gum resin) was tapped. The tourist industry in turn occupied what had become an 'abandoned space' after the demise of *chicle* production in the 1960s.

The island of Cozumel, which lies opposite Playa Del Carmen geographically, was one of the first pioneer tourist zones on the coast, and provides an early illustration of the way in which tourist economies can develop successfully in highly demarcated spaces, such as tropical islands. The Grand Hotel Louvre on Cozumel, owned by Refugio Granados, had been constructed in the 1920s. Advertised in the Revista de Quintana Roo, in 1929, the owners publicised its merits in the following terms:

> Tourists, tourist, tourists, travellers and travel agents! If you want a well-ventilated room and are demanding of the very best in attention, come to the Gran Hotel Louvre. In addition it has a magnificent restaurant attached. Set meals and a la carte meals are available in a constantly changing menu. Expert chef. Calle Juarez with Zaragoza. Proprietor Refugio Granados. (Dachary and Burne 1998, 394)

Between the late 1920s and 1940s two other hotels were built on Cozumel, the Yuri and Playa, but at this time most visitors to what are today major Mayan archaeological sites on the mainland, still slept in improvised cabins. The majority of tourists still left Cozumel by boat, landed on the mainland coast at Tankah, stayed briefly at the most important copra estate near-bye, and then either cut a path in the jungle to Tulum, or took a boat along the coast.

It was another century before modern tourism arrived in Cozumel, with the construction of Hotel Playa and the patronage of an influential American, William Chamberlain. From about 1952 onwards Chamberlain enticed numerous foreigners to the area, and constructed the first tourist *cabanas*, which he named 'Hotel Mayalum'. This was also the first recorded attempt to link the region and its coastal tourist attractions to the cultural life of the Maya, the historical antecedents of the 'Maya World', the brand name for most of this zone today.

In the mythology of pioneer coastal tourism, the main protagonists in Cozumel were adventurous Americans and a medley of rather unusual Mexican businessmen. On the 13th of February 1948 a Panamanian merchant vessel, the 'Narwhal', under Captain J. Wilson Berringer, with a crew of ten, transporting bananas from Guatemala to Mobile, Alabama, was cast onto the reefs off the island. The owner of the boat, Charlie Fair travelled from New York to Cozumel to take charge of the rescue and supervise the paperwork. Here he soon made contact with Carlos Namur, one of the few local people to speak English. Namur, who is now celebrated in the museum of Cozumel as a 'founder and tourist pioneer', booked the American into the Hotel Playa, and Charlie Fair was so entranced with the island, and his stay there, that he almost forgot the circumstances of his arrival, and wrote to his friends recommending they join him.

By 1957 an article on the island had appeared in the American glossy magazine, '*Holiday*', and the first eight tourists arrived on a new flight from Merida to Cozumel. Unfortunately their 'host', the indefatigable Carlos Namur, was himself in the United States at the time, and the tourists had to be put up with local families, some of them on the second floor of the building occupied by the harbour-master. Sharing this accommodation only excited their interest more, and since several of the tourists were journalists, they soon made good copy of their visit to tropical Mexico. Soon afterwards, in the 1960s, the French filmmaker Jacques Cousteau discovered the reefs nearby, and added some media celebrity to the island.

In Mexico Cozumel had blazed a modest trail, as a tourist destination, followed by Islas Mujeres, where relatively small hotels and guest-houses began to cluster around the modest central square, and provided important facilities for discriminating groups of Mexicans and Americans anxious to avoid large-scale tourism. By 1975, 90,000 tourists were visiting Islas Mujeres annually. Behind much of this growth were powerful new political interests, later to play a part in the development of Cancun, and linked to the person of President Luis Echevarria, whose godfather was a leading businessman on the island.

During the 1960s, 14 new hotels were built in Cozumel, with a total of 400 beds, an apparently modest figure in the light of subsequent developments. But by the end of the decade, 57,000 tourists had visited the island, two thirds of them foreigners. This remarkable success prompted some of the inhabitants to examine their own histories more carefully. It was soon revealed that almost the entire population was made up of 'pioneers', or 'founders' (forjadores). Refugees from the Caste War had in fact repopulated contrary to the prevailing view, created by global tourism, that the Mexican Caribbean lacked any 'identity' of its own. Unlike the rebel Maya who held the mainland, the 22 families of refugees who arrived in Cozumel in 1848, felt themselves to be the only surviving 'Mexicans' on the peninsula.

Cozumel had played an important advance role in tourist development because, apart from its roster of former chicle entrepreneurs, who were interested in putting their capital into a profitable new business, it also boasted an international airport, originally built during the Second World War for United States reconnaissance. Cozumel had traditionally been a staging post for the natural resources of the region; now it was a natural watering hole for foreign tourists, moving in the opposite direction. Unlike Cancun, however, the pioneers and founders of Cozumel had been its own indigenous bourgeoisie (Dachary and Burne 1998).

'Tropical Paradise': The Consumption of Space by Mass Tourism

The Mexican Caribbean coast was largely absent from mainstream history until Cancun was built, and the coast rediscovered almost a century later. Today a myth has developed around Cancun that probably explains why so much of its history is still unwritten. One of the principal tourist guides to the area says:

> Cancun, until very recently, was an unknown area. Formerly it was a fishing town but over a period of thirty years it evolved into a place that has become famous worldwide. It is located in the south-east of Mexico with no more "body" to it than the living spirit of the Mayas, a race that mysteriously disappeared and who were one of the great pre-Columbian cultures in Mexico. The only thing that remained was the land transformed into a paradise on earth. (*Everest Tourist Guide* 2002)

This extract reveals all the major myths about the area: Cancun was uninhabited when it was 'discovered'; it embodied the spirit of the ancient Maya (who had mysteriously disappeared); and the few remaining mortals who survived had the good fortune to be in possession of 'paradise'. These three myths guide much of the 'Maya World' tourist discourse today. That is: space was devoid of culture, Indians were devoid of ancestors, 'natives' were possessed of 'paradise'.

The development of Cancun, beginning in the 1970s, made earlier tourist incursions seem very modest indeed. In the view of some observers Cancun was chosen because the Mexican Caribbean was like a political tinderbox, liable to explode at any time. Cancun was not simply a gigantic tourist playground, in this view, it was an 'abandoned space' on the frontier, which needed to be 'settled, employed and occupied'. Even in 1970 almost half of the population of Cancun was from outside Quintana Roo; as the zone developed it pulled in people from all over southeast Mexico.

Before work even started on the vast physical infrastructure of Cancun, the Mexican Fund for Tourist Infrastructure (*Infratur*) and the *Banco de Mexico* completed an unusually complete feasibility study of the tourist potential of the region. The study reported that the withdrawal of Cuba from the tourist scene had left a vacuum that Mexico was in a weak position to exploit, since so much of its Caribbean coast was undeveloped. The danger was that other places such as the Bahamas, Puerto Rico, Jamaica and the Virgin Islands, would fill the vacuum. The study suggested that two sites should be given priority for Mexican investment: Cancun, in the Caribbean and Ixtapa-Zihuatanejo on the Pacific. The early development of Cozumel gave the development of Cancun an advantage, and the reasons why the Yucatan Peninsula should be favoured were spelled out in the document. It possessed an army of under-employed or irregularly employed workers, since the demise of henequen and chicle, and these workers lived close to some of the most beautiful marine environments in the Caribbean. Rapid tourist development would bring them both together.

Cancun could only be developed if all the available land was acquired by the project. The task of land acquisition, much of it is the form of lakes and marine lagoons, proved to be a mammoth operation. Unfortunately the man who was its guiding light, Carlos Nadir, died before his work could be completed. The project was divided into five sub-projects, separating the tourist zone from the new city. A bridge was built connecting the island of Cancun with the mainland, and the harbour of Puerto Juarez. At the same time an international airport was constructed which could handle incoming flights from Europe and North America, as well as Mexico.

The second part of the project involved a massive drive to 'sanitize' the zone, eradicating mosquitoes like those that had bothered Stephens and Catherwood, as well as most other forms of wildlife, and providing a secure supply of fresh water by constructing twenty enormous holes in the porous rocks. Yucatan has no rivers. This was followed by the electrification of the new zone, linking it with the grid in Yucatan, and opening up a vast new telecommunications network. Finally, the whole area was subjected to building and construction on a scale hitherto unknown in the Caribbean.

About two-thirds of the capital for the development of Cancun, initially 142 million dollars, was provided by the Mexican state, with help from Inter-American Development Bank loans. The scale of this investment, and the risks borne by the Mexican Government, virtually assured complementary private investment of a similar magnitude. Cancun began to function as a tourist resort in 1974 with fewer than 200 hotel beds. By 1980, when the project's first phase was completed, there were 47 hotels, 4,000 beds and almost 300,000 tourists staying in Cancun. The coast was passing from a forest enclave, linking tropical forest products with the consumption of hardwoods and chewing gum in the United States, to a tourist emporium, bringing people from far away to utilize their consumer power on the Mexican Caribbean coast.

The collapse of oil prices in 1981 forced a massive devaluation of the Mexican currency the following year and, as a consequence, more efforts were made throughout the 1980s and 1990s to earn additional foreign exchange from tourism. Environmental concerns, although frequently voiced, did little to hold back the pace of tourism on the Caribbean coast, nor the gradual destruction of the coastal habitat. Pollution became a growing problem, and Cancun spawned slums, which spread northwards, and sewage, which turned the lagoon on which the city was constructed, into a diseased sewer, alive with algal blooms, and exuding a terrible stench. Ecological problems were mirrored by a growth in criminal activity, including the large-scale laundering of drug money through inflated resort development. Drug barons moved into Cancun in the late 1980s, and one of them, Rafael Aguilar Guajardo, was famously gunned down in Cancun in April 1993.

By the early 1990s Cancun had lost much of its initial appeal, even to tourists. It had developed too quickly, and at too much cost, and the developers feared that however much lip service was paid to the environment, it was evident that mass tourism, especially from the United States and Europe (which was increasingly the market for Cancun's resort owners), was moving elsewhere. Cancun had been the principal example of what has been described as an '... archipelago of artificial paradises' in tropical Mexico (Loreto and Cabo San Lucas in Baja California,

Ixtapa near Acapulco, Puerto Escondido on the coast of Oaxaca) but Cancun had always been the jewel in the Mexican tourist crown (Dachary and Burne 1998).

Gradually foreign tourists began to follow the Mexican tourists, the backpackers and beachcombers, south of Cancun to the coastal area opposite Cozumel, where local 'tourist pioneers' established themselves in the 1970s, in places like Akumal. Most of the tourists however did not travel so far south, and they arrived eventually at Playa del Carmen. As we have seen. Here the 'pioneers' were of more recent provenance, like Ted Rhodes quoted in a tourist magazine. They were also instilled with 'Green' ambitions:

> Ted Rhodes is a local developer and pioneer for ecologically sound technologies, who is attempting to combine state-of-the-art technology, while enjoying the benefits of eco-tourism. He's only been in the Playa area since 1995, but is in the process of planning and developing six major projects … carrying disdain for the use of the word "eco", which he feels has been an over-abused term for a less than fully understood concept. Ted describes his ventures as "raw jungle converted with the hand of Mother Nature, to create a positive impact, using Mother Nature's rules … He works with the natural elements of the land, employing natural building materials from agriculture to culture, including water treatment which respects the composition and inhabitants of the land …". (*Playa Magazine*, August 1999, 8)

Comments like those of Ted Rhodes have received attention because they encapsulate the difficulty with which advocates of more sustainable tourism have to grapple. It is clear that much of the development of Mexico's Caribbean coast has been at the expense of conservation objectives, whether marine turtles, mangroves or coral reefs. The natural environment is fragile and needs protection. Nevertheless the economy of the region is highly dependent on tourism, and any suggestion that the environment is under threat rebounds against tourism. The response has been to provide a new 'eco-tourist' discourse that appears to pay attention to the concerns of the environmentalist and concerned tourist. Coastal development has been 're-branded' as 'eco-friendly', 'natural' and 'sustainable'. However, these new ways of repackaging development pay scant attention to the history of the area, which shows every sign of social and political conflict and little consideration for long-term sustainable development.

Reinventing Histories and the Politics of Space

Chacchoben is the name given to a new 'heritage' village, built deep in the forest of southern Quintana Roo. It is a construction of the tourist industry, the local peasant community and the state government of Quintana Roo, built on the site of an original settlement of *chicleros.* The location of Chachoben is important because it signals the development of one of the most ambitious tourist frontiers in

Latin America. A six-lane highway is being built, linking the existing road south to the largely undeveloped coast, to Mahahual and on to Xkalak, almost one hundred miles. Here a new generation of tourist pioneers is establishing itself, around diving and game fishing. These 'pioneers' threaten to leave when the tourist 'armies' descend, as they fear they will. The electric grid has only just arrived; the pylons were erected in April 2003. Meanwhile, fishing communities like Xkalak, on the coast, which was destroyed once by Hurricane Janet in 1955, are being gradually rebuilt, in preparation for the arrival of cruise ships, expected to dock in the port of Majahual nearby. A new generation of 'itinerant' tourists is setting foot on a stretch of the Mexican Caribbean coast formerly only known to pirates, *chicleros*, copra plantation workers and Mayan fishing families. An 'abandoned space' is being reclaimed and occupied by new visitors, who leave their 'mother ship' for only a few hours at a time, as tour coaches take them inland to spend their dollars in villages like Chacchoben.

Conclusion

This chapter has argued that the way in which space is 'consumed' needs to be set in historical context. Investigating an area such as the Mexican Caribbean we meet several different historical narratives of space, and competing interpretations. The vast urban coastal development known today as 'Playa Del Carmen' to the international tourist industry, is also known to some by its original Mayan name, 'Xiaman-Ha', and by others as 'Solidaridad' (or solidarity) the revolutionary epithet used by the local administrative authority. The space represented by Playa can be translated into several spaces, each with their own history and mythology.

It is also clear that if we investigate the connections between this coast and the outside world it is clear that they have existed even when the area was looked upon as having been 'abandoned', for example in the early colonial period after the fall of the Mayan empire and the hegemony of the Spanish. The Mexican Caribbean was important to pirates and privateers, mercenaries like Henry Morgan, who performed illegal acts on behalf of the British crown. Indeed, it is difficult to draw convincing lines of 'legality' in the white sand of the Caribbean, since natural resources were often exploited without license, and certainly without the consent of indigenous populations. The histories and politics of space are reinventions, for different generations and for different groups of people. I have argued that, taken together, these represent 'layered' histories, which serve to sanctify place through the activities of people whom some refer to as 'pioneers'. In many cases spaces are 'discovered' by successive generations of travellers and visitors, as the uses to which they are put are transformed and, occasionally, memorialized. So the burgeoning tourist economies today find 'pioneers' in Cozumel, Akumal and Playa, whose early efforts (as we have seen) serve to sanctify current activities and lend the gravitas of history to market opportunities today. Similarly, the tourists from North America and Europe who visit the new 'Eco-Parks' being built on the coast, 'discover' the marine environments and 'Mayan' cultures in new, hybridized forms. And the 'eco-tourists' who visit the Biosphere Reserve of Xian-Ka'an, under the direction of wildlife experts and specialist companies, 'discover' a space that has been drawn on the map, to provide cartographic evidence of global conservation (and a legitimacy that is often at odds with what happens there on the ground).

September 11th was an event that the world will not easily forget – indeed, its impacts are still felt in Mexico, as well as in most of the rest of the world. But the spaces of production and consumption on the Mexican Caribbean, or the 'Mayan Riviera' as it is represented today, are only the latest spatial imaginaries in a long history of changes surfaces and appearances. They illustrate that the fortunes of New York are closely linked with those of the Mexican Caribbean, and those of the Caribbean are linked with generations of people elsewhere, especially in Europe and North America, whose daily lives depended on connections that they were usually only dimly aware of.

References

Andrews, A. and Jones, G. (2001), 'Asentamientos Colonials en la Costa de Quintana Roo', *Temas Antropológicos* 23: 1, 231–246.

Blomley, N. (2003), 'Law, Property and the Geography of Violence: The Frontier, the Survey and the Grid', *Annals of the Association of American Geographers* 93.

Bridge, G. (2002), 'Resource Triumphalism: Post-industrial Narratives of Primary Commodity Production', *Environment and Planning A* 33, 759–766.

Bridge, G. and Jonas, A. (2002), 'Governing Nature: The Reregulation of Resource Access, Production, and Consumption', *Environment and Planning A* 34: 756–766.

Cesar Dachary, A. and Arnaiz Burne, S. M. (1998), *El Caribe Mexicano: Una Frontera Olvidada* (Chetumal: University of Quintana Roo).

Dachary, A. C. and Arnaiz Burne, S. M. (1998), *Cozumel: Los Anos de Espera* (Merida: Fundacion de Parques y Museos de Cozumel).

Encyclopedia de Quintana Roo. (1998), Volume Three (Cancun: Mensa Publicaciones).

Everest Tourist Guide to Cancun and the Riviera Maya. (2002), Cancun.

Cancun and Riviera Maya. (2002), *Everest Tourist Guides* (Spain: Leon).

Jones, G. (2003), 'Imaginative Geographies of Latin America' in Swanson, P. (ed.), *The Companion to Latin American Studies* (London: Edward Arnold).

Lefebvre, H. (1991), *The Production of Space*, trans. by Nicholson-Smith, D. (Oxford: Blackwells).

Martins, L. (2000), 'A Naturalist's Vision of the Tropics: Charles Darwin and the Brazilian Landscape', *Singapore Journal of Tropical Geography*, 21.

Playa Magazine. (1999), August, Playa Del Carmen.

Primack, R. B., Bray, D., Galletti, H. A. and Ponciano, I. (eds) (1998), *Timber, Tourists and Temples: Conservation and Development in the Maya Forest of Belize, Guatemala and Mexico* (Washington DC: Island Press).

Ramos Diaz, M. (1999), 'La Bonanza del Chicle en la Frontera Caribe de Mexico', *Revista Mexicana del Caribe* 7: 172–193.

Redclift, M. R. (2001), 'Changing Nature: The Consumption of Space and the Construction of Nature on the "Mayan Riviera"' in Cohen, M. and Murphy, J. (eds), *Sustainable Consumption* (pp. 165–183) (New York: Elsevier).

Redclift, M. R. (2004), *Chewing Gum: The Fortunes of Taste* (New York: Taylor and Francis).

Redclift, M. R. (2006), *Frontiers: Histories of Civil Societies and Nature* (Cambridge, MA: MIT Press).

Reed, N. (2001), *The Caste War of Yucatan* (California: Stanford University Press).

Salvatore, R. D. (1996), 'North American Travel Narratives and the Ordering/Othering of South America (c. 1810–1860)', *Journal of Historical Sociology* 9: 1, 85–110.

Simonian, L. (1995), *Defending the Land of the Jaguar: A History of Conservation in Mexico* (Austin: University of Texas Press).

Stephens, J. L. (1988), *Incidents of Travel in Yucatan, Vol. II* (Mexico City: Panorama Editorial).

Chapter 5
Recognition and Redistribution in the Renegotiation of Rural Space: The Dynamics of Aesthetic and Ethical Critiques

John Wilkinson

Introduction

In *From Farming to Biotechnology* (1987) we argued that, from the middle of the nineteenth century through to the advent of biotechnologies, the industrialization of the agrofood system had successively reduced the rural to the residual elements resistant to the technological transformation of natural processes. We argued furthermore that technology was a moving frontier and that with the emergence of biotechnologies even these residual elements faced the threat of a more radical erosion. The corollary of this physical displacement of agricultural by industrial processes was a withering away of rents generated in the rural space. The values attached to food, however, primarily under the pressure of convenience considerations, only partially adapted to the processed and reconstituted offerings of the industrial food system. Alongside, transformation technologies, preservation strategies appealed to this persistent valorization of 'pre-industrial' food in their attempts to reproduce the original organoleptic qualities of the agricultural product.

Rural actors and, in varying degrees, States and multilateral bodies adopted a variety of strategies in their attempts to redefine values and therefore rents around precisely those rural activities threatened by industrialization. Alternative production-consumption networks emerged around organics, 'origin' products, local production and fair trade, and, in developing countries, integrated rural development projects oriented to the small farmer, peasant sector. Nevertheless, the dominant response was that of a variety of price support policies to smooth over rather than challenge the transition to an industrialized agriculture. The combined potential of the different alternative networks only emerged with the shift in consumer dynamics as from the '70s and '80s of the last century. Product differentiation and market segmentation transformed the 'natural' and 'origin' values of these networks into superior quality products and processes. The emergence of large-scale retail attuned to consumer demand and not identified with any particular commodity supply structure consolidated this shift, transforming the products of alternative networks into mainstream markets. The new constellation of interests around demand issues was also felt in the surprising strength of opposition to the application of biotechnologies to food production. This in turn led to a demand for the reorganization of rural spaces on the principle of identity preservation, a measure reinforced by parallel concerns with food safety.

While the alternative networks which we have mentioned above, and the newer movements such as *Via Campesina* and Slow Food, are often considered as forming a single category we argue in this chapter that they may have very different implications from the standpoint of rural spatial dynamics and rent generation. We distinguish between those networks based on alternative, in principle, universal values and those which anchor values in the uniqueness of localities and cultures. We argue that these can be understood as incorporating either an 'aesthetic' or an 'ethical' critique of the dominant agrofood system, drawing on the conceptual framework of Boltanski and Chiapello (1999). Alternatively, they can be understood to be based respectively on strategies oriented either to recognition or redistribution, using the analytical framework of Fraser (2005). In each case the dialectic of market and social movement has different consequences for the organization of rural space and the appropriation of rural rents.

The incorporation of alternative networks into mainstream channels is often seen to be at the cost of undermining the values of locality and/or the participation of subaltern rural social groups. We argue, however, that a broader understanding of the dynamics of social movements points rather to a dialectic in which mainstreaming is repeatedly counteracted with the emergence of new demands reposing the values of locality and alternative production consumption systems. We draw on Michel Callon's (1998) notions of 'framing' and 'overflowing' to capture how market institutions inevitably generate a problematic of 'externalities'. We also incorporate Brunori's (1999) discussion of struggles over the symbolic capital associated with special quality products to help understand the significance of such new social movements as Slow Food, various forms of sustainable production systems, and the resurgence of large-scale, even global, trust-based direct sales in both organics and fair trade made possible by the internet.

These frameworks of analysis, we argue, allow for a more sanguine perspective on the reconstitution of values and rents associated with locality and the differentiation of rural space. Globalization and liberalization have, however, redefined the scale of special quality markets. A notable example here has been the establishment of an international regime, via the TRIPs, for geographical indication products. In response to the globalization of these markets, proposed indications by developing countries often now involve huge territories – the whole of the pampas region for Argentine beef, the country as a whole for Colombian coffee, and a region straddling two countries in the case of basmati rice (Sinergi 2006). Similar pressures arising from the magnified demands of the global market are leading to the revision of existing geographical indication demarcations in Europe, as notably in the case of Champagne (Bertrand 2008). At the same time, in the context of globalization quality food and non-food demands led to the mobilization of new spaces in developing countries, dedicated to what have become known as 'non-traditional' exports – fruits, seafood, vegetables, cut flowers. Differently from traditional tropical exports (coffee, cocoa, cotton), these new spaces have often been organized around dedicated production poles, frequently on the basis of irrigation. In addition, they are primarily specialized, labour intensive, short cycle production systems. As such, they are less integrated than the traditional cash crop into peasant production systems, and tend to rely heavily on hired, especially female labour (Barrientos and Dolan 2006).

Globalization has, on the other hand, been accompanied by sustained high growth in large emerging nations which has given new life to the commodity production and trade which predominated in and among developed countries in the post second world war period. Here again the transition to an urban animal protein diet in these emerging countries is redefining the geographical map as whole new territories are drawn into oils, grain and meat production, particularly in the Southern Cone countries but also on the African continent and in regions of Central Europe (Wilkinson 2008). The energy crisis and the promotion of biofuels have massively reinforced this refocusing on commodity production and have generated a tension between the goals of fuel and food production. Global warming in its turn is putting a new premium on the demarcation of territorial spaces particularly in the case of regions, now often viewed primarily as biomes, considered to be ecologically fragile. Rather than 'origin' products, therefore, we now have labels certifying that products are 'not from' different sensitive regions. The proto-typical case here would be 'green' soy from Brazil, which is guaranteed as 'not from' the Amazon forest region. With the development of energy crops there is pressure for the adoption of large-scale zoning regulation to ensure that biofuels do not threaten food production spaces (Oxfam 2008).

We suggest that a useful approach to the evaluation of the potential of quality based production consumption circuits for regenerating localities in the face of green revolution style commodity production is to distinguish between those initiatives which are based on particularistic aesthetic claims and those which rely on more universal, ethical objectives. In addition, we have argued that this potential should not be reduced to a discussion of alternative networks and their subsequent absorption within a trajectory of mainstreaming. Rather, these movements should be understood in the more dialectical perspective of a continuous process of 'framing' and 'overflowing' to adopt the vocabulary used by Callon to capture the interplay between markets and the externalities they generate. In the rest of the chapter we explore these arguments in greater detail and conclude with a necessarily tentative outline of the impact of these combined trends for the redesigning of rural spaces.

Alternative, but often Complementary, Strategies for the Appropriation of Rural Space and Occupation of Food Markets

In recent years, an increasing number of countries have seen the resurgence of agrarian reform movements, involving the demand for a redistribution of primary assets, a demand whose respectability has been enhanced by the widely influential contributions of Amartya Sen and Rawls on issue of justice. Agrarian reform may take the form of full-scale social movements, as in the case of the Brazilian Landless Movement (MST) and the *Via Campesina.* It has also become legitimated, although in attenuated form, in the Land Credit programmes of the World Bank and other multilateral organizations (Leite 2006). Nevertheless, while the axis of this movement is clearly based on claims of redistributive justice, once achieved, the products and services which emerge from the agrarian reform settlement may become oriented to niche-market segmentation based on special qualities associated with this movement. There is in fact a constant interplay between the universal claims for redistribution and the particular strategies for market recognition.

A different response has focused on redefining the status of 'traditional' production and practices, which are now repositioned as special quality products, reflecting their growing rarity value as the rural is overlaid by urban/industrial encroachment. The classic reference here is the *appellation d'origine* movement in France and Italy, which was extended to the whole of the European Union and later incorporated, as geographical indications (GI), into the global TRIPs legislation (Barjolle and Sylvander 2000). At the same time, it can take the form of rural tourism and development strategies whose justification lies in the natural and cultural specificities of different localities. The key notion here is that of territory understood as a unique outcome of nature and culture whose value can be expressed in particular products and services (Casabianca et al. 2005).

The defence of traditional products and practices, particularly in developing countries, has often taken the form of the promotion of sustainable production strategies, involving indigenous, exotic, and non-timber forest products (Boisvert 2002). Global NGOs and religious bodies have been notably active in the development of these markets, both through alternative and mainstream circuits. Here, however, the particular is defended in terms of the universalistic criteria of sustainability applicable, in principle, equally to medium and large-scale production systems. The environmental certification movement is not directed specifically at small-scale producers, nor are these latter the principal beneficiaries (Thompson 2003).

The same is broadly true of the organics movement initiated as an alternative movement of small producers. Differently from *appéllation d'origine* products, however, the principles of organic production are not locationally or category specific. As an alternative model, organics has both universalistic and normative pretensions in relation to industrialized agriculture. Nevertheless, in the process of its institutionalization it has become widely adopted as a component of mainstream segmentation strategies (Guthman 2004). Organics, therefore, became desembedded from its roots in small farming to the extent that it became defined in terms of universalistic production processes rather than specific origin criteria. The agroecological movement, more closely identified with (idealized) peasant practices, can be seen as a response to this mainstreaming of organics (von der Weid and Altieri 2002), as also the resurgence of informal trust-based local or direct (virtual) organic markets (Raynolds 2004).

A strategy, which has gained considerable momentum in recent years as markets and trade have become the hallmarks of development strategy, is the fair trade movement whose aim is also to redistribute value back upstream in the supply chain to the rural producers. The claim here, however, is not for special quality status but for justice defined as remuneration for the rural activities of the agrofood/agroindustrial chain which generates sufficient income to cover production costs and allows for the reproduction of the farming unit in conditions of citizenship (Raynolds and Long 2007). In this sense, it complements land reform movements in favour of a redistribution of assets, generated now in the sphere of trade. While its implementation in the form of alternative trading circuits may take the form of niche markets, paralleling artisan and 'tradition' products, the universalistic claims based on fairness challenge the principles of (and serve as a norm for) conventional trade, and are complemented by a social movement for the reform of the global trading system (Wilkinson 2004).

The terms in which these movements are negotiated and justified oscillate between demands for recognition and redistribution (Fraser 2005), involving a mix of particularist and universalist mobilization strategies. In the context of neo-liberal reforms, however, it is more to the market than the State that the demands of collective action are focused. Once a market perspective is adopted even strategies based on radical critiques of dominant values suffer the threat of appropriation as mainstream marketing adopts the values of counter cultures and social movements. Endogenisation of the aesthetic critique (Boltanski and Chiapello 1999) via market segmentation is evident in the adoption of values associated with nature/tradition/artisan production by food industry and retail leaders. A parallel endogenisation of the ethical critique can be seen also in the extension of corporate social responsibility to include strategies of ethical trading. On the other hand, values expelled by the marked often resurface in the form of new social movements, as in the case of the Slow Food Movement (Fonte 2006), which can be understood as a resurgence of the aesthetic critique or the Solidarity Economy movement which repositions the ethical critique (Laville 2007).

Our analysis highlights the tension between critique and cooption, between social movement and markets, between alternative and mainstream trading circuits. At the same time, it draws attention to the distinct, but often complementary, strategies elaborated by subaltern groups in the rural context based respectively on aesthetic and social critiques with their different forms of justification and negotiation. We apply the concepts of recognition and redistribution (Fraser 2005), which have been developed to discuss issues of race, gender and social inequality, to the different types of demands underpinning rural mobilizations. We will also draw on the French convention approach focusing on justified and justifiable patterns of social action, especially in the more recent formulation of Boltanski and Chiapello (1999). Before discussing in detail the implications of these different strategies we situate them within the broader framework of transformations in the global agrofood system

The Quality Turn, New Patterns of Regulation, Globalization and the Return to the Commodity

Four factors have transformed the dynamic of the agrofood system since the beginning of the '90s. In the first place, there has been a generalized shift in the rules of the game as regards investment, intellectual property and trade. While some countries moved precociously on these questions, the nineties marked the alignment of national economies to the precepts of the WTO and particularly the TRIPS. Internal market regulation was relaxed, external tariffs were eliminated or drastically lowered, intellectual property rights were extended to seeds and food, and business law reform accorded national status to the subsidiaries of transnationals. Given the uneven development of different national economies and the more general polarization between developing and developed countries, this process of institutional adjustment has placed trade at the centre of international negotiations and conflict (Wilkinson 2004). The more so, given the special status of the agricultural sector, of vital interest to most developing countries, and the sector for which most lenience has been shown to the persistence of the protectionist systems historically in place in developed countries, particularly the European Union and the Unites States. This set of institutional reforms has been justified by the defence of trade rather than the domestic market as the terrain of growth and development strategies. It is not surprising therefore, that the nature of this trade should become the privileged focus of global negotiation and conflict, and that the fair trade movement has received increasing attention from mainstream actors (Stiglitz and Carlton 2006).

The second major factor responsible for the paradigm shift in the dynamic of the agrofood system, and one greatly facilitated by the above reforms, is the rapid transnationalization of leading agrofood actors, among which we may also include NGOs such as Greenpeace, Actionaid, Oxfam, Rainforest Alliance (Gereffi et al. 2003). Foreign direct investment (FDI) is the catchphrase resuming this process and all sectors of agrofood have been implicated. A simplified historical view of agrofood internationalization would see successive waves involving first the global traders, followed by agricultural inputs and machinery in the wake of the 'green revolution', and then, as domestic markets assume relevance, investments by the final food industry. Though there is some explanatory value to this evolutionary schema, recent FDI has involved all sectors of the agrofood chain. The biotechnology revolution has launched a global restructuring of the seeds and chemicals industries. The fragmentation of production systems and out-sourcing, leading to the emergence of global production chains/networks, has also affected all sectors of agrofood. Nevertheless, the factor most characteristic of this movement has been the transnationalization of the retail sector, which fits well with the evolutionary schema presented above. While French supermarkets, particularly Carrefour were precocious in their international presence, it is from the nineties that we see the global operations of Tesco (Britain) and Wal-Mart (US), the world's leading retailers, being put into place (Reardon et al. 2001).

As in the case of the global production networks referred to above, this transnationalization of the retail section should not be seen simply as an adjustment to the new conditions of foreign markets, particularly those of developing countries, in the light of the institutional reforms described earlier. While the dynamic of these 'local' markets is crucial, this transnationalization implies the construction of global retail circuits in which the traditional boundaries between 'domestic' and 'export' markets lose their significance. Not only is retail involving itself more directly in international trade, it is also constructing supply systems where both domestic and export goods are subjected to the same logistical and marketing criteria. Particularly in the case of developing countries, retail is simultaneously replacing traditional outlets and conditioning access to its shelves on increasingly uniform global quality and logistical criteria. The farmer no longer knows whether his produce, once purchased by the retailer, will be marketed locally or absorbed into global supply chains. For this to occur, however, global criteria on product and process standardization need to be defined and implemented (Reardon et al. 2008).

In the third place, agrofood studies have identified a fundamental break in the dynamic of the world food system as from the '70s of the last century, marking the crisis/end of a food regime based on commodity staples (Friedman and McMichael 1989). French studies have analysed this same period as one marked by a radical shift to an economy based on quality (INRA 1995). Product differentiation and market segmentation now became the dominant competitive strategies and product innovation began to undermine the traditionally long life cycles of leading branded products. Initially this took the form of what has been called 'delayed innovation' (Galizzi and Venturini 1996) with modifications being introduced only in the final industrial phase (very often packaging). This plethora of new products, nevertheless, posed special problems for food regulation where the public sector had been traditionally responsible for monitoring food safety and minimum quality standards. To deal with this new situation, the focus of regulation was shifted from products to processes and at the same time responsibility for implementation was increasingly passed on to the private sector. The adoption of HACCP and ISO systems relating to industrial processes was the prime expression of this shift in regulation practices (Marsden et al. 2000).

The quality turn, however, also corresponded to major shifts in demand provoked both by demographic and occupational factors and also by new thinking on issues of public health. The absorption of women into the workforce led industry to assume more and more activities originally performed in the home kitchen and the demands for convenience were met with a move from products to (ready to [h]eat) meals (Goodman and Redclift 1991). This in turn led to an exponential increase in the demand for ingredients placing a premium on the development of logistics to adapt the supply chains to 'just-in-time' food production. On the other hand, the greying of the population in the industrialized countries and the adoption of a preventive rather than an interventionist approach to public health placed food at the centre of health and wellness strategies (Wilkinson 2002). In its turn, quality as a competitive strategy became more focused on substantive content claims and firms gave greater attention to the development of 'special/superior quality' products (Sauvée and Valceschini 2003). For reasons peculiar to the history of the food industry, approximation to the qualities of the original agricultural product became a key benchmark of industrial food quality.[1]

Quality strategies, therefore, could no longer depend on the control over industrial processes but had to extend to the whole supply chain and to the conditions of agricultural production. Leading food firms and retail have all developed their special quality criteria creating new barriers to entry for their would-be suppliers. Individual firm strategies have, at the same time, been combined with sector initiatives to ensure new minimum private standards, the prime example being the imposition of GLOBALGAP standards for agrofood products entering European markets (Busch and Bingen 2006). Successive food scares, in their turn, have led to tighter public measures for ensuring food safety in European markets. Deepening the shift from product to process regulation, the strategy adopted by the European Union was the imposition of traceability procedures, first in the most critically affected meats sector, but then as a norm for quality control in all supply chains. Under this system the agricultural input must be monitored and identifiable from planting or rearing right through to its sale to the consumer (Green and Hy 2002).

And fourthly, the North-South axis of agrofood trade and investment is rapidly being complemented by South-South flows to such an extent that a leading country like Brazil now exports a greater value of agrofood products to other southern countries than it does to the traditional markets of the North. These South-South flows have been accompanied by a shift from differentiated to commodity products as large emerging countries begin to reproduce the transition to an animal protein diet that characterized Northern countries in the early and middle years

1 The industrialization of the food system never completely substituted the original agricultural product given the biological restrictions both of photosynthesis and digestion. As a result, industrial strategies which had been primarily based on transformation of the agricultural product (milk to cheese) focused their strategies on preservation technologies (freezing, chilling) reproducing in so far as possible the original characteristics of the agricultural product. Industrial food quality, therefore, became increasingly associated with the degree to which it reproduced the quality of the agricultural input. The mainstreaming of organics can be understood in this light. For an overall account of the industrialization of the agrofood systems see Goodman et al. (1987).

of the last century. The sustained demand from these countries, led by China, is leading to a large-scale expansion of commodity production in many developing countries of Latin American, Africa and Asia. Foreign direct investment on the part of developing countries, especially China, India and Brazil is responsible for a rapidly increasing proportion of this expansion (Rama and Wilkinson 2007). The North-South axis, for its part, is being redefined by the entry into global trade of new non-food commodites – biofuels – as the industrialized world seeks to respond to the twin crises of fossil fuel energy and climate change. These two tendencies are redefining the geographies of developing countries as natural comparative advantage threatens to push aside the cultural advantages underpinning competition on the basis of quality, understood as locality (Fonte 2006). The constructed values of place are once again threatened by the competitive attractiveness of space.

In the following two sections we will discuss geographical indications and fair trade as paradigmatic examples respectively of strategies of recognition or aesthetic critique and redistribution or ethical critique showing how each is inversely threatened in the context of globalization and mainstream strategies. On the one hand, the particularist values of origin products are challenged either by territorial competition as supply outstrips demand or by de-territorialzation as demand strains the established demarcations of supply. On the other, the universalist values of fair trade suffer pressures to reformulation in terms of particularist, niche, 'habitat' strategies. While the high quality commodity segment is largely inaccessible to the small holder, sustainable principles can provide a special quality status – as in the case of bird friendly or shade grown coffee.

Geographical Indications as Aesthetic Critique and Strategy of Recognition

The heyday of Fordist industrialization was built on a sustained shift from a rural to an urban society in which backs were generally turned to the values of the countryside. The shift itself was made possible by the industrialization of agricultural activities, which simultaneously marginalzed diversified production systems and rural agroindustries. Local markets based on these latter persisted but were viewed as residual, not forming part of the strategies of the food industry which was more focused on the emergent mass markets where scale and price were the key criteria. Growing affluence and nostalgia for the now distant rural world provided a new lease of life for artisan food products geared to the expanding middle class, occupying a niche not immediately contested by the new segmentation strategies of the food industry. In this sense, *appellation d'origine* products represent the clearest expression of a strategy based on the aesthetic critique, counterposing the craft traditions of local food production to the homogenous mass products of the industrial food system. As such, they include the demand for recognition of the uniqueness of the historically constructed values of a particular locality, which becomes the object of *sui generis* intellectual property rights.

In this period, the *appellation* model spread from wine to cheeses and meats and a range of new products for which specific ecosystem and cultural claims could be made. Seen as a typically, southern European conception and practice, legislation guaranteeing collective rights to 'origin' products was surprisingly put into place by the European Union, and Northern European countries, such as Britain, after initial resistance turned with enthusiasm to this new mechanism for valorizing the local (Sylvander 1995; Tregear 2003). Later, and again in spite of persistent opposition from anglo-saxon countries (USA, Australia), *appéllations*, now as geographical indications, were incorporated into the global TRIPS agreement (Hughes 2006).

A market strategy based on the recognition of explicitly anti-industrial values, seemed for a while to offer the best strategy for renewing traditional communities. That these special quality products benefited from a *sui generis* form of collective intellectual property reinforced the idea that there existed a sharp frontier between these markets and those of the dominant agrofood system, apparently guaranteeing an autonomous space of market recognition for the rural artisan sector (Bérard and Marchenay 2004).

Two developments, however, which we have noted in the previous section, are seriously eroding the barriers erected between the *appellation d'origine* and the industrial food worlds and threaten the long-term viability of a distinct sector of artisan food production. On the one hand, public and private regulations increasingly require the adoption of 'industrial' practices as the condition of market access. Even in the most prestigious and traditional regions of artisan production such as the Emilia Romana, and Parma districts in Italy, *denominazione de origine* producers of globally reputed cheeses and hams now have to adopt HACCP and ISO practices (Wilkinson 2005). Their adoption will tend to involve changes in the traditional practices defining the artisan character of the protected activity. If this is not the case, additional costs are incurred which put further strain on the smaller farmers involved in these activities.

On the other hand, in their segmentation strategies, leading food firms are themselves now appealing to the values of traditional and 'natural' production and are no longer content to leave 'artisan quality' in the hands of the traditional farming sector. It should be said that the *appellation d'origine* sector was never immune to pressures from dominant agrofood actors. The sector was protected from the impact of scale economies only to the extent that its *cahier de charges* stipulated processes and practices resistant to scale-up. Lax definitions of production processes led to rapid industrial concentration in the case of some successful brands. In others, agricultural activities might allow for scale economies, leading to a concentration of farms, while the artisan character of processing activities was preserved. Yet again, financial concentration showed itself to be compatible with decentralized production activities (INRA 2000).

The novelty of the most recent period, however, has been the direct competition by the industrial food and retail sectors for control of this market segment, which as with luxury foods, tends to grow much faster than the average for the food industry in the industrialized countries. Industrial and retail marketing strategy are now replete with product claims to natural/traditional methods of cultivation and processing (stone-milled, wood-smoked). Unilever provides an interesting case because its strategy reveals a possible marketing weakness in *appéllation* products – the rigidity of their production processes in the light of market segmentation. Unilever has developed a line of traditional olive oils including ingredients not permitted for *appéllation* products, such as aromas and special tastes, which allow for the targeting of specific consumer segments and settings (www.unilever.com). Whether this proves to be a market advantage over *appellation* products remains to be seen, but it is clear that there is no longer a reserved space for GI products.

This is even more evident in the case of supermarkets where *appéllation* products have to compete in the same gondolas not only with mass-produced products, but also with other 'artisan' products based on brand or more generic, reputational appeals (farmer status etc). Studies in the late '90s suggested that *appéllation* products were able to command an average 15–20 per cent price premium (INRA 2000) but heavy targeting of cheeses, meats and oils by agribusiness is now reducing margins for all but the most highly reputed products. Supermarkets have gone further, however, developing their own special quality supply chains where the marketing strategy is based on the quality of the original product, marking a radical break from the earlier food industry strategy of 'delayed innovation' discussed earlier. In a similar form to *appéllation* products, these supply chains involve the adoption of specific production practices which are subject to the guarantees of certification. The British supermarket, Tesco has developed a special quality line for fresh vegetables with the suggestive title, *Nature's Choice*, involving the certification of a branded supplier, Castle Brand. In France, Carrefour has promoted a similar quality brand.

At the same time, this valorization of the agricultural/rural origins of the product has opened up opportunities for innovative responses from rural actors marketing territorial values without recourse to the *appéllation* procedures (Sauvée and Valceschini 2003). It is interesting that in these cases – the above author's cite the Milieukeur pig brand in Holland and the Asturian meat brand – the owners of the brand are not the direct producers as such. In the former case, we are dealing with a consortium involving consumer and environmental associations, the public sector and representatives of the pig producers. In the latter, it is rather the regional authority which is the repository of the brand whose execution and monitoring is delegated to an inter-professional body. It is clear here that strictly economic values and organization are no longer considered sufficient to guarantee what are not exclusively commercial, but equally civic valorization strategies. These special quality supply chains, dominated by civic organizations, have rather the flavour of social movements, where the demarcation lines between markets and society become diffuse. In the same way that Parmesan cheese producers have adopted the values of the organic movement for which they receive a higher premium, these special quality supply chains draw on the values of sustainability and consumer rights. The valorization of particularist traditions, therefore, is increasingly premised on their alignment with new universalistic sources of value attribution, even at the risk of diluting authenticity. Recognition is itself, therefore, now subject to qualification with the danger that the *origin* products and the traditional producers become increasingly subject to urban quality values.

While food industry and retail strategies work to undermine the uniqueness of place through symbolic appropriation of locality and tradition in marketing strategies they also undermine the specificity of origin through the proliferation of 'traditional' quality labels and brands. On the other hand, the very success of geographical indication products in the context of the shift to global markets puts enormous strains on the specificity of the local supply base. One response would be to reaffirm the uniqueness of the product, maintaining or even reinforcing the rigour of spatial demarcation and production norms. Such a strategy, however, in the context of global competition is necessarily exceptional, with the product occupying a luxury but marginal niche. More frequently, the opportunities and the demands of global markets are leading either to pressures for a redefinition of the original demarcated territory or for more flexible rules on the origin of raw materials which compose the final product (INRA 2000). In either case, the aesthetic critique and the strategy of recognition run serious risks as the unique features of place become diluted.

The Slow Food Movement has developed an original version of the *appéllation d'origine* approach based on the uniqueness of all localities and local practices which are in their turn focused on the renovation of local cuisines (Miele and Murdoch 2003) However, in generalizing this concept, the basis of justification has shifted to the universalistic values of biodiversity (both nature and culture). As Slow Food expands, however, its social movement characteristics, which have allowed it to eschew the State dependent strategy and exclusiveness of GIs, must now confront the market test of retail mainstreaming, with as yet unclear implications for its particular strategy of quality communication and underwriting (Fonte 2006). In developing countries, as we have seen, the claims of 'terroir' status tend to be pitched from the outset on a scale aligned with the goal of occupying global markets (Wilkinson and Cerdan 2008).

Fair Trade and the Universalist Thrust of the Ethical Critique

The fair trade movement is based on very different criteria, directed at the generalized injustice characterizing food industry and retail relations with the primary producers (small farmers, rural workers) of global, commodity, food and agroindustrial production chains. While the medium, long term results of fair trade practices may, through allowing for savings and subsequent investments, provide the conditions for eventual product diversification, the immediate claims for value redistribution are not based on special quality, tradition or territory but on minimalist and universal demands for redistributive justice.

The current success of the fair trade movement in terms of market expansion, media coverage and occupation of the political agenda is perhaps related to the principle that agreement on what is fair is easier than agreement on what is good and is also compatible with different conceptions of the good. It therefore defines the preconditions on the basis of which the pluralist pursuit of the good can then be pursued (Arnsperger and Parijs 2000). Translated into the terms of our discussion, the aesthetic strategy behind GIs can be judged by the market whereas the justice claims of the fair trade movement imply new rules and regulations for the functioning of markets. The strategy for promoting these goals may be voluntary systems of market regulation, and the fair trade movement has been primarily independent of the State and market oriented, but the overall objective is for the transformation of the terms of trade (Tallontire and Vorley 2005). Hence, the anti-WTO mobilizations are an intrinsic part of the fair trade movement.[2]

The fragility of the fair trade movement lies in its focus on the paradigm of the commodity market whose prices, although susceptible to manipulation behind the scenes by the economic power of oligopoly markets or by deliberate collusion, are those of the stock exchange built on the logic of supply and demand. It is precisely these markets that are currently being undermined by the network economy which has emerged consequent on the quality turn. Increasingly traditional commodity markets are giving way to the networked coordination of supply systems in which prices are both determined by the buyer and largely hidden from public scrutiny (Busch 2004). Paradoxically, within this logic, fair trade becomes another component of the quality turn accelerating the shift from commodity markets to managed supply systems.

The response of the leading actors in the food industry directly affected by the challenge of fair trade (Nestlé, Sara Lee) has been twofold. On the one hand, they have tried to reposition the ethical critique on to the aesthetic terrain where, as we have seen above, they have demonstrated considerable capacity for absorbing critique. In this sense they argue that the solution lies in the adoption of quality criteria able to capture price premiums (May et al. 2004). These strategies proposed for the traditional commodity sector can be understood as an extension of the broader strategy to promote 'non-traditional' exports mentioned earlier. This approach would be in line both with dominant agrofood tendencies and with the different strategies directed at re-appropriating value within the rural segments of

2 In fact levels of State involvement in fair trade vary from country to country. France and Belgium have taken initiatives for its regulation. In Britain, fair trade has been adopted as the policy of the International Cooperation Ministry (DFID). In Europe generally the public sector has been mobilized as a collective consumer. In the USA, on the other hand, public involvement is absent. Traceability in Europe and biosecurity in the USA may lead to the inclusion of fair trade circuits in more general regulatory measures. In developing countries uneven levels of public involvement also prevail with Mexico developing a private national fair trade system, whereas the Brazilian movement in this direction has had heavy public sector participation.

the product chain, and is broadly convergent with territorial valorization strategies. In the case of coffee, the principal fair trade product, this would involve organic, shade grown, bird friendly and locational/origin criteria, in addition to 'intrinsic' quality characteristics (taste, aroma etc.). On the other hand, these leading firms have moved to absorb the ethical critique posed by the fair trade movement, and particularly the fair trade Labelling Organization (FLO) whose criteria involve commitment to stipulated price and working conditions. In this sense, they have adopted halfway social labels (Utz Kapeh) or developed, as in the case of Starbucks, their own socially audited supply systems (Grodnik and Conroy 2007).

Perhaps the most notable feature of the recent period has been industry's move on to the offensive on the ethical front, within which we can include issues of sustainability. Many leading firms are now committed to triple bottom line and voluntary, social accountability, auditing systems (ISEAL, Price Waterhouse, the Ethical Trade Alliance (ETI)).[3] Within the fair trade movement this has taken the form of the unexpected certification of Nestlé (coffee) and Dole (bananas) by the FLO, which marks a sharp shift from earlier opposition, particularly in the case of Nestlé which had insisted in the 'quality' option. It should be clear, however, that certification in the case of Nestlé refers to only a tiny fraction of its total coffee trading activities (1 per cent) and in the case of Dole only to specific plantations. Within the framework of Boltanski and Chiapello's argument (1999), the later '80s and early '90s saw a resurgence of social critique consequent on the disarming/absorption of the aesthetic critique contained in the social movements arising from the '60s. Today, they argue, we are witnessing an attempt to construct a new system of justification based on the criteria of the network/connectionist economy. The channelling of social and environmental critiques via the construction of new systems of governance based on largely private (but with public endorsement) certification and auditing regulatory systems represents the alignment of the agrofood system with this emerging network, 'world of projects'. Whether this world becomes a 'city', in the terminology of Boltanski and Thévenot's *De la Justification* (1991), within which new hierarchies and inequalities become legitimated, depends on the way markets, social movements and State are redefined in this process.

Critique and Appropriation or the Permanent Dialectic of Markets, Networks and Movements

The State has traditionally been the focus of social movement demands but in the network society its centrality is now called in question at two levels. On the one hand,

3 ISEAL = The International Social and Environmental Accreditation and Labelling Alliance; Price Waterhouse has now specialized in social auditing and has developed a fair trade auditing system with the French fair trader, Altereco. The ETI is a British association involving the leading food industry and retail players. See Smith and Barrientos (2005).

as we argued above, both the institutional reforms and the transnationalization of corporate operations, which have accompanied globalization, have weakened the authority of national States. On the other, as we also saw in our earlier discussion, governance on a global scale has been largely assumed by private actors in the form of voluntary regulatory systems which determine access to and the rules of participation in different markets. In this respect, States are largely assuming an underwriter role, with non-elected multilateral organizations (WTO, World Bank) increasingly responsible for the institutional design and monitoring of the system as a whole.[4] Social movements and their organizations have reflected this change as they campaign against corporations and the WTO and promote alternative global forums for international law and regulation (Kyoto Protocol, Convention on Biodiversity). These actions are combined with the direct negotiation with transnational corporations over the rules governing specific markets or production and labour practices which often involve the development of competing labelling and certification systems – IFOAM, FSC, WWF, FLO.[5]

In the '80s, economic options were still discussed in terms of market versus hierarchy with intermediary organizational forms being seen as essentially unstable and temporary (Williamson 1985). Granovetter and the new economic sociology contested this view arguing that economic organizational forms could vary and that intermediary forms of coordination were possible to the extent that they were anchored in social networks (Granovetter 1985). Since then, however, hybrid organizational forms have steadily replaced both market and hierarchy (public and private) and the network form has established itself as the dominant pattern of economic coordination (Busch 2004). While the anchoring of economic ties in social networks remains a key support for economic transactions (Uzzi 1996), the network form presents itself rather as an endogenous response to economic coordination characterized by high levels of uncertainty and rapidly changing market conditions. Differently from Williamson's (1985) supposition that frequent transactions in contexts of asset specificity (sunk costs) tend towards vertical integration, it is now argued that network forms of cooperation are structures in which trust is produced, tested and promoted endogenously (Sabel 1993; Wilkinson 2005).

In addition, the network form is more open to the collaboration of erstwhile 'non-economic actors', whether in the negotiation of global regulation often conducted within the framework of 'global policy networks' (Benner et al. 2004) involving

4 In the light of the increasing importance of large 'emergent' nations and the current financial and economic crisis which has undermined the credibility of neo-liberal policies and the globalized market national and regionally-based policies are gaining force and with them a new legitimacy of the State.

5 IFOAM – International Federation of Organic Movements; FSC – Forestry Stewardship Council; WWF – World Wide Fund for Nature. The Marine Stewardship Council (MSC) is an interesting example of a joint-venture between NGOs and transnationals, in this case the World Wide Fund for Nature and Unilever.

multilateral organizations, transnationals and NGOs, or in the innovative supply chains which we discussed above where civil society actors play a key role. Within the rural studies literature, the notion of 'alternative networks' is used to refer to commercial circuits developed within a social movement dynamic, particularly in the case of organics or fair trade (Goodman 2004; Raynolds 2004). What is needed, however, is a broader theorization of the permanent tension, or 'dialectic without synthesis' (Simmel 1997), between markets and social movements in the case of transactions heavily laden with values.

Alternative networks often take the form of unmediated producer–consumer relations as in the case of direct sales or local market fairs, particularly so for organic produce. Fair trade, on the other hand, being based on long circuits, has led to the transformation of NGOs into economic actors either as traders or as the organizers of retail outlets. The dedicated fair trade shops are a hybrid form using voluntary 'movement' labour, thereby lowering costs (particularly important in the high wage Northern economies) but often leading to market insensitivity (Wilkinson 2007).

In other cases, a more business approach has been adopted, a notable example being the Café Direct sales in England, a firm created by four fair trade organizations – Oxfam, Traidcraft, Equal Exchange, and Twin Trading. Café Direct had a turnover of 13 million pounds sterling in 2004 and launched an Ethical Public Shares Offering (EPO) which would give investors 56 per cent of the shares, with producer groups holding 5 per cent and the four original owners 39 per cent (Barrientos et al. 2007). A further example of the movement directly adopting a business format would be the Day Chocolate Company founded at the end of the '90s. Fifty-three per cent of the firm's shares are owned by the NGO Twin Trading, 33 per cent by the Ghana cocoa farm cooperative, Kuapa Kokoo and 14 per cent by the Body Shop International. Another leading NGO, Christian Aid has preference shares in the company and the charity Comic Relief is partner in the launching of one of its brands (DFID 2001).

Recently a leap into the world of virtual trade has been taken by the International Fair Trade Association (IFAT), which has established a joint venture with the eBay electronic auctioning site for the internet sale of artisan products. IFAT is organizing a tour of producer organizations by the software company responsible for the production of eBay's electronic catalogues. With the active support of IFAT's social movement network otherwise isolated producer groups will now be able to negotiate their goods directly on the internet (www.worldofgood.ebay.com).

In a more indirect way, social movement characteristics become an integral part of market dynamics. Mainstream market growth of fair trade products, at around 20–30 per cent per year, makes this segment one of the most dynamic sectors of the food industry and is overwhelmingly movement driven (Krier 2005; FTF 2006) Movement campaigns, whether directly political, as in the anti-WTO mobilizations, or in the form of campaigns (fair trade weeks, fair trade towns, organizations of producer group tours), directly promote the market and are geared to influence consumer support, rather than for instance the traditional campaigns for donations. To the extent that most fair trade has now become mainstream a cynical view would see this campaign activity as a form of unpaid publicity, whose principal beneficiary is the supermarket. A blistering indictment of supermarket opportunism and an exposé of the dilemmas of mainstreaming can be found in Oppenheim (2005) and a similar critique in the French context is made by Jacquiau (2006).

Callon (1998) has developed an analysis of the relation between market and society as one of permanent 'framing' and 'overflowing'. To the extent that a market is defined and regulated, externalities are created which subsequently become the terrain of renegotiation. This is particularly acute in the case of environmental and social externalities and we would argue that Callon's approach is convergent with the 'dialectical' relation we have posited between markets and social movements as a key feature of contemporary economic activity. Brunori (1999) has developed a similar argument specifically relating to new quality markets in agrifood. The shift to quality markets means that agribusiness is permanently searching for new ways of linking consumption and production in terms of immaterial rather than functional characteristics.[6] The alternative networks of social movements, such as organics, fair trade and regional products, have what Brunori aptly describes as 'high symbolic density' which presents itself as an attraction to mainstream business to the extent that this can be transformed into economic power in the form of premium pricing. The danger is that the alternative products become absorbed as commodities within the differentiated markets of the quality economy.

Brunori distinguishes between different resources as the basis of empowerment – social, technological, economic and symbolic. Mainstreaming involves the subsumption, particularly, of social and symbolic to that of economic power. The social movement response, in its turn, depends on its ability to reintegrate these different resources so that economic power is not able to become independent of social and symbolic power. In practical terms, this would mean campaigning for higher standards once the original standards have been transformed into conventional criteria, establishing new connections (GMO-free products), integrating an increasing number of alternative characteristics (organic + fair trade + sustainability + local distinctiveness), or developing new distribution channels.

In the course of our discussion we have pointed to a number of ways in which the social movement has responded to mainstreaming along the lines suggested by Brunori. The Slow Food movement has tied in local production to the goal of biodiversity, fair trade in the South has raised the demand for the application of fairness to local markets and the adaptation of fair trade criteria to local conditions, and IFAT has launched fair trade through internet auctions. This line of argument has also been taken up in the context of Northern markets (Jaffee et al. 2004). The issue of integrating the different strands of the alternative

6 This distinction, however, is often difficult to apply as the 'immaterial' may often refer to emerging functional characteristics, such as health.

movements remains a key terrain of struggle since mainstream actors, have taken the lead in fusing social and environmental criteria, albeit in diluted form, as in the Ethical Trade Initiative and Triple Bottom Line accounting. The shift from single to multi-issue transnational social movement organizations in the recent period is consistent with this dynamic (Smith 2005). Organizational inertia may be a limiting factor here as the social movement organizations are not immune to the effects of bureaucratization (Campbell 2005). It is probable, however, that the issue of alternative forms of product/process guarantees to those of mainstream certification (participatory certification, systems of conformity, see Fonseca 2005) will provide the social movement response to what are seen to be exclusionary conditions of market access.

To the extent that aesthetic, social and environmental values are increasingly channelled through the market, or rather through the vehicle of economic transactions, there is in principle no limit to the process of 'framing' and 'overflowing', involving permanent negotiation and conflict between mainstream and social movement actors. At the same time, both social movement and mainstream actors are mobilized around issues of governance. Given the weakened role of national States and the predominance of private regulatory arrangements, it is unlikely, however, that a politicization of the social movement will lead to the alternating cycles of consumer and political mobilization identified by Hirschman (1970) for an earlier period. In the context of globalization and transnationalization it is more likely that the three facets of social movement activity – alternative circuits, mainstreaming and political campaigning –continue to develop synchronically.

Conclusions

The initial phase of globalization was consolidated under the aegis of liberalization and the shift to the quality economy. In developed countries this strengthened the turn to origin based products and in developing countries opened up the perspective for 'non-traditional' exports each of which in different ways anchored values in specific forms of territorial advantage. The aesthetic critique based on the values of territory and *terroir* prospered in developed countries becoming institutionalized first in the European Union and later in the TRIPS agreement. In the developing country context, on the other hand, the opportunities of non-traditional exports were insufficient to compensate the collapse of traditional commodity markets and gave rise to the ethical critique of the fair trade movement, which increasingly took on board the convergent universalistic demands of sustainability. Both these movements, as we have seen, suffer from pressures to cooptation and adaptation as dominant actors also contest the values of tradition and equity. Nevertheless, as the market arena gives way to networks and as these networks become permeated with broader society demands the appropriation associated with mainstreaming has been accompanied by the resurgence of a range of new social movements

– Slow Food, the solidarity economy, agroecology, direct sales now by internet, and the creation of a global organization for GI products, Origin.

Three new factors are profoundly changing the dynamic of agrofood globalization and with it the design of production-consumption spaces – the demand for agricultural commodities by the large emerging economies, the promotion of biofuels and the actual and projected effects of climate change. The first two give a new lease of life to the commodity rather than the quality economy – foodstuffs increasingly from South to South and biofuels restoring the traditional North/South commodity flows. The third factor is more ambiguous since it simultaneously points to the need for increasing levels of local food and energy sufficiency while at the same time it is clear that an increase in floods and droughts and shifts in regional specialization will reinforce the key role of commodity trade.

In this new context we are seeing a swing in which the aesthetic, particularist critique of the dominant agrofood system in the name of the uniqueness of rural traditions and localities plays now an accompanying role to a dominant ethical, universalistic critique around notions of sustainability whose territorial implications are the subject of intense debate. On the one hand, policies based on 'food miles' favour the promotion of local food systems, although the local here is less the 'terroir' and more a function of putative energy and climate change benefits. As a result support for fair trade itself has now increasingly to confront the test of the ecological footprint. On the other hand, the resurgence of the global commodity economy transforms the scale in which territories are conceived. In the Northern countries, set-aside lands are brought into play for biofuel production, whereas in the South the over-riding issue is the protection of large eco-systems - the tropical humid forests and the swamplands providing carbon sinks – from the expansion of the commodity frontier. These may be defended via private certification and traceability systems validated at the point of purchase or, depending on different national State capacities and configurations, by public zoning regulations. Given the pressures from both the emerging and the Northern countries for food/feed and biofuel the central territorial tensions in developing countries will increasingly be around local/domestic versus export production (World Development Report 2007). While artisan quality has been central to family farm and peasant-based strategies in many developing countries and regions it is likely that in the coming period greater attention will be given to their role in basic food production. In this context it is likely that the aesthetic critique, in the form of GI products, will in the developing country context be assimilated into the more entrepreneurial high quality niche markets.

References

Allaire, G. and Boyer, R. (eds) (1995), *La Grande Transformation de l'Agriculture* (Paris: INRA/Economica).

Arnsperger, C. and van Parijs, P. (2000), *Éthique Économique et Sociale* (Paris: La Decouverte).

Barjolle, D. and Sylvander, B. (2000), 'Some Factor of Success for "Origin Labelled Products"' in *Agrifood Supply Chains in Europe* (Paris: INRA).

Barrientos, S., Conroy, M. E. and Jones, E. (2007), 'Northern Siocial Movement and Fair Trade', in Raynolds, L., Murray, D. and Wilkinson, J., *Fair Trade. The Challenges of Transforming Capitalism* (London: Routledge).

Barrientos, S. and Dolan, C. (eds) (2006), *Ethical Sourcing in the Global Food System* (London: Earthscan).

Barrientos, S. and Smith, S. (2007), 'Mainstreaming Fair Trade in Global Production Networks' in Raynolds, L., Murray, D. and Wilkinson, J., *Fair Trade. The Challenges of Transforming Capitalism* (London: Routledge).

Bérard, L. and Marchenay, P. (2004), 'Local Products and Geographical Indications: Taking Account of Local Knowledge and Biodiversity', *International Social Science Journal*, UNESCO.

Bertrand, J. (2008), 'Champagne Growing Area Extended', *Agence Press*, 3rd March 2008.

Boisvert, V. and Vivien, D. (2005), 'The Convention on Biodioversity. A Conventiionalist Approach', *Ecological Economics* 53: 4.

Boltanski, L. and Chiapello, E. (1999), *Le Nouvel Ésprit du Capitalisme* (Paris: Gallimard).

Bolstanski, L. and Thévenot, L. (1991), *De La Justification* (Paris: Gallimard).

Benner, T., W. H. Reinicke and Witte, J. M. (2004), *Multisectoral Networks in Global Governance: Towards a Pluralistic System of Accountability* (Oxford: Blackwell).

Brunori, G. (1999), *Alternative Trade or Market Fragmentation? Food Circuits and Social Movements* (Italy: University of Pisa).

Busch, L. (2004), *The Changing Food System: From Markets to Networks* (Trondheim).

Busch, L. and Bingen, J. (2006), 'Introduction: A New World of Standards' in Bingen, J. and Busch, L. (eds), *Agricultural Standards: The Shape of the Global Food and Fiber System* (Springer).

Callon, M. (1999), *The Laws of the Markets* (Oxford: Blackwell).

Campbell, J. L. (2005), 'Where do we Stand? Common Mechanisms in Organizations and Social Movement Research' in Davis, G. F., McAdam, D., Scott, W. R. and Zald, M. N. (eds) *Social Movements and Organization Theory* (New York: Cambridge University Press).

Casabianca, F., Sylvander, B., Noel, Y., Beranger, C., Coulon, J. B. and Roncin, F. (2005), 'Terroir et Typicité', Paper presented to the Congrès International des Terroirs Viticoles, 2006, Montpellier, France.

DFID. (2001), http://www.dfid.gov.uk/Media-Room/Case-Studies/2007/A-Divine-story-DFIDs-contribution-to-a-fair-trade-success/.

Fonseca, F. (2005), 'Certification and Acreditation: Alternative Certification and a Network Conformity Assessment Approach', *The Organic Standard* 38, Grolink AB, www.organicstandard.com.

Fonte, M. (2002), 'Food Systems, Consumption Models and Risk Perception in Late Modernity', *International Journal of Sociology of Food and Agriculture* 10: 1. Available at: Available at: http://www.otago.ac.nz/nzpg/csafe/ijsaf/archive/vol10/Fonte.pdf.

Fonte, M. (2006), 'Slow Food's Presidia: What do Small Producers do with Big Retailers?' in Marsden, T. K. and Murdoch, J. (eds), *Research in Rural Sociology and Development, Vol. 12. Between the Local and the Global: Confronting Complexity in the Contemporary Agri-Food Sector.*

Fraser, N. (2005), *Qu'est-ce que La Justice Sociale?* (Paris: Éditions Découverte).

Friedmann, H. and McMichael, P. (1989), 'Agriculture and the State System', *Sociologia Ruralis*, 29, 93–117.

FTF. (2006), http://www.fairtrade.org.uk/press_office/press_releases_and_statements/archive_2006/dec_2006/fairtrade_fortnight_2007.aspx.

Galizzi, G. and Venturini, L. (1996), 'Product Innovation in the Food Industry' in Galizzi, G. and Venturini, L. (eds), *The Economics of Innovation: The Case of the Food Industry* (Physica Verlag).

Gereffi, G., Garcia Johnson, R. and Sasser, E. (2003), 'The NGO Industrial Complex', *Foreign Policy*, July.

Goodman, D. and Redclift, M. (1991), *Refashioning Nature: Food, Economy and Culture* (London: Routledge).

Goodman, M. (2004), 'Reading Fair Trade: Political Ecological Imaginary and the Moral Economy of Fair Trade Goods', *Political Geography* 23: 7, 891–915.

Goodman, D., Sorj, B. and Wilkinson, J. (1987), *From Farming to Biotechnology* (Oxford: Blackwell).

Granovetter, M. (1985), 'Economic Action and Social Structure: The Problem of Embeddedness', *American Journal of Sociology* 91: 3, 481–510.

Green, R. and Hy, M. (2002), 'La Traçabilité: Un Instrument de la Securité Alimentaire', *Agroalimentaria* 15: 15.

Grodnik, A. and Conroy, M. E. (2007), 'Fair Trade Coffee in the United States: Why Companies Join the Movement' in Raynolds, L., Murray, D. and Wilkinson, *J Fair Trade. The Challenges of Transforming Capitalism* (London and New York: Routledge).

Guthman, J. (2004), *Agrarian Dreams: The Paradox of Organic Farming in California* (Berkley: UC Press).

Hirschman, A. O. (1970), *Exit, Voice and Loyalty* (Cambridge, MA: Harvard University Press).

Hughes, J. (2006), 'Champagne, Feta and Bourbon: The Spirited Debate about Geographical Indications', *Hastings Law Journal* 58: December, 299; www.Hughes_44.doc.

INRA. (2000), *The Socio-Economics of Origin Labelled Products in Agri-food Supply Chain: Spatial, Institutional and Co-ordination Aspects* (eds),

Sylvander, B., Barjolle, D. and Farfini, F., Actes et Communications 17–1, 2 vols. (Paris: INRA).

Jacquiau, C. (2006), *Les Coulisses du Commerce Équitable* (Paris: Mille et Une Nuit).

Jaffe, D., Kloppenberg, J. and Monroy, M. B. (2004), 'Bringing the Moral Charge Home', *Rural Sociology* 69.

Krier, J. (2005), *Fair Trade in Europe 2005: Facts and Figures on Fair Trade in 25 European Countries* (Brussels: FINE and the Fair Trade Advocacy Office).

Laville, J.-L. (2007), *L'Économie Solidaire* (Paris: Hachette Littératures).

Leite Pereira, S. (2006), *Agrarian Reform, Social Justice and Sustainable Development*, Issue Paper 4, International Conference on Agrarian Reform, Porto Alegre, Brazil, FAO.

Marsden, T., Flynn, A. and Harrison, M. (2000), *Consuming Interests: The Social Provision of Foods* (London: UCL Press).

May, P. H., Mascarenhas, G. C. C. and Potts, J. (2004), *Sustainable Coffee Trade: the Role of Coffee Contracts* (Montreal: International Institute for Sustainable Development-IISD).

Miele, M. and Murdoch, J. (2002), 'The Practical Aesthetic of Traditional Cuisines: Slow Food in Tuscany', *Sociologia Ruralis* 42: 4.

Oppenheim, P. (2005), 'Fair Trade Fat Cats', *The Spectator*, November.

Oxfam International. (2008), *Another Inconvenient Truth*, Oxfam Briefing Paper 114 (Oxford).

Rama, R. and Wilkinson, J. (2007–2008), 'Foreign Direct Investment and Agri-Food Value Chains in Developing Countries: A Review of the Main Issues', *Commodity Market Review* (Rome: FAO).

Raynolds, L. (2004), 'The Globalization of Organic Agro-Food Networks', *World Development* 32: 5, 725–743.

Raynolds, L. and Long, M. (2007), 'Fair/Alternative Trade: Historical and Empirical Dimensions' in Raynolds, L., Murray, D. and Wilkinson, J., *Fair Trade. The Challenges of Transforming Capitalism* (London and New York: Routledge).

Raynolds, L., Murray, D. and Wilkinson, J. (2007), *Fair Trade. The Challenges of Transforming Capitalism* (London: Routledge).

Reardon, T., Codron, J.-M., Busch, L., Bingen, J. and Harris, C. (2001), 'Global Change in Agrifood Grades and Standards: Agribusiness Strategic Responses in Developing Countries', *International Food and Agribusiness Management Review* 2: 3, 432–435.

Reardon, T., Timmer, C. P. and Berdegué, J. (2008), 'The Rapid Rise of Supermarkets in Developing Countries' in McCullough, E. B., Pingali, P. L. and Stamoulis, K. G. (eds), *The Transformation of Agri-Food Systems* (London: Earthscan).

Sabel, C. F. (1994), 'Learning by Monitoring: The Institutions of Economic Development' in Smelser, N. and Swedberg, R. (eds), *The Handbook of Economic Sociology* (Princeton: Princeton University Press and Russell Sage Foundation).

Sauvée, L. and Valceschini, E. (2003), 'Agro-alimentaire: La Qualité au Coeur dês Relations Entre Agriculteurs, Industriels et Distributeurs' in *Demeter 2004* (pp. 181–227) (Paris: Armand Colin).

Simmel (apud Arnsperger, C. and Parijs, P.) (2000), *Éthique économique et social* (Paris: La Découverte).

SINERGI. www.origin-food.org.

Smith, J. (2005), 'Globalization and Transnational Social Movement Organizations' in Davis, G. F., McAdam, D., Scott, W. R. and Zald, M. N. (eds), *Social Movements and Organization Theory* (New York: Cambridge University Press).

Smith, S. and Barrientos, S. (2005), 'Fair Trade and Ethical Trade: Are there Moves Towards Convergence?' *Sustainable Development* 13, 190–198.

Stiglitz, E. J. and Charlton, A. (2005), *Fair Trade for All* (Oxford: Oxford University Press).

Sylvander, B. (1995), 'Conventions on Quality in the Fruit and Vegetables Sector: Results on the Organic Sector', *Acta Horticulture* 340, 241–246.

Tallontire, A. and Vorley, B. (2005), *Achieving Fairness in Trading between Supermarkets and Their Agrifood Supply Chains* (London: UK Food Group).

Thompson, A. (2003), 'Behind the Label: How Well is the Forest Stewardship Coubncil Protecting Trees?', *The Environmental Magazine*, Jan–Feb.

Tregear, A. (2003), 'From Stilton to Vimto: Using Food History to Re-think Typical Products in Rural Development', *Sociologia Ruralis* 43: 2, 91–107.

Uzzi, B. (1996), 'The Sources and Consequences of Embeddedness for the Economic Performance of Organizations: The Network Effect', *American Sociological Review* 61, 674–698.

Weid von der J. M. and Altieri, M. (2002), 'Perspectivas de Manejo de Recursos Naturais com Base Agroecológica par Agricultores de Baixa Renda, Século XXI', in Albuquerque Lima, D. and Wilkinson, J. (orgs.). *Inovação nas Tradições da Agricultura Familiar* (Brasília: Paralelo 15).

Wilkinson, J. (2001), 'Building Trust. A Review of Richard M. Locke', *Econômica* 3: 2.

Wilkinson, J. (2002), 'GMOs, Organics and the Contested Construction of Demand in the Agrofood System', *International Journal of Sociology of Agriculture and Society* 11, 9–21.

Wilkinson, J. (2004), 'The Food Processing Industry, Globalization and Developing Countries', *Journal of Agricultural and Development Economics* 1: 2, 184–201, *eJADE* (FAO) www.fao.org/es/esa/eJADE.

Wilkinson, J. (2006a), 'Network Theories and Political Economy. From Attrition to Convergence?' in Marsden, T. K. and Murdoch, J. (eds), *Research in Rural Sociology and Development, Vol. 12. Between the Local and the Global: Confronting Complexity in the Contemporary Agri-Food Sector* (Amsterdam: Elsevier).

Wilkinson, J. (2006b), *Fair Trade Moves Center Stage* (Rio de Janeiro: Edelstein Publications) www.edelsteincenter.org.

Wilkinson, J. (2007), 'Fair Trade. Dilemmas of a Market Oriented Global Social Movement', *Journal of Consumer Policy* 30: 3, 219–239.

Wilkinson, J. (2009), 'Globalization of Agribusiness & Developing World Food Systems', *Monthly Review* 61: 4.

Wilkinson, J. and Cerdan, C. (2008), *A Brazilian Perspective on GIs*, Paper presented to SINER-GI Final Conference, Geneva, 2008.

Williamson, O. (1985), *The Economic Institutions of Capitalism* (New York: Free Press).

World Development Report. (2007), *Agriculture for Development* (Washington: World Bank).

PART II
Consumption in Space and Place

Chapter 6

Ethical Campaigning and Buyer-Driven Commodity Chains: Transforming Retailers' Purchasing Practices?

Alex Hughes, Neil Wrigley and Martin Buttle

Introduction

This chapter explores how ethically-sensitive consumption is challenging the organization of retailer-driven supply chains. In particular, the effects of ethical campaigning on changing retailers' purchasing practices in global supply networks are examined. Retailers' supply chains have been the subject of intense scrutiny by ethically-inclined consumers, social scientists, the media and civil society activists for over two decades now. The adverse effects of western corporate buying power upon labour conditions, in particular those in the global South, have been of great concern to these groups and have been highlighted in numerous critical articles and trade justice campaigns. Since the 1990s, an extensive ethical trading agenda has developed trans-nationally to address these political-economic issues, with codes of conduct concerned primarily with labour standards (though in some cases also addressing environmental issues) set up by a range of stakeholder groups to foster greater worker welfare at sites of production (Barrientos and Dolan 2006). However, while labour codes have become commonplace, many activists more recently have argued that these standards rarely attend to the powerful commercial buying practices that put producers under so much economic pressure in the first place.

Many of the labour problems experienced at sites of export production, including extremely low wages, excessive overtime and poor worker safety, are frequently created by the downward pressure exerted on producers by powerful buyers on behalf of large retail and brand name companies. These buyers continually place suppliers under pressure by altering product specifications at short notice, increasing orders, shortening delivery times and reducing the prices paid to producers (Ethical Trading Initiative 2008). Such pressure on suppliers, linked strongly to demands emanating from the sphere of consumption such as fast and disposable fashion and demand for low cost goods, lies at the heart of problems concerned with improving workplace standards. And yet, retailers' powerful purchasing practices have rarely been addressed in the codes of conduct set out since the 1990s by ethical trading organizations. This gap between the work of ethical trading initiatives and the organization of global supply chains in practice is now being addressed by non-governmental organizations (NGOs), trade unions, consultancies and leading-edge retailers, predominantly (though not exclusively) in the UK. As a result of targeted campaigning and critical journalism, the UK's Ethical Trading Initiative (ETI) has been organizing a 'Purchasing Practices Project' since 2005 that seeks to tackle directly the commercial buying practices that lie at the root of many key ethical trading problems. This initiative, and the broader ethical purchasing practices agenda into which it fits, is a new area of development in the governance of global supply networks that critically connects spaces of consumption with the spaces of production in the contemporary global economy.

This chapter seeks to interrogate the ways in which this 'purchasing practices' agenda is shaping UK-based retailers' responsible sourcing strategies. First, the scene is set with regard to the landscape of globalizing trade and the growing extent to which the governance of global supply networks has become driven by retail and brand-name corporate buyers. Second, the emergence of an ethical trading agenda constructed to counter this buying power is discussed, with attention paid to its embeddedness in a broader field of ethical consumption. The following sections focus, in turn, on: (i) the development of a purchasing practices agenda by activists and ethical consultancies in the UK; and (ii) the corporate response to this activism in terms of retail participation in the ETI Purchasing Practices Project since 2005 and internal strategic thinking on commercial buying practices. Attention is also paid to some of the reasons why this specific programme of work on ethical trade has developed more strongly in the UK context of retail and consumption than it has elsewhere in the world. In so doing, it is suggested that the spatiality of the purchasing practices agenda is connected to broader spaces of consumption that form the subject of this book. Overall, the chapter addresses the influence of consumption trends in both shaping the nature of retail buying power and, in the form of ethical campaigning and advocacy work, attempting also to counter this power. The convergence of these two contrasting and clashing articulations of consumer interest—demands for desirable goods at cheap prices on the one hand, and moral concern about distant conditions of work on the other—lie at the heart of the challenge facing retailers with respect to their purchasing practices. And it is this strategic challenge and its relationality to the spaces and politics of consumption that forms the focus of our chapter.

The empirical material discussed is from a wider research project sponsored by the Economic and Social Research Council in the UK between 2005 and 2007. This study aimed to evaluate UK and US differences in retailers' approaches to ethical trade in the food and clothing sectors, influenced by contrasting national contexts of ethical campaigning. It involved 96 in-depth interviews with UK and US ethical trade practitioners working for leading corporate food and clothing retailers and those involved in campaigning for, and developing, ethical trade on behalf of multi-stakeholder organizations, NGOs and trade unions. This chapter draws specifically on the UK-based interviews. These include 20 interviews with organizations on the advocacy side of ethical trade, incorporating NGOs, trade unions, the media and the ETI. On the corporate side, 26 representatives from retailers, auditors and consultants were interviewed, including seven of the top ten UK food retailers and nine of the leading ten clothing retailers in terms of market share. Significantly for this chapter on the purchasing practices agenda, this sample of key informants includes the NGO and two ethical consultancy firms leading the campaigning and pressure on retailers in this area. It also includes six of the eight corporations, one of the three unions and four of the five NGOs signed up to the ETI's Purchasing Practices Project. Interview material is also backed up by reports and web-based sources on the issue, some of which have been published since the ESRC-funded research was conducted.

Buyer-driven Commodity Chains and Retailer Power

Since the 1980s, social scientists have acknowledged the increasingly powerful ways in which trends and practices of consumption influence the organization of production. Most significantly, the shift in economic power from production to services and consumption was recognized by commentators in the very large and wide-ranging literature on post-Fordism and flexible specialization (see, for example, Amin 1994 and Harvey 1989). More specifically, studies of commodity chains and networks concerned with relationships between the production, circulation and consumption of goods and services have recognized and sought to theorize the influence of consumer trends and their articulation through the strategies of powerful corporate buyers. One of the most influential commentaries on this development comes from Gary Gereffi and the school of thought on global commodity chains (GCCs). Gereffi (1994a, 1994b) provides a characterization of the 'modes of organization' governing the commodity chain. Distinguishing between producer-driven and buyer-driven chains, Gereffi draws attention to the importance of the buyer-driven chain in driving trade in consumer goods such as garments, footwear, toys, consumer electronics, household products and hand-crafted items. Emphasis is placed upon the power of commercial capital in allocating economic resources and wealth within—and therefore defining the shape of—GCCs. The most recent incarnation of GCC theory, re-labelled the global value chains (GVC) approach, also asserts the significance of buyers on behalf of large retail and brand name companies in governing the organization of trade, often through what are referred to as 'relational' forms of networked co-ordination (Gereffi et al. 2005), as does the more spatially-aware global production networks (GPN) analytical framework associated with the Manchester School of economic geography (Coe et al. 2008). There is not space here to review the theoretical merits of the large literature that has evolved on commodity chains and networks. However, the attention paid to the role of retail and consumption in these networks is significant to note.

In terms of detailed empirical research on retailers and buying power, informed by these theoretical approaches, studies have highlighted the significance of the forces wielded by retailers in the supply chain. The shift in the balance of power from manufacturing suppliers to retailers in both the food and clothing industries of the UK, for example, was noted by economic geographers in the 1990s (Bowlby et al. 1992; Bowlby and Foord 1995; Crewe and Davenport 1992; Doel 1996; Hughes 1999). In the UK, following the abandonment of resale price maintenance in 1964 (Doel 1996) and the subsequent development of retailer buying power, the largest of the 'big capital' retail chains became increasingly able to dictate the terms and conditions of trade to their manufacturing suppliers, and listing fees were demanded from manufacturers by retailers before new products were accepted onto supermarket shelves. Retailers also extracted increasingly favourable pricing terms from their manufacturing suppliers, often involving non-cost-related discriminatory discounts. They extended their buying power still further by negotiating favourable terms of payment from these suppliers (Hughes 1996; Wrigley 1992).

From the mid-1990s onwards, academics also have documented the increasing globalization of retailers' supply chains within a political-economic context of neoliberalism, retail FDI liberalization, and rapidly expanding trans-national retail activity (Coe 2004; Coe and Hess 2005; Reardon et al. 2003; Wrigley 2000; Wrigley et al. 2005). Work by Barrett et al. (1999, 2004) and Dolan and Humphrey (2004) draws attention to the ways in which UK supermarket chains forge increasingly direct relationships with suppliers of high-value horticultural produce from African producers, for example, forming 'fully-integrated' distribution channels. Retailers develop tight specifications for the commodities they source, placing growers under pressure to produce crops that meet these specifications. Binding contractual agreements are rarely set up with these overseas suppliers, and instead buyers spread their supplies of each product category over a wide range of global sources. This means that the retail chains can pull out of supply relationships at any time. Moreover, retailers operate on high profit margins, above all in the area of non-traditional commodities, with large-scale profits rarely flowing back to producers. This serves to perpetuate poor conditions of work at sites of production, including a lack of job security, pressure to work overtime at particular times of peak demand, and the employment of large numbers of temporary workers (Kritzinger et al. 2004).

Similar developments are observed in the global clothing trade, although in contrast to the direct supply relationships forged between retailers and producers of food products supply channels for clothing and footwear are shown to involve vast networks of sub-contracting relationships through which retailers orchestrate demands (Crewe 2004; Crewe and Davenport 1992; Hale 2000; Tokatli 2008). Consumer demands linked to trends recently associated with 'fast fashion' are, in particular, shown to place export producers under extreme stress through downward pressure on prices and constantly changing orders (Tokatli 2008; Weller 2007). This, in turn, creates similar problems for labour conditions as discussed for the food industry. It is against this backdrop of retailer power in the supply chain that ethical trading programmes have been called for and developed. Furthermore, the aforementioned commercial buying practices themselves have been the focus of attention by recent campaigners in the purchasing practices agenda that forms the focus of this chapter.

Campaigning and Ethical Consumption in the UK: Setting the Scene for the Corporate Purchasing Practices Agenda

In the context of globalizing supply chains driven by retailers, both critical journalism and direct campaigning orchestrated by civil society organizations have led to the development of ethical trading programmes concerned with labour and environmental standards at sites of production (Freidberg 2004). Supermarkets and high street retailers in the UK that have been targeted by campaigners and journalists include the leading four food retailers—Tesco, Sainsbury, Asda and Morrisons—and clothing retailers such as Marks and Spencer, Next, Debenhams and brands owned by Philip Green.

Media exposure and civil society campaigning on the subject of ethical trade is suggested by cultural and political geographers to be strongly connected to new waves of ethical consumption and associated emotions of care. With relevance for this chapter, consumers' purchasing decisions are seen to be shaped increasingly by consideration of companies' ethical reputations and 'credence factors' (Dolan and Humphrey 2004; Ethical Consumer 2007). At the heart of trends in ethical consumption, in which citizens as consumers are encouraged increasingly to make responsible purchasing decisions in the context of complex networks of relations, lie emotions concerned with ethics of care (Barnett et al. 2005). And 'caring at a distance' is seen to shape various forms of ethical consumption, including campaigns for trade justice (Barnett et al. 2005; Bryant and Goodman 2004; Clarke et al. 2007; Dolan 2005; Goodman 2004). Clarke et al. (2007) emphasize the powerful role played by intermediaries, including the media and NGOs, in raising awareness of the often global, social and political significance of consumers' purchasing decisions. They also note the 'politics of shame' (238) simultaneously directed by these organizations at corporate actors, including retailers, in the course of playing their 'brokering role' (240) between consumers on the one hand and corporations on the other. In the absence of significant corporate responsibility and labour codes before the mid-1990s, NGOs in the UK such as Oxfam, Christian Aid and Catholic Agency for Overseas Development (CAFOD) drew consumers' attention to the very poor labour conditions at sites of export production in retailers' ever more globalizing supply chains (Freidberg 2004; Hughes 2001). Such campaigning was frequently reported in broadsheet newspapers and was supported by radio and television documentaries (Wrigley and Lowe 2002).

As a result of the work of intermediaries such as the media and NGOs, and in the absence of state regulation of trans-national production networks, corporate responsibility in retailers' supply chains has been implemented widely through labour and environmental codes of conduct involving minimum standards for workers and environmental protection at sites of production (Barrientos and Dolan 2006; Christopherson and Lillie 2005; Jenkins 2002). Multi-stakeholder organizations involving institutional collaboration between companies, NGOs and sometimes trade unions have been particularly influential in guiding corporate approaches to labour standards (Barrientos and Dolan 2006; Jenkins 2002). In the UK, the ETI, for example, influences the codes and procedures used by many high-profile retailers and brand manufacturers in the governance of ethical supply chains. The ETI was formally established in 1997 and now represents one of the largest and most strategically significant civil initiatives of its kind. It consists of corporate, NGO and trade union members and is supported by the UK government's Department for International Development (DfID).

Lying at the heart of the ETI is the base code of conduct, which the organization encourages all corporate members to use in the implementation of responsible business standards in the context of their own-brand supply chains. This base code consists of nine provisions that build directly on core ILO conventions. These provisions, exclusively concerning labour standards, are: employment is freely chosen; freedom of association and the right to collective bargaining are respected; working conditions are safe and hygienic; child labour should not be used; living wages are paid; working hours are not excessive; no discrimination is practised; regular employment is provided; and no harsh or inhumane treatment is allowed (Crewe 2004; Hughes 2005). However, the terms of trade between retail buyers and their suppliers are absent from the base code, lying untouched by this form of voluntary supply chain governance.[1]

Since the establishment of the ETI, journalists and campaigning groups in the UK have continued their focus on retailers' supply chains, with a developing critical spotlight on retailers' ethical trading programmes themselves. Television documentaries have continued with, for example, BBC *Panorama* programmes in 2000 and 2008 examining retailers' labour codes in global supply chains for clothing. Newspaper reports also have maintained their critical focus on supermarkets and high street clothing retailers with respect to their supply chain management (Mathieson 2006; Mathieson and Aglionby 2006a, 2006b; Pratley and Finch 2005). With significance for an emerging ethical trading agenda focusing on purchasing practices, the trading impacts of fast and disposable clothing fashion at the low cost end of the market have been the focus of media reports (e.g. *The Guardian* 16/07/07). Campaigns aimed at retailers also have continued, exemplified by Action Aid's *Rotten Fruit* report (Wijeratna 2005) illustrating how

1 The ETI now has 'Principles of Implementation' that provide general guidelines for companies in the execution of their ethical trading strategies, including buying and awareness-raising (see http://www.ethicaltrade.org/Z/lib/base/poi_en.shtml; accessed 22/07/08). However, these principles were not in place before the purchasing practices agenda was set by campaigners and consultancy firms.

workers on South African fruit farms supplying Tesco are being exploited despite the retailer's claims to be trading responsibly. Oxfam's *Trading Away Our Rights* report (Oxfam 2004), discussed below, emerged also as one of the key vehicles for focusing the debate on the power and negative impacts of retailers' purchasing practices.

Challenging Retailers' Purchasing Practices: The Case of the UK Retailers

UK-based trade justice work and the focus on corporate purchasing practices

Extensive campaigning and critical media coverage on the subject of global supply chains and labour issues since the mid-1990s means that the majority of UK retailers, many of whom are ETI members, now have well-established ethical trading strategies and labour codes. However, by the early part of the twenty-first century the very buying practices that put so much pressure on suppliers and working conditions in the first place remained curiously untouched by these strategies and codes. Reflecting on the terrain of the ethical trading debate at this time, one NGO campaigner explains that:

> Nobody had talked about purchasing practices before. And when you went to companies, they just reeled off all the things they were doing around codes. And then you said, "What about purchasing practices?" and they just dried up because nobody had their corporate speech on this yet. And we realised that this was an issue. It was a new thing. (Interview, 19/09/05)

Critical commentators in the UK, including academics, campaigners and ethical consultancies, recently have drawn attention to the fact that retailers' purchasing practices—involving downward pressure on prices, altering product specifications at short notice, increasing orders and shortening delivery times—are actually *undermining* corporate ethical trading programmes. As one campaigner puts it:

> The right brain and the left brain are not connecting in the sense that the ethics is a wish instead of a corporate responsibility, but it is being undermined by the practices of the commercial buyers for whom ethics are not, in some cases, issues at all. So you've got a disconnect there, which is pretty fundamental. (Interview, 04/10/05)

This failure of labour codes and ethical trading programmes to engage with practices connected to retail buying power itself reflects a larger point made by Blowfield (2005) that the corporate responsibility movement falls short of addressing the fundamental values of capitalism concerning profitability, free trade, the freedom of firms to disinvest, private property, commodification and, of direct relevance to the purchasing practice debate, market power. In Blowfield's view, the vast majority of corporate responsibility initiatives only engage with the more negotiable aspects of corporate activity. Retailers' commercial buying practices have, until recently, fallen outside of the range of supply chain activities monitored by ethical trade and codes of conduct. One trade union commentator suggests that:

> Purchasing practices goes to the heart of the capitalist machine. It touches raw nerves like pricing and practice … On the one hand you've got audit and compliance efforts, so there are people from one side of the organization going in to examine the negative effects and trying to rectify things. And then there is somebody from the other side of the organization going in saying we want it this price and we want it fast. You put those two heavy impositions on a supplier and that's bound to have an effect. (Interview, 14/11/05)

This reflection draws attention to the organizational conundrum facing retailers attempting to incorporate responsible purchasing practices into their corporate ethical trading agendas. That is, that the teams responsible for executing ethical trading programmes and organizing the auditing of labour codes are frequently *detached from* the teams concerned with corporate buying.

The key problems associated with the downward pressure exerted on suppliers by retail buying practices, and the organizational disconnect between the aims of ethical trade on the one hand and the profit-maximizing objectives of commercial purchasing on the other, were placed in the critical spotlight both by the ETI at its biennial conference in 2003 and by three important reports published in the UK in 2004. While a small minority of leading-edge corporations based outside the UK were beginning to address the impacts of their buying practices, for example Nike and Gap in the USA, this sharp focus on purchasing practices in the UK ethical trade debate was unprecedented. So when the three reports—*Buying Your Way into Trouble*, *Taking the Temperature* and *Trading Away Our Rights: Women Workers in Global Supply Chains* by Acona Limited, Impactt Limited and Oxfam respectively—were published, shock waves were felt by ethical consumers, the media and the corporate community alike. While trade unions like the International Textile, Garment and Leather Workers Federation (ITGLWF) and NGOs such as Banana Link and Christian Aid also drew corporate attention to the adverse effects of retail buying, it was these three reports that provided the crucial vehicles for placing purchasing practices more centrally on the ethical trade agenda.

The reports written by Acona Limited and Impactt Limited are similar in that they represent pieces of published research and strategic recommendations for companies by ethical consultancy firms, both of whom have mainstream corporations as their core clientele and corporate responsibility as their focus. In Acona's case, corporate sustainability is linked to the wider management of business risk. Impactt Limited, who at the time of writing has been operating for just over ten years, has an agenda more specifically focused on the ethical management of supply chains. Both companies, though, represent part of the significant 'communities of practice' (Wenger 1998) involved in creating and disseminating knowledge on the ethical economy. For Impactt, buying practices had been part of their agenda for corporate change prior to the publication of their *Taking the Temperature* report, featuring on their private training programmes offered to retail clients, and they continue to be recognized as a key ethical trading challenge for retailers. For Acona, their report accompanied other pieces of commissioned research on ethical trade, including a study of homeworkers.

While Impactt's 2004 report touches on buying practices as one of many ethical trading issues, the Acona report on *Buying Your Way into Trouble* places purchasing practices centre stage. The report was written by the consultancy firm on behalf of Insight Investment, the asset management of UK financial services company, HBOS (Halifax Bank of Scotland), and its intended audience is managers responsible for purchasing within large brand name and retail corporations. As one of the Acona authors suggests,

> The strength of our report was that we were trying to build a business case for [ethical] purchasing practices. (Interview, 09/05/06)

The publication interrogates the nature of corporate buying practices and their supply chain impacts based on the consultancy's own experience of working with clients and also thirteen in-depth interviews, including those with companies in whom Insight Investment has a shareholding. The report focuses on the adverse effects of problematic buying cultures within firms and highlights the fact that buying cultures frequently undermine ethical trading strategies within those very same firms. Illustrating the tension between commercial buyers and ethical trading managers, some buyers were reported to refer to ethical trading personnel as the 'sales prevention team' (Acona 2004, 32). Drawing attention to the organizational separation of commercial and ethical trade teams within companies, one of the authors of the report recounts that:

> In one of the companies we interviewed for our report, we managed to get the Director of Marketing and the Head of CSR [Corporate Social Responsibility] in the room together, and at the end of the meeting—it was an hour long meeting—the Director of Marketing turned to his colleague and said, "That's the longest I've ever spent in a room with you". And they'd been working for this company for five or six years. (Interview, 09/05/06)

A key recommendation made by Acona for companies is therefore to adopt a more integrated approach to managing efficient supply chains and simultaneously maintaining ethical standards. To do this, they suggest that companies carefully manage their 'critical path'[2] and build ethical supply chain management into a more sustainable buying culture.

The Oxfam (2004) report on *Trading Away Our Rights: Women Workers in Global Supply Chains* also attends critically to the subject of purchasing practices, but within a broader frame of reference concerned with the rights of female workers at sites of export production. The report covers both clothing and agricultural sectors and represents a significant piece of research and advocacy work on the part of this large and highly influential NGO. Its author, Kate Raworth, explains that there were three key aims of this work—changing the terms of the ethical trade debate, changing policy and changing practices.

> The biggest goal was to change the terms of the debate, so that people could no longer talk about codes of conduct without looking at the commercial side. (Interview, 19/09/05)

The report was based on extensive research in 12 countries, including interviews with hundreds of women workers, as well as factory and retail managers and bureaucrats. It presents coverage of the severe pinch points felt by export producers in both clothing and agriculture supplying large retail firms and highlights, in particular, the very poor conditions of work and lack of labour rights experienced by female employees in developing countries. In this context, the purchasing practices of large retail firms, including Tesco, are demonstrated to create insecurity at sites of production. The author of the report sought the expert advice of the buying teams based within Oxfam itself, as the organization also operates its own ethical business using global supply chains. This experience fed into the report in order that realistic recommendations could be made. Amongst several key recommendations, retailers are urged to integrate respect for labour rights into their overall supply chain management and purchasing practices. Consumers also form a target group, being asked to 'insist that retailers and brands ensure that their sourcing and purchasing practices support, rather than undermine, workers' rights' (Oxfam 2004, 8). Such targeting of consumers as well as companies and governments is a critical move and illustrates the ethical 'brokering role' (Clarke et al. 2007, 240) played by this NGO between consumers and corporations on the subject of purchasing practices. The key findings of Oxfam's work were communicated to the public through the dissemination of the report on Oxfam's website and press releases broadcast through the media, including broadsheets such as *The Guardian*. Significantly, as discussed above, the message that retailers' purchasing practices continue to undermine ethical trading efforts is continually being reinforced in the media through recent newspaper articles and television documentaries concerning the negative impact on producers and workers of

2 The 'critical path' is a management term referring to the series of inter-linked, time-critical tasks that must be performed in order for the buying process to proceed. It typically includes: design, planning, sourcing, sampling, production and logistics.

extreme downward pressure on prices connected with fast and disposable fashion (for example, *The Guardian* 16/07/07; BBC *Panorama* documentary screened on 23/06/08).

Oxfam took its ethical brokering role still further by taking the findings and recommendations of its *Trading Away Our Rights* report directly to the UK's ETI—the multi-stakeholder organization sitting at the centre of developments in ethical trade and shaping many retailers' ethical supply chain strategies (Jenkins 2002; Hughes et al. 2007). Kate Raworth recounts:

> About a year ago after the report came out, the first thing we did, we went and did a presentation of the report to the company caucuses—the garment one and the fresh produce one [of the ETI]—which was good because they said "yeah". It was the first time they had heard this story. (Interview, 19/09/05)

This strategic connection made by Oxfam with the ETI, and the aforementioned consultancy reports and other campaigning trade unions and NGOs, have been critical in taking the purchasing practices agenda directly to the retailers. The following section of this chapter addresses the extent to and way in which the retailers have responded to the calls for more ethical purchasing practices, both through the ETI and with respect to their own corporate strategies.

The response of UK retailers to campaigning on corporate purchasing practices

In the first decade of the twenty-first century, as a response to the campaigning and reports discussed above, both the ETI and its UK retail members began to acknowledge the importance of addressing purchasing practices as a *part of* their ethical trading efforts. As one high street retailer puts it, 'I think it's going to be the hot topic. It's certainly not going to go away' (Interview, 20/06/06). Acknowledging that purchasing practices are a major challenge for ethical trading management, the Head of Quality Assurance working for a prominent UK chain of department stores notes that:

> There's no way I'd be able to sit here and honestly say to you guys or anybody else that our purchasing practices are completely in line with our ethical trading principles or policies. That would be a nice thing to say, but I'd be lying if I did. We know that we're not perfect. We don't make [buying] decisions at the right time. (Interview, 03/05/06)

A key way in which campaigners and ethical consultancies have persuaded UK retailers to take purchasing practices seriously is through the ETI, as highlighted above. Both the 2003 and 2005 ETI biennial conferences organized sessions concerning the challenge of marrying 'the commercial' with 'the ethical'. The next large meeting of the organization to be held in October 2008 also has the issue centrally on its agenda. The whole ethos of the ETI centres on the importance of learning in the development of best practice for ethical trade. Central activities of the organization are conferences, seminars, workshops, research and, importantly, experimental projects that each involve a selection of the ETI's members from the NGO, trade union and corporate communities.

From April 2005, the ETI set up an experimental project on purchasing practices, which remains active at the time of writing. This project aims to discover how retailers' buying practices can more effectively support working conditions at sites of production. Eight companies are signed up to the project-Asda, Debenhams, Gap (based in the USA), Inditex, Marks and Spencer, New Look, Next and WH Smith. These companies are each partnered by a participating NGO or trade union in order to work collaboratively on assessing and improving their purchasing practices with respect to ethical supply chain management. Trade unions involved are the ITGLWF, Prospect and the Transport and General Workers Union. NGOs working on the experimental project include some of the most prominent organizations in the debate—CAFOD, Homeworkers Worldwide, Oxfam, Traidcraft and Women Working Worldwide. As reported publicly on the ETI's website:

> Project group participants are involved in the following activities: (i) analysing the critical path (the pipeline of decisions and actions from design to delivery); (ii) engaging suppliers to investigate the impact of key purchasing decisions; (iii) raising buyers' awareness of working conditions; and (iv) collaborating with other companies to explore the impact of multiple buyers sourcing from one supplier and the role of agents.[3]

This experimental project arguably represents the most strategically significant ethical trading work on the purchasing practices issue in the UK and beyond, as it draws in some very high profile companies and organizations and represents a sustained attempt over time to address the challenge of ensuring that retailers' everyday buying practices do not undermine the efforts of ethical trade and labour codes. Demonstrating the extent to which the project and its objectives have captured the attention of a broader range of ETI members, the organization also held a members' meeting on 29 November 2007 on the subject of purchasing practices with over 100 people in attendance incorporating more than 60 corporate representatives, including retailers' buying staff. The meeting was organized around the discussion of three case studies concerning, in turn, the impacts of downward pressure on prices, late orders and the lack of corporate integration between the activities of commercial and ethical teams (ETI 2008; Interview, 09/05/06). The ETI also now has 'Principles of Implementation' that provide general guidelines for companies in the *execution* of their ethical trading strategies, including buying and awareness-raising.[4] though these are not monitored standards. In the remainder of

3 http://www.ethicaltrade.org/Z/actvts/exproj/purchprac/index.shtml (accessed 09/03/08).

4 http://www.ethicaltrade.org/Z/lib/base/poi_en.shtml (accessed 22/07/08).

the chapter, we discuss some of the ways in which the ETI and retail engagement with the purchasing practices debate represent a successful step forward in the field of ethical trade. The limitations and challenges of the ETI's Purchasing Practices Project (and the wider issue of changing buying practices) are then highlighted, including the challenge for the retailers of integrating more ethically-sensitive purchasing practices into their everyday management of supply chains.

Evaluating the challenges of the ETI Purchasing Practices Project and the changing corporate strategies of ethical buying practice

With the exception of the USA's Worker Rights Consortium, which is evaluating the impacts of buying practices on ethical production in the supply chains specifically for college apparel, the ETI is quite unique in its acceptance of purchasing practices onto the ethical trading agenda. Most other companies and multi-stakeholder organizations working on ethical trade focus on the monitoring of labour codes with little consideration for the terms of trade and corporate demands affecting them (Hughes et al. 2007). Reflecting on their success in engaging the ETI, and acknowledging the organization's willingness to take the issue on board, the author of Oxfam's *Trading Away Our Rights* report comments that:

> I don't see another organization that is where the ETI is on purchasing practices ... So if the ETI has actually got companies around the table talking about this, I think that is fantastic ... Give them credit because I think they could have said, "Sorry, but that is outside the terms of reference. We were set up to promote learning and the implementation of codes of labour practices, and that has nothing to do with purchasing practices". But they didn't. They actually changed the ETI to reflect this. (Interview, 19/09/05)

One of the explanations for this willingness to take on such a significant challenge relates to the ETI's developmental philosophy, where its members are comfortable with experimentation and raising standards to which they can aspire rather than setting rigid codes that are measurable and immediately achievable. Remarking on their companies' decisions to get involved with the ETI's Purchasing Practices Project at its inception back in 2005, the following ethical trading managers working for two major UK high street retailers commented as follows:

> We wanted to get involved with it. We recognized, actually, that probably what we do in here and how we operate probably does have an effect on suppliers' ability to be able to comply with the code. (Interview, 16/02/07)

> My belief is that it's better to be in at the beginning and be there to start looking at your own processes and things now because the pressure is going to come at some point, whether it's in a year or two years or three years, and it's better to be on the front foot and have already taken a look and be starting to move

> in the right direction than be playing catch-up in three years time. (Interview, 04/05/06)

This latter retailer is one of the companies leading the way in terms of developing more ethically-sensitive purchasing practices, and the manager commented on a very successful working relationship with their partner NGO on the ETI project. Overall, there is clear recognition that the ETI, along with campaigners, consultancies and leading-edge retailers, have made significant headway in setting the corporate purchasing practices agenda and developing new work in this area. However, there are several key challenges that the ETI and its members face both in the execution of the experimental project and in the translation of this strategic agenda into significant improvements in everyday supply chain management. These organizational challenges can be categorized broadly as: (i) maintaining appropriate levels of corporate and non-firm involvement in the ETI project; (ii) challenges associated with consulting suppliers on the impacts of purchasing practices; (iii) the challenge for retailers' ethical trading managers of gaining internal support for ethical purchasing practices from their top-level management and buyers; and (iv) the limitations associated with inter-firm collaboration on sensitive areas of retailers' business such as product prices. These challenges are discussed briefly in turn.

Developing an experimental project on the theme of retailers' purchasing practices is a radical move by the ETI. While the organization has been successful in initiating this programme of work and attracting interest from many firms, maintaining high levels of active participation in the project itself represents the first key challenge. Progress has occasionally been slow and several companies and NGOs have been forced to pull out due either to problems gaining the support of their top-level management or to a lack of resources to fund their involvement (personal interviews). The following comments made by a trade union and a corporate participant respectively towards the start of the project's life reflect these problems:

> The company I am working with is really dragging its feet … The issue is, I assumed that the company involved in this project had cleared the way with senior management to be involved and it seems like what we are having to do is build a case to take it to management to get their approval just to start, and we've been going months now. (Interview, 14/11/05)

> Initially there were a lot of corporate members sitting round the table wanting to be involved. We're at a point now where there are two members who are actually actively participating in the project and three who are still hanging on in there but are stuck for whatever reason. And all the rest have fallen by the wayside … I think that one of the reasons is the challenge to business to take a long hard look at itself and look at the way that you operate … It really pushed

> organizations maybe a bit quicker than they were ready to go and start asking some searching questions. (Interview, 04/05/06)

Another necessary component of the ETI Purchasing Practices Project is consultation with suppliers in order most effectively to understand the impact of buying practices on producers and working conditions. One ethical trading manager working for a leading UK supermarket chain participating in the project explained the importance of supplier consultation when investigating his company's critical path (or supply chain) and referred to the need for a 'two-way conversation' when it comes to working with suppliers on this issue (14/06/05). However, the Director of the ETI points out that such dialogue with suppliers represents a radical departure for retailers from their usual way of operating in the supply chain:

> You have to do what most retailers have not done very much in the past, and that is to ask suppliers what they think. And suppliers are not used to being asked what they think, just being told what to do. So that does take time. (Interview, 12/09/05)

Ethical campaigning on purchasing practices has therefore forced a response from retailers that challenges established corporate approaches to the organization of buyer-driven chains (Gereffi 1994a, 1994b) and instead involves retailers experimenting with forms of collaboration with producers concerning the effects of their buying power on labour conditions.

In addition to the tests associated with developing a more dialogic relationship with suppliers external to the retail firms is the challenge of gaining support from other employees *within* the retail corporation itself. In other words, while retailers' ethical trading managers may be fully on board with the purchasing practices agenda, they still need to convince both top level management and their extensive buying teams of its importance before any improvements can be made. We have already highlighted the problems experienced by some ethical trading managers in persuading their line managers and company directors to agree to corporate participation in the ETI experimental project. But there is the additional challenge of getting buyers themselves both to appreciate and to respond to the need for more ethical purchasing methods. As the person responsible for ethical trade at the headquarters of a major UK chain of department stores explains:

> It is about getting what could be the lowest paid people in [our company] on a buying team to realize that if they don't send that e-mail to that supplier that night, then we lose our production slot, which then means that the factory are going to—if we still want the same delivery date—are going to have to work overtime. And it's all these knock-ons, and that for me is what "purchasing practices" is looking at. (Interview, 03/05/06)

One way of improving buyers' awareness of ethical issues is for retailers to develop training programmes. Several large retail chains have worked with NGOs and consultancy firms to develop short courses on the subject for their buying teams (Hughes 2006; Interview with UK supermarket chain, 08/08/06). However, the extent to which these educational initiatives result in improved purchasing in practice is yet to be assessed. In short, though, there are some leading retailers who are working to improve their purchasing practices internally. The challenges of ethical trade, though, have traditionally also involved significant collaboration *between* retail corporations. The ETI has been instrumental in fostering the sharing of ideas and best practice between companies. But as the purchasing practices agenda cuts straight to the heart of fundamental capitalist strategies, the sharing of ideas on this theme appears to be more problematic. Sensitive areas of retailers' business such as pricing and payment terms are central to the organization of purchasing and are therefore important topics of debate within the ETI experimental project. This poses a conundrum for retailers participating in the project who, on the one hand, wish to work collaboratively on the improvement of buying practices, but on the other hand need to protect their competitive position by ensuring that sensitive aspects of their supply chain strategies are kept confidential. The following quotes from a leading UK supermarket chain and an ethical consultancy firm respectively reflect this challenge:

> One of the difficulties I think we have is that the way companies buy things is the source of their competitive advantage and we couldn't share that if we wanted to from an OFT [Office of Fair Trading] point of view … I think they'd be fairly worried if there seemed to be some sort of collaboration on the way we buy things, with our competitors. So there was that issue to cut through and the group [the ETI Purchasing Practices Project], I don't think, really ever did. (Interview, 10/05/06)

> Some individual ETI member companies are doing good stuff on purchasing practices. I think it is a mistake to try and do purchasing practices as a group project, which is what they're doing, because I don't see how you can do it. Because surely, if you're going to improve purchasing, if you're going to make this social and beneficial improvement, if you're going to make it so better buying is better for the company, it's something you don't share with your competitors, so it has to be an individual thing. (Interview, 03/05/06)

There are therefore significant challenges associated with the purchasing practices agenda that place limits on the extent to which retailers' buying practices can be adapted to meet the important demands of ethical trade. The evidence is that campaigners and ethical consultancies have successfully set an agenda concerning the need to create more ethically-sensitive purchasing practices. This agenda has been accepted by the UK's central multi-stakeholder organization for ethical trade, the ETI, and leading retailers are experimenting collaboratively with ways of improving their buying practices and developing fairer relationships with suppliers. However, it would appear that changes to retailers' purchasing practices beyond the ETI's project are currently embryonic and small in scale. Some civil society organizations are pessimistic about the extent to which buying practices will be altered, while other commentators take a positive view that even small-scale modifications in purchasing represent positive change for the better. Taking a pragmatic view, an ethical trading manager working for one UK clothing retailer involved in the ETI Purchasing Practice Project acknowledges the challenges, but also recognizes the value of steady change:

> How I'm approaching it is through awareness raising … [saying to buyers] "Do you realise that if you change that design and that specification and you ask for re-sampling and you don't move that delivery date, that probably means 36-hour working in the factories?" … And sometimes you can do something about it, sometimes you can't. The way I always tackle these issues is to think that if I can even get some thought about it, if it stops one, two or three instances of that happening, thank goodness for that. It's better than saying, "Well it's a big problem, it's impossible" because then you make no change whatsoever … You chip away at it. I've just realised over the years, change is slow and sustainable change is slow. (Interview, 20/06/06)

Other retailers, who have co-operative and ethical values more firmly rooted in their long-standing corporate philosophies, suggest that responsible purchasing practices are best incorporated into all aspects of everyday business decisions. Such a view, represented here by the Head of Corporate Social Responsibility for a UK department store chain, appears to dovetail with the recommendations made by Acona and Oxfam in their 2004 reports:

> All of our responsible sourcing activity is done in conventional business practice—having the right buyers, selecting the right suppliers, developing the right products, paying a fair price … Democratic management runs all the way through [our] business and because of that, because you're working for a business which actually you own, the way in which you make those management decisions, the way in which you work and interact with other partners, with other suppliers, with customers, is quite a different business model. (Interview, 02/05/06)

Conclusion

There is evidence to suggest that the field of ethical consumption, part of which includes the influence on consumers' purchasing decisions of corporate reputations regarding the treatment of workers in supply chains, is evolving in terms of the ways in which it is challenging the organization of production networks. This chapter has addressed the critical 'brokering role', to use Clarke et al's (2007, 240) term, played by civil society organizations operating between spaces of consumption and worlds of trade and production. This brokering role during the 1990s involved NGOs, trade unions and critical journalists articulating consumer concerns about labour conditions in supply chains and effectively pressuring companies to develop codes of conduct for labour and environmental conditions at sites of production. However, reflecting Blowfield's (2005) point that the corporate social responsibility movement touches and modifies only the more negotiable aspects of capitalism, these codes of conduct tended to ignore the terms of trade that disadvantaged suppliers and their workforces so severely in the first place. The next step for ethical trade at the beginning of the twenty-first century has therefore been for UK-based ethical brokers like NGOs to, in Oxfam's words, 'change the terms of the debate' so that retailers' market power and associated purchasing practices have become the subject of scrutiny and potential change. This has generated a challenge for retailers in that they now have to respond to two opposing forces coming from the sphere of consumption—the need to provide desirable goods and services at the lowest possible prices on the one hand, and the pressure also to treat their suppliers fairly in terms of their logistical, financial and social demands. Evidence in this chapter suggests that civil society organizations, the media and cutting-edge ethical consultancy firms based in the UK have successfully carved out a challenging new ethical trade agenda on the subject of retailers' purchasing practices.

In terms of explaining the relationship between this action on the part of NGOs and ethical consultancies and the broader field of consumer politics, Miller's (2000) notion of the 'virtual consumer' is insightful. In capturing the characteristics of the contemporary global economy, Miller (2000) highlights the prevalence of the 'virtual consumer' over 'real', embodied consumption practices and consumer preferences, at least in terms of how economic action in the productive and distributive spheres is justified and rationalized. In the field of buyer-driven commodity chains, just as discourses of fast and disposable fashion, for example, justify and rationalize clothing retailers' strategies of aggressive purchasing practices, so it is that representations of ethical consumers' concerns also underpin the powerful advocacy strategies of the civil society organizations that seek to challenge them. As such, while it is not always direct action on the part of consumers themselves that challenges the corporate organization of commodity chains, representations of consumers' ethical concerns about distant workers frequently form at least a part of the justification for the campaigns that are articulated by organizations such as NGOs. While the responses from retailers to the specific campaigns concerning corporate purchasing practices are embryonic and are currently only scratching the surface of the majority of mainstream retailers' buying strategies, a purchasing practices agenda has at least been set and is being executed by the influential ETI.

With respect to the geography of campaigning on the purchasing practices issue, the specificity of this agenda in the UK can be noted. A broader UK–US comparison of retailers' ethical trading programmes, resulting from our comparative project, reveals that such an agenda is more wide-reaching and influential in the UK than it is in the USA. Indeed, discussion with ethical trade practitioners reveals that purchasing practices are more centrally on the ethical trading agenda in the UK than they are elsewhere in the world, though more rigorous research is needed to investigate this view. With respect to the USA, while Nike, Gap (influenced by it membership of the UK's ETI) and the Worker Rights Consortium have made significant headway in integrating ethical purchasing practices into mainstream supply chain management, purchasing practices are not a part of the majority of US-based retailers' social compliance strategies. This UK–US contrast is linked, first, to a difference in campaigning contexts where the USA has tended to be more campus-based and the UK more activist-based (Hughes et al. 2007; personal interviews), the latter allowing for a clearer brokering role to be played by organizations like Oxfam. Second, while the US ethical trading movement has driven a much stronger compliance-monitoring, audit-driven methodology for governing global supply chains, the UK agenda for ethical trade has been arguably softer and more experimental in approach (Hughes et al. 2007, 2008). This more developmental context for ethical trade in which UK-based retailers are situated allows companies to accept the challenge of pursuing a cultural change in purchasing practices that defies simple measurement and contrasts with audit-based approaches that have been labelled as 'quick fix' solutions (Clean Clothes Campaign 2005). As a key NGO commentator in the ethical trade debate comments:

> The British companies are happy to make commitments which they regard as being aspirational commitments. American companies are uncomfortable to sign up to something that they can't nail to a number or something that can't be categorically proven. (Interview, 30/08/06)

With significance for this volume on consuming spaces, the geographical specificity of the ethical purchasing practices agenda, facilitated in part by particular civil society organizations, can be theorized using the GPN approach associated with economic geographers. One of the key concepts defining the GPN approach is that of embeddedness. While societal forms of embeddedness capture the ways in which firms' production chains are influenced by the broad-level social and institutional contexts in which firms are based, and network forms of embeddedness refer to connectivity and relations between different agents in the commodity chain across different spatial scales, the notion of territorial embeddedness represents the ways in which a firm's production networks are '… "anchor[ed]" in different places (from the nation-state to the local level) …' (Henderson et al. 2002, 452, original emphasis). This 'anchoring' at the national or local level includes the influence of particular political organizations such as NGOs and trade unions, which are able to exert 'collective' and 'countervailing' power on corporations and thereby challenge their strategies (Henderson et al. 2002). The ethical campaigning and consultancy work on retailers' purchasing practices, conducted by UK-based organizations, can therefore be viewed as being part of a particular 'ethical complex' (Freidberg 2004) that is territorially embedded in the specific ethical trading and ethical consumption debates within the UK context. This concurs with Clarke et al's (2007) point that ethical consumption is produced through the specific actions and practices of key agents and linked to '... the strategic choices made by organizations and activist groups to mobilize 'the consumer' in particular ways, faced with various opportunity structures and the availability of different bundles of resources' (246). These 'opportunity structures' and 'bundles of resources' vary between different geographic contexts.

In sum, a purchasing practices agenda has emerged in the UK context for ethical trade that has been powerfully shaped by campaigning and ethical consultancy groups playing a brokering role between spaces of consumption and global supply chains. This agenda has developed specifically in the UK, embedded in a context of experimental and activist-led approaches to ethical trade (Hughes et al. 2008). Two key arguments can finally be made from this analysis. First, research into the governance of commodity chains and networks should continue to take the role of consumption seriously (Coe et al. 2008), incorporating the influence of ethical issues. Second, when the producer-consumer relation is being interrogated, its spatiality should be addressed. The emergence of a purchasing practices agenda in the field of ethical trade is developing in specific ways in particular places through the ethical brokering role played by key organizations representing the consumer and workers' interest. Recognizing such geographical specificity is important for understanding the ever-changing ways in which consumption influences the dynamism of the capitalist space economy.

Acknowledgements

We are very grateful to the editors of this volume for their valuable comments on an earlier draft of this chapter. We also acknowledge the support of an Economic and Social Research Council grant (RES-000-23-0830), which funded the study reported. And finally, we wish to thank all the interviewees who participated in that study for their insights and their time.

References

Acona. (2004), *Buying Your Way into Trouble: the Challenge of Responsible Supply Chain Management* (London: Insight Investment).

Amin, A. (1994), *Post-Fordism: A Reader* (Cambridge MA: Blackwell).

Barnett, C., Cloke, P., Clarke, N. and Malpass, A. (2005), 'Consuming Ethics: Articulating the Subjects and Spaces of Ethical Consumption', *Antipode* 37, 23–45.

Barrett, H. R., Browne, A. W. and Ilbery, B. W. (2004), 'From Farm to Supermarket: The Trade in Fresh Horticultural Produce from Sub-Saharan Africa to the United Kingdom', in Hughes, A. and Reimer, S. (eds), *Geographies of Commodity Chains* (pp. 19–38) (London and New York: Routledge).

Barrett, H. R., Ilbery, B. W., Browne, A. W. and Binns, T. (1999), 'Globalization and the Changing Networks of Food Supply: The Importation of Fresh Horticultural Produce from Kenya into the UK', *Transactions of the Institute of British Geographers* 24: 2, 159–174.

Barrientos, S. and Dolan, C. (eds) (2006), *Ethical Sourcing in the Global Food System: Challenges and Opportunities to Fair Trade and the Environment* (London: Earthscan).

Blowfield, M. (2005), 'Corporate Responsibility: Reinventing the Meaning of Development', *Affairs* 81, 515–534.

Bowlby, S. R. and Foord, J. (1995), 'Relational Contracting between UK Retailers and Manufacturers', *International Review of Retail, Distribution and Consumer Research* 5, 333–61.

Bowlby, S., Foord, J. and Tillsley, C. (1992), 'Changing Consumption Patterns: Impacts on Retailers and their Suppliers', *International Review of Retail, Distribution and Consumer Research* 2, 133–50.

Bryant, R. and Goodman, M. (2004), 'Consuming Narratives: The Political Ecology of 'Alternative' Consumption', *Transactions of the Institute of British Geographers* 29, 344–366.

Christopherson, S. and Lillie, N. (2005), 'Neither Global nor Standard: Corporate Strategies in the New Era of Labor Standards', *Environment and Planning A* 37, 1919–1938.

Clark, G., Gertler, M. S. and Feldman, P. P. (eds) (2000), *The Oxford Handbook of Economic Geography* (Oxford: Oxford University Press).

Clarke, N., Barnett, C., Cloke, P. and Malpass, A. (2007), 'Globalizing the Consumer: Doing Politics in an Ethical Register', *Political Geography* 26, 231–249.

Clean Clothes Campaign. (2005), *Looking for a Quick Fix: How Weak Social Auditing is Keeping Workers in Sweatshops* (Clean Clothes Campaign).

Coe, N. M. (2004), 'The Internationalization/Globalization of Retailing: Towards an Economic Geographical Research Agenda', *Environment and Planning A* 36, 1571–1594.

Coe, N. M., Dicken, P. and Hess, M. (2008), 'Global Production Networks: Realising the Potential', *Journal of Economic Geography* 8: 3, 271–295.

Coe, N. M. and Hess, M. (2005), 'The Internationalization of Retailing: Implications for Supply Network Restructuring in East Asia and Eastern Europe', *Journal of Economic Geography* 5, 449–474.

Crewe, L. (2004), 'Unravelling Fashion's Commodity Chains', in Hughes, A. and Reimer, S. (eds), *Geographies of Commodity Chains* (pp. 195–214) (London and New York: Routledge).

Crewe, L. and Davenport, E. (1992), 'The Puppet Show: Changing Buyer-Supplier Relations within Clothing Retailing', *Transactions of the Institute of British Geographers* 17, 183–197.

Doel, C. (1996), 'Market Development and Organizational Change: The Case of the Food Industry', in Wrigley, N. and Lowe, M. (eds), *Retailing, Consumption and Capital: Towards the New Retail Geography* (pp. 48–67) (Harlow: Longman).

Dolan, C. (2005), 'Fields of Obligation: Rooting Ethical Sourcing in Kenyan Horticulture', *Journal of Consumer Culture* 5, 365–389.

Dolan, C. and Humphrey, J. (2004), 'Changing Governance Patterns in the Trade in Fresh Vegetables between Africa and the United Kingdom', *Environment and Planning A* 36, 491–509.

Ethical Consumer. (2007), 'Ethical Consumerism on the Up and Up', *Ethical Consumer* EC105, 9: March/April.

Ethical Trading Initiative. (2008), *Purchasing Practices: Case Studies to Address Impacts of Purchasing Practices on Working Conditions* (London: ETI).

Freidberg, S. (2004), 'The Ethical Complex of Corporate Food Power', *Environment and Planning D: Society and Space* 22, 513–531.

Gereffi, G. (1994a), 'Capitalism, Development and Global Commodity Chains', in Sklair, L. (ed.), *Capitalism and Development* (pp. 95–122) (London: Routledge).

Gereffi, G. (1994b), 'The Organization of Buyer-driven Global Commodity Chains: How US Retailers Shape Overseas Production Networks', in Gereffi, G. and Korzeniewicz, M. (eds), *Commodity Chains and Global Capitalism* (Westport: Greenwood Press).

Gereffi, G., Humphrey, J. and Sturgeon, T. (2005), 'The Governance of Global Value Chains', *Review of International Political Economy* 12, 78–104.

Goodman, M. (2004), 'Reading Fair Trade: Political Ecological Imaginary and the Moral Economy of Fair Trade Foods', *Political Geography* 23, 891–915.

The Guardian. (16/07/07), 'Asda, Primark and Tesco Accused over Clothing Factories', p. 1 and 9.

Hale, A. (2000), 'What Hope for "Ethical" Trade in the Globalized Garment Industry?', *Antipode* 32, 349–356.

Harvey, D. (1989), *The Condition of Postmodernity* (Oxford: Backwell).

Henderson, J., Dicken, P., Hess, M., Coe, N. M. and Yeung, H. W.-C. (2002), 'Global Production Networks and the Analysis of Economic Development', *Review of International Political Economy* 9, 436–464.

Hughes, A. (1996), 'Forging New Cultures of Retailer-Manufacturer Relations', in Wrigley, N. and Lowe, M. (eds), *Retailing, Consumption and Capital: Towards the New Retail Geography* (pp. 90–115) (Harlow: Longman).

Hughes, A. (1999), 'Constructing Competitive Spaces: On the Corporate Practice of British Retailer-Supplier Relationships', *Environment and Planning A* 31, 819–840.

Hughes, A. (2001), 'Multi-stakeholder Approaches to Ethical Trade: Towards a Reorganization of UK Retailers' Global Supply Chains?', *Journal of Economic Geography* 1: 4, 421–437.

Hughes, A. (2005), 'Corporate Strategy and the Management of Ethical Trade: The Case of the UK Food and Clothing Retailers', *Environment and Planning A* 37: 7, 1145–1163.

Hughes, A. (2006), 'Learning to Trade Ethically: Knowledgeable Capitalism, Retailers and Contested Commodity Chains', *Geoforum* 37: 6, 1007–1019.

Hughes, A., Wrigley, N. and Buttle, M. (2007), 'Organizational Geographies of Corporate Responsibility: A UK–US Comparison of Retailers' Ethical Trading Initiatives', *Journal of Economic Geography* 7, 491–513.

Hughes, A., Wrigley, N. and Buttle, M. (2008), 'Global Production Networks, Ethical Campaigning, and the Embeddedness of Responsible Governance', *Journal of Economic Geography*, 8, 345–367.

Impactt. (2004), *Taking the Temperature: Ethical Supply Chain Management* (London: Impactt).

Jenkins, R. (2002), 'The Political Economy of Codes of Conduct', in Jenkins, R., Pearson, R. and Seyfang, G. (eds), *Corporate Responsibility and Labour Rights: Codes of Conduct in the Global Economy* (pp. 13–30) (London and Sterling: Earthscan).

Kritzinger, A., Barrientos, S. and Roussouw, H. (2004), 'Global Production and Flexible Employment in South African Horticulture: Experiences of Contract Workers in Fruit Exports', *Sociologia Ruralis* 44, 17–39.

Mathieson, N. (2006), 'Green Savaged over Arcadia's Ethical Standards', *The Observer* 5 March, 2.

Mathieson, N. and Aglionby, J. (2006a), 'The True Cost of Cheap Clothing', *The Observer* 23 April, 4.

Mathieson, N. and Aglionby, J. (2006b), 'Exposed: Life at Factory that Supplies our Fashion Stores', *The Observer* 23 April, 1.

Miller, D. (2000), 'Virtualism- the Culture of Political Economy' in Cook, I., Crouch, D., Naylor, S. and Ryan, J. R. (eds), *Cultural Turns/Geographical Turns: Perspectives on Cultural Geography* (187–215) (Harlow: Prentice Hall).

Oxfam. (2004), *Trading Away Our Rights: Women Working in Global Supply Chains* (Oxford: Oxfam International).

Pratley, N. and Finch, J. (2005), 'Shop tactics', *The Guardian* 6 January, 2–3.

Reardon, T., Timmer, C. P., Barrett, C. B. and Berdegue, J. (2003), 'The Rise of Supermarkets in Africa, Asia, and Latin America', *American Journal of Agricultural Economics* 85: 5, 1140–1146.

Tokatli, N. (2008), 'Global Sourcing: Insights from the Global Clothing Industry- the Case of Zara, a Fast Fashion Retailer', *Journal of Economic Geography* 8: 1, 21–38.

Weller, S. (2007), 'Fashion as Viscous Knowledge: Fashion's Role in Shaping Trans-national Garment Production', *Journal of Economic Geography* 7: 1, 39–66.

Wenger, E. (1998), *Communities of Practice: Learning, Meaning and Identity* (Cambridge: Cambridge University Press).

Wijeratna, A. (2005), *Rotten Fruit: Tesco Profits as Women Pay a Huge Price* (London: Action Aid).

Wrigley, N. (1992), 'Antitrust Regulation and the Restructuring of Grocery Retailing in Britain and the USA', *Environment and Planning A* 24, 727–49.

Wrigley, N. (2000), 'The Globalization of Retail Capital', in Clark, G., Gertler, M. S. and Feldman, P. P. (eds), *The Oxford Handbook of Economic Geography* (pp. 292–313) (Oxford: Oxford University Press).

Wrigley, N., Coe N. M. and Currah, A. D. (2005), 'Globalizing Retail: Conceptualizing the Distribution-based TNC', *Progress in Human Geography* 29: 4, 437–457.

Wrigley, N. and Lowe, M. (eds) (2002), *Reading Retail: A Geographical Perspective on Retail and Consumption Spaces* (London: Arnold).

Chapter 7

The Cultural Economy of the Boutique Hotel: The Case of the Schrager and W Hotels in New York

Donald McNeill and Kim McNamara

Introduction

The rise of the so-called 'boutique hotel' has been one of the biggest stories in hotel development since the late 1980s (Riewoldt 1998; Watson 2005). As a reaction to the rapid spread of standardized, easily replicated corporate hotel formats in the post-war period, a number of independent hotel entrepreneurs began to focus on an individualized guest experience, based on high quality service combined with a distinctive design ambience. Although there are numerous hotels that had been established with that ethos, the popularization of the format – and the apparent coining of the term 'boutique' – was undertaken by Ian Schrager in New York in the 1980s and 1990s. Hoteliers became interested in bold, extravagant designs to appeal to an increasingly differentiated and reflexive set of hotel consumers. This has now achieved a worldwide reach, to such an extent that the major hotel corporations are now rolling out 'boutique' brands, despite the apparent contradiction that this entails.

In this chapter, we consider the significance of the boutique hotel within broader debates about consumption sites (cf. McNeill 2008). Hotels act as a site for the processing, welcoming, and orientation of travellers, but also as a destination and site of social exchange among city dwellers. The attention paid to its design and configuration is thus one of the key decisions to be made by developers and operators, ranging from the functional to the luxurious. The chapter focuses on the economy and design ethos in the production of hotel space, and identifies the assumptions made about who will be consuming it, and how. This is fundamental to the evolution of the boutique form: the calculated risk of increasing the design quality of a hotel's components is that it is reflected in higher revenues for the hotel's operators.

To illustrate this, we consider some recent developments in the production of design or boutique hotels in New York, notably the range of Ian Schrager operated hotels (known as the Morgans Hotel Group), and their corporate competitor, Starwood's W brand. We stress *production*, rather than merely *design*, because the financing of hotel development is based upon a calculated analysis of the role of public areas within the overall hotel programme. As hotels are commercial developments, any non-revenue driving spaces are regarded with suspicion by investors in hotels that fall below the luxury segment. We explore the nexus between their production (their financing, programming, management, and design) and consumption (their commoditization through nightly room rates, their patronage by non-guests, their mix of lobby-related functions such as bars and cafes).

Towards a Cultural Economy of the Boutique Hotel

The hotel has now arrived as a significant interest for social historians seeking to explain the formation of modern business districts and urban economies, from their centrality in the historical development of urban tourism as a business (Cocks 2001), to their establishment as a key part of modern business districts (Davidson 2005), to their role in diffusing American cultural norms overseas (Wharton 2001), to their significance in both colonial 'outposts', and then again by post-colonial tourist economies (Peleggi 2005). However, we want to suggest here that hotels contribute to their urban economies in fairly complex ways. While the standard corporate hotels were established to service the so-called Fordist guest, with apparently conservative tastes in everything from food to music to design, the growing importance of fashion and media 'creatives' within cities has posed a challenge to how urban economies are read (Edwards 2006).

The definition of the boutique hotel is contested. As Watson (2005, 12) summarizes:

> the word carries a sense of being small, originally referring to a small shop or business selling items such as clothes or jewellery. It conveys a suggestion of quality, that the items are perhaps crafted, limited edition, or even bespoke, and certainly luxurious and fashionable.

This focus on an experiential relationship is certainly central to contemporary consumption debates, particularly in terms of the commercialization of hospitality within cities (Bell 2007). This can also be framed through the cultural economy approach that is now well-established in sociology and geography, which theorizes the urban economy as 'the instanciation of much of the matter that makes markets' (Amin and Thrift 2007, 158). This approach emphasizes 'the absolute centrality of passion in mobilizing and sustaining drive in contemporary capitalism, including the libidinal energies that motivate entrepreneurs, speculators and investors … and the compulsion of consumption that now so energizes human need, pleasure and identity' (Amin and Thrift 2007, 147). The capitalist economy is thus less 'economistic' than it might appear: 'passion of varying intensity and form is shot through the producer-consumer relations on which capitalist success is delicately balanced (147–8).

Tapping into such passions presents a challenge to hotel designers, particularly when guests split their consumption preferences depending on, for example, their work and leisure subjectivities, or variations in work and personal budgets. According to Postrel (2004):

> Aesthetic identity – *I like that, I'm like that* – is more specific and personal than "That's attractive" in some universal sense … The subjective value of a particular form varies from person to person. A young female Westin guest transported to the clubby, pin-striped Sheraton room might appreciate its attractiveness but still feel slightly out of place. (107)

Boutique hotels have, therefore, fitted into a rather different consumption framework than the standardized hotel offers of the Fordist period. They are certainly implicated in the changing geographies of the 'global fashion city' (Breward and Gilbert 2006), allowing the likes of Donatella Versace or Sonia Rykiel to enter into agreements with hotel developers to showcase their products (Gross 2004). This reflects the growing understanding of the cross-branding inherent in the massification of luxury: Armani, Bulgari, and Camper have taken the hotel concept further by establishing a limited number of branded hotels in fashion centres such as Milan, Paris, London and New York. The logic of this is to 'let customers indulge in the fantasy by moving into rooms that are live-in ads for sheets and towels, home decorating fabrics and furniture, bath and beauty products' (Gross 2004).

However, even without an association with an established brand, the boutique hotel market has utilized design in novel ways in order to rework the traditional rules of hotel development. The basic formula used in determining hotel profitability is RevPAR, Revenue per available room, where revenue is the result of occupancy multiplied by average nightly room rate (PKF Consulting 1996). Conventionally, hotel development formats follow an understanding of minimum room size standards, which has evolved within the sector through the twentieth century. Additional facilities (such as a bar, business centre, and fitness suite) will also drive revenue. Crucially, room prices reflect location (centrality and amenity), which in turn has a strong correlation to land prices when the development was undertaken, along with orthodox supply and demand conditions within the local hotel market. What entrepreneurs such as Schrager intuited was that by filling larger numbers of attractively designed small rooms than the convention, a competitive edge can be achieved based on the design premium.

This could be achieved by rethinking the city centre hotel offer in its entirety, which in turn required lateral thinking in terms of the conventional arithmetic used by investors in determining the potential yield of hotels. This involved maximizing the design quality of the hotel overall, playing on the romantic attachments and dramatic potential of these areas, using this to compensate for a smaller than average room size, based upon an assumption of a reduced length of time spent by the (younger than average) guests in their room.

Here, the lobby assumes a pivotal role. Unlike the lobbies of apartment buildings and commercial office blocks, hotel lobbies 'must do all the impressing, greeting, dispersing, and securing' of these other building types with a more diverse public in mind (Berens 1997, xiii). By programming them as social, theatrical spaces, Schrager and Starck were able to create a surplus value over and above the basic offer of a hotel: the selling of a bedroom, now with an 'urban living room' added (Slesin 1990). In the private spaces of the hotel, the bedrooms, this aesthetic judgement of the design object can be taken slowly and at leisure. In the public space of the lobby, however, there is a conscious 'aesthetics of strangeness' which some may find confronting. As with a stage, one is self-conscious if crossing the lobby floor – it is purposely designed to be so. Creating this sensation in the hotel is thus a commercial risk:

> Cutting-edge design is rarely seen in hotel lobbies. When done well this design style creates exciting and visionary places, defining its own clear version of individualistic sophistication while not referring to other times or far off places. More modern than today could ever be, it invents its own vocabulary and stakes out its own design frontier. Critical to cutting-edge design is the creation of a theme that will not grow old and tiresome before it is time for the first guests to check in. (Berens 1997, 3)

The bedroom was also packaged in a way that appealed to a differing form of consumption choice. It is now accepted that interior design decisions go beyond the merely functional, and that the purchase of furniture is an important act of home-making (Reimer and Leslie 2004). By extension, the boutique hotel is often seen as an experimental place where avant-garde or unusual design can be experienced as part of the consumption package, without requiring long-term commitments to the design artefact in question. Indeed, the boutique hotel may be the home's 'other' space: for some consumers, it may provide a forum for escape, changed identity and performance characteristics. Thus for Danny Miller, 'shopping is not about possessions per se, nor is it about identity per se. It is about obtaining goods, or imagining the possession or use of goods' (1998, 141; in Reimer and Leslie 2004, 204). As the concept evolved, the boutique hotel bedroom thus became a catalogue of home products that can be tried and tested. Luxury Egyptian cotton sheets can be slept on for a night, bathroom faucets can be turned on and off, and compared against those in one's own home.

The production of the boutique hotel (as a fixed element of the built environment) follows that of mobile commodities, such as a car, through a process that Callon et al. (2003) call 'the economy of qualities'. Both objects have 'a life, a career':

> Seen from the angle of its conception and then production, it starts off by existing in the form of a set of specifications, then a model, then a prototype, then a series of assembled elements and, finally, a car in a catalogue that is ordered from a dealer and has characteristics that can be described relatively objectively and

> with a certain degree of consensus. Once it is in the hands of its driver the car continues moving, not only on roads but also, later, for maintenance purposes to workshops, then to second-hand dealers. At times it becomes again an object on paper, which takes its place alongside other cars in the guide to second-hand car prices in specialized magazines. (61)

The idea of the hotel as having a career, or biography is one that is already well accepted by hotel historians (e.g. Jim 2005, who develops the idea from Kopytoff 1986). The life of the hotel is one of great fluidity and churn, given the rapid turnover of guests and visitors, their impact on the spaces in terms of wear and tear, and the engaging similarities that it has with more conventional biographies of human subjects. The early development of the product owed much to the dynamism of creative entrepreneurs without formal business experience (Currid 2007; Eikhof and Haunschild 2006). Once in the hands of its guests, its identity continues to evolve, before succumbing to the same fall in quality of any durable consumer object, and the requirement of maintenance. And, when its owners – the investors – seek to rebalance their portfolios, the hotel again becomes an 'object on paper', this time listed not in the features of lifestyle magazines, but in the commercial listings of real estate agencies.

The hotel's life cycle is in turn closely related to the specificities of local land markets, the urban economic sectors that dominate central business districts, and the physical structure of existing buildings. The story of how New York generated its own version of the boutique hotel format must also take into account its history of sceptical loan institutions, origins in a loosely regulated nightlife sector, and an evolving demographic that meant that the hotel was consumed as much by locally embedded social elites as much as the out-of-town sleepers. Both the production and consumption of the boutique hotel good are thus defined by place specific criteria.

Ian Schrager's Boutique Hotels

It is interesting that the boutique hotel took such a vigorous hold in Manhattan, and perhaps mirrored a shift during the 1970s in attitudes to the city as a world fashion centre (Rantisi 2006). By the mid-1980s, the city had – arguably – become 'the premier global creative hub of fashion, music, art, design, and these worlds collided in the insomniac, coke-fuelled, disco-lit world of nightlife' (Currid 2007, 36). One of the driving forces of this moment was Schrager and his then-business partner Steve Rubell, who opened the nightclub Studio 54, credited with the establishment of disco as a musical form. The club was notorious for its strict entry policies, prefacing a climate where 'nightlife became increasingly exclusive, letting in only those creative producers who were also linked to the cultural marketplace or the culture of cool, while wannabes were left on the wrong side of the velvet rope' (Currid 2007, 37). The club's success was short-lived. In 1980, Rubell and Schrager were convicted of tax evasion, and were sentenced to prison for three years. It was in prison that the pair 'worked on plans to open a new kind of hotel' (Dougary 2003), one based on the same kind of principles as their night-club: a sense of avant-gardism, scene-setting, and high design, with an emphasis on youth. Their first hotel, Morgans, on Madison Avenue, opened in 1984, and was, according to some commentators, the very first boutique hotel (Stern et al. 2006, 1087). The hotel was designed by the famed French interior designer Andrée Putman, with deliberately darkened corridors and Robert Mapplethorpe photographs hanging in the rooms. After the initial success with Morgans, Schrager opened the Royalton on West 44th (1988), and despite being rocked by Rubell's premature death in 1989, the Paramount, on West 46th (1990) and then the Hudson, on 58th Street and 9th Avenue (1999). The hotels would be of different types: the Royalton would run to 169 rooms, the Hudson to over 1000. Yet each would be key hotels in establishing the aesthetic and concept of his further ventures into the hotel business.

Schrager employed the French product designer, Philippe Starck, to work on his post-Morgans hotels, and would go on to collaborate with him as he opened hotels in other major cities, such as the Mondrian (Los Angeles), Clift (San Francisco), Delano (Miami) and St Martin's Lane (London). The collaboration with Starck would be the defining style for Schrager hotels in the early years – a modern, quirky and sometimes surrealist aesthetic – the most extreme of which would be showcased in the public areas. The Royalton highlighted this approach. It was built as a bachelor's residence in 1897, but became run-down, and was re-imagined by Starck and Schrager as a 'high style home away from home' (Stern et al. 2006, 1088). The lobby's narrow 180 foot long space featured a royal blue carpet, likened to a model's runway, on which guests must strut up to check in (Solomon 2005). By the early 1990s it had attracted the reputation as lunch headquarters of the nearby Condé Nast publishing house, 'the nexus for the elegant editors of the high-profile magazines and Seventh Avenue designers and others in the fields of fashion, beauty, design and celebrity journalism' (Dullea 1992). The restaurant, 44, and the bar, thus operated at the nexus of what Currid (2007) calls the 'fluid economy' that underpins creative production. Such fluidity 'allows creative industries to collaborate with one another, review each other's products, and offer jobs that cross-fertilise and share skill sets, whether it is an artist who becomes a creative director for a fashion house or a graffiti artist who works for an advertising agency' (7).

The Hudson, which opened in October 2000, aimed at a younger clientele, and exemplified the trade-off between small rooms and dramatic public areas, as well as Schrager's acumen at carrying out complex property deals (Weiss 1997). The building had an unusual history. Anne Morgan, daughter of the American financier J. Pierpont Morgan, financed it as a women's residence. In 1941, a portion of it became the Henry Hudson Hotel (Stern et al. 2006, 1090). Other floors housed a variety of uses over the years, including a local public service television station and student housing for St Luke's-Roosevelt Hospital Centre. Schrager bought the building in 1997, and saw its potential as a 'modern YMCA, provocative and original – like an urban spa' (Gray 1998). The Hudson's 'lifestyle' concept was aimed at young people, boasting room rates starting at $95 a night. In reality, this price was an introductory deal, and rates would soon rise as high as $245, the hotel's service standards and room size meeting with negative critical notice (Flack 2007).

As Meryl Gordon notes, in a 2001 profile of the hotelier for the *New York Times Magazine*, Schrager was clear about his design rationale: 'Since we're doing highly stylized things ... we've been able to jettison traditional ideas of opulence'. This allowed for particular cost-cutting features. The bathrooms at the Royalton were lined in Vermont slate at $1 a foot, rather than with imported marble at $8 a foot. The bathroom fittings at the Hudson and Paramount – both ex-single room occupancy (SRO) hotels – were rechromed rather than replaced. The extra light fittings for each room were run off an extra cable from the existing ceiling light, rather than being rewired (Gordon 2001). As well as cost-cutting on fittings, Starck's design experience was used to draw attention away from the tiny rooms (12x14 feet in the singles), through the use of an elongated chair and 'Brancusi-style' lamp fitting in order to 'distort the proportions of the rooms, creating a surreal quality overall' (Stein 1991, 72 in Stern et al. 2006, 1088–9).

A key aspect of the Schrager formula was the creation of a 'scene' in the lobby. In many ways, this involved harking back to early modern conceptions of these spaces:

> The urban hotel lobby, as a mediator between our public and private worlds, is where the life lived in public is important. To see and be seen is the essence of urban style. Lobbies provide the backdrop for scenes of urban drama. People are primary. In some lobbies it is difficult to determine which demands more attention, the people or the décor. Indeed, effective designs endeavour to make the guests look good by creating spotlights of activity and complimentary settings. These lobbies are dependent upon human activity and interaction for the design to come to life. Good design strives to create an atmosphere that encourages conversation, socializing, and the urban constant, serendipity. (Berens 1997: xv)

For Schrager, this fitted with a marketing approach that shunned advertising and used publicists to charm journalists and organize parties, relying on an additional element of word of mouth to expand the hotel's name. It also involved the use of a casting director to recruit staff, who were chosen for 'coolness' rather than hotel experience. The bar and restaurant were prominent features, located at a central, highly visible point adjacent to the lobby. This, combined with the lobby design, was an important mechanism in creating the hotel's distinctiveness in the city. The Hudson, particularly, created a significant buzz in Manhattan's social scene. The bar and restaurant became fashionable destinations, to such an extent that some guests complained that they were unable to check-in due to the bar's door policy. *New York Times* architecture critic Herbert Muschamp described the Hudson as being for creative types 'with more dash than cash … a refuge … for all those who crave what is lacking in the city outside' (Muschamp 2000). In this sense the Hudson 'transposes the street dynamic into a compact indoor version' (Muschamp 2000). To achieve this, the hotel becomes a controlled and highly calculated space – much like the controlled (and exclusive) space of Studio 54. The lobby, for example, is only accessible by way of an escalator up from street level, where guests travel through a yellow Perspex tunnel to the lobby area. Here, guests are met by deliberately dissonant materials and furnishings such as a standing bear coffee table, dark wood panelled walls, and a kitsch chandelier. Other aesthetic quirks include a giant watering can in its 'private park', an outdoor escape for its guests.

Yet while admiring the Schrager approach, Muschamp sounded a timely note of alarm: 'How much design can people tolerate? How many people are willing to pay for a room in which all but a few inches of floor space is covered by the mattress? How do you create a raving, Dionysian bar scene without repelling the Apollonian creative types you would also like to attract?' (Muschamp 2000). This would be a problem faced by a Schrager as he sought to soften the approach of his hard-line door-staff, and encourage the cool kids to hail taxis for guests. As he admitted, 'But if you think we can get away with attitude because we have the coolest place in town, it won't last' (in Gordon 2001).

The 'W' and the Corporatization of the City

After the initial successes of Morgans, the Royalton and the Hudson, Schrager's hotels faced a challenging economic climate. While the terrorist attacks of 9/11 caused a generalized, if temporary, crisis in the city's economy, the competitive advantage enjoyed by the boutiques had already begun to be eroded by the response of the major hotel groups, and by other boutique chains such as Kimpton. The major hotel groups such as Starwood, Intercontinental, or Marriott contain a range of hotel brands dedicated to a specific demographic, carefully reflected in the design of its space, public and private. Starwood were the quickest of the corporate groups to respond to the boutique trend. Their market research identified a major popular demographic cleavage in American consumer society, between baby-boomers and Gen-Xers. According to Levere (2005), these groups (respectively, the 78 million Americans born between 1946 and 1964, and the 58 million Americans born between 1965 and 1980) are the key markets that hotels have to pitch their offer to. They were also aware of the growing trend for using lobbies for business meetings, often among non-guests seeking neutral, informal – and free – meeting rooms (Welch 2006).

In 1998, Starwood launched their W chain, a high end brand with a strategy of creating individual hotels in major world cities. Driven by their CEO, Barry Sternlicht, the W chain looked closely at Schrager's market but entered at a higher level. Starwood opened five Ws in Manhattan alone, each with a distinctive design rationale. Their offer rested on a number of clearly articulated and branded components. Each hotel has a lobby shop selling carefully selected design artefacts, along with some W branded merchandise; each W has a strongly articulated design, but with a different designer to maximize the feel of each hotel's individuality; the rooms are equipped with high-tech gadgetry and were early guarantors of wi-fi access; guests are allowed to choose from a range of pillow types from a menu; the lobby is purposely designed to act as a focal point and city destination, as with the Schrager prototypes, and are often designed to open into the bar area.

The Ws quickly followed the Schrager example in creating 'destination' bars, aimed at trend-conscious city residents as much as out-of-town guests (Elliot 1998; McKinley 2000). Indeed, Starwood lured Randy Gerber, who had previously run Schrager's bars at the Paramount, Morgans, and Mondrian, to set up at the W in Lexington Avenue (Gordon 2001). In 2001, a few months before September 11, a journalist from the *New York Times* followed Schrager and two of his managers on a reconnaissance mission to the Starwood's W Lexington Hotel: 'Like the commander of a SWAT team, Schrager barks out orders. "We're going in for two minutes", he said. "I want to know: What's the vibe? What's the security like? Who's the crowd? Do they have a DJ?"' (Gordon 2001). This vignette is of interest, of course, in its distance from the usual conventions of good hotel-keeping: the vibe appears to matter more than the service standards. And yet, such factors are clearly important in an increasingly complex market. The Californian Joie de Vivre Group's self-promotional *The Secrets to Boutique Success* (Joie de Vivre n.d.) pins down why the W has been successful:

> Boutique hotels create loyalty by tapping into psychographic niches. No longer can a hotel say that its ideal guest is a 48-year-old businessman from Chicago as that demographic may not be as telling about this man as the words that executive would use to describe himself. W Hotels realized this and created a whole ad campaign that described both their hotel and their target demographic of customer: "witty, warm, welcoming, wired and wonderful" (or as some of their detractors suggest, "wannabe") … Most of us wouldn't be overtly flattered to be called the kinds of words we'd use to describe America's best-known hotel chains: bland, predictable, ubiquitous, safe, and consistent. (Joie de Vivre, n.d., n.p.)

This would be a successful strategy for Starwood, and reflected the corporatization of the boutique approach. Schrager was squeezed not only by the corporates, but also by the upswing of boutique hotels across New York, a fact reflected in his group's difficulty in coping with the post-9/11 downturn:

> There are … dozens of boutique hotels that have never been the setting for a fragrance kickoff or a model's birthday party but that offer the same basic stylish alternative to conventional hotels. With their new-age soaps and wenge wood cabinetry and Phillipe Starck-like lighting, they amount to a new hip

> homogeneity, as uniform in their own way as any standard hospitality chain. (Hamilton and Rich 2003)

Schrager and his partners in the Morgans Hotel Group (owned by the investment fund, NorthStar Capital) were seen to have over-expanded and over-borrowed, relying on 'Schrager's ability to deliver publicity, an instant reputation for a new hotel as a hot spot and *a cachet that could be translated into high room rates*' (Hamilton and Rich 2003, emphasis added). This 'economy of prestige' was thus seen to be financially unsustainable, and – additionally – the Schrager formula was being adopted and employed by competitors with a more sober business model. When Schrager resigned from Morgan's Hotel Group in 2005, its owner, Edward Scheetz, renovated the Starck lobby, replacing it with a 'clubby' look, 'mostly dark metal and leather' (Bernstein 2007). However, despite Schrager's absence, the new look seems to softly pay homage to his club days, the Royalton's website even welcoming viewers with black and white paparazzi shots, reminiscent of Studio 54 itself.

It is tempting to read this as a metaphor for Manhattan more generally. As numerous authors have lamented (e.g. Currid 2007; the contributors to Hammett and Hammett 2007), the avant-gardism that fuelled Studio 54 and the earlier Schrager hotels has dissipated as the central city has 'suburbanized', with the creative community of the 1970s and 1980s inadvertently fuelling gentrification. In this context, we could see Schrager as one actor from that era who has survived, redefining himself and his hotel product as the times have changed.

Conclusions

To conclude, we have described the renewed significance of boutique hotels within urban economies, and how the innovative product of Schrager has been copied by major hotel groups and brands, as they attempt to emulate the boutiques' success. We can suggest that the Schrager 'moment' arrived at a time of rapid social change, certainly in terms of consumer tastes. The popularity of the early post-war 'Fordist' hotels was clear: 'The visual signs sought by the middle-class, suburban consumer were not those of prestige, history, and community, but rather those of hygiene, accessibility, and privacy' (Wharton 2001, 175). In a similar way, boutique hotels have tracked shifting consumer sentiment through their self-conscious expression of urbanity, appealing to consumers with avant-garde tastes, and rapidly diversifying into a wide range of significantly differing hotel offers.

To sustain competitiveness within such a knowledgeable market is difficult, however. We could see hotels as being one of a small category of building types that require constant maintenance and regular reinvention to ensure profitability (Graham and Thrift 2007). This is by no means a singular process of creative destruction. Lobbies, for example, lend themselves well to certain kinds of high design, as one of the key activities undertaken there – sitting – is a timeless process.

Chairs become 'design classics' as a result; but 'there is unlikely to be a classic telex-machine design' given the rapid technological changes in office systems, for example (Julier 2000: 71). On the other hand, hotel interior architecture can be subject to the fickleness of consumer taste. As noted above, the Royalton would be refurbished in 2007 under new owners, and the Starck lobby ripped out. Even Starwood's W found itself alienating some of its baby boomer clientele, with its over-loud music, 'destination' bars, and challenging contemporary design. The group responded with yet another brand, a toned down variation of the W known as Aloft, which seeks to retain the 'boutique' aesthetic of Ws yet remain differentiated from the standardized, familiar, conservative designs of its Sheraton hotels.

At its most basic level, the boutique hotel is a reflection of a creative cultural economy with regards to what consumers demand in terms of space and location. This is embedded in the notion of the 'boutique' itself. As Watson (2005, 12) notes, the translation of 'boutique' into hotel culture 'is at best a little garbled':

> Some people claim that Schrager's own hotels, which have several hundred rooms, are not boutique – after all, how small is small? I think Schrager meant that the hotel should evoke a personal relationship with the guest, offering him/her a unique and surprising experience which cannot be exactly repeated in any other environment.

What we see, then, is an innovation in terms of the production of space within a particular land economy, which is meshed with consumer demand for a range of goods which the purchase of the commodity (the hotel room) allows access to, including the immediate public spaces of the hotel itself, as well as the nightclubs, bars, and restaurants of its surrounding neighbourhood. It is important to think of the consumption of the boutique hotel as a distributed action, one which is related to the temporal nature of the enjoyment of the good itself. This is made up of several components, both routine (as in taking a shower or sleeping) and unique (as in going to a 'scene' nightclub or describing the experience to jealous peers). Unlike the mass produced commodities described by Callon et al. (2004), the early days of the boutique hotel were notable as being more than just a singular product with a stable consumer market. Instead, the process of trial and error involved in its design was a reflection of

> the co-construction of supply and demand … On the one hand, it leads to a singularization of the good (so that it is distinguished from other goods and satisfies a demand that other goods cannot meet). On the other hand, it makes the good comparable to other existing goods, so that new markets are constructed through the extension and renewal of existing ones. (Callon et al. 2004, 65; after Chamberlin 1946)

If we understand how this translated into the corporate version of the boutique offered by chains such as the W, we can gain some insight into the global diffusion, or translation, of the boutique hotel concept into the mainstream hotel business.

References

Amin, A. and Thrift, N. (2007), 'Cultural-economy and Cities', *Progress in Human Geography* 31, 143–161.

Bell, D. (2007), 'The Hospitable City: Social Relations in Commercial Spaces', *Progress in Human Geography* 31: 1, 7–22.

Berens, C. (1997), *Hotel Bars and Lobbies* (New York: McGraw-Hill).

Bernstein, F. A. (2007), 'Royalton Shake-up, from Top to Lobby', *New York Times* 11 November, www.nytimes.com, accessed 15 November 2007.

Breward, C. and Gilbert, D. (eds) (2006), *Fashion's World Cities* (Oxford: Berg).

Callon, M., Méadel, C. and Rabeharisoa, V. (2004), 'The Economy of Qualities', in Amin, A. and Thrift, N. (eds), *The Cultural Economy Reader* (pp. 58–79) (Oxford: Blackwell). Reprinted from Economy and Society (2002) (31: 2, 194–217).

Chamberlin, E. H. (1946), *The Theory of Monopolistic Competition: A Reorientation of the Theory of Value*, 5th Edition (Cambridge MA: Harvard University Press).

Cocks, C. (2001), *Doing the Town: The Rise of Urban Tourism in the United States, 1850–1915* (Berkeley: University of California Press).

Currid, E. (2007), *The Warhol Economy: How Fashion, Art, and Music Drive New York City* (New York: Princeton University Press).

Davidson, L. P. (2005), 'Early Twentieth-Century Hotel Architects and the Origins of Standardization', *The Journal of Decorative and Propaganda Arts* 25, 72–103.

Dougary, G. (2003), 'The Hip Hotelier – Journeys 2003 – Interview – Ian Schrager', *The Times* 11 January, www.timesonline.co.uk, accessed 28 October 2004.

Dullea, G. (1992), 'The, Uh, Royalton Round Table', *New York Times* 27 December, www.nytimes.com, accessed 15 October 2004.

Edwards, B. (2006), 'Shaping the Fashion City: Master Plans and Pipe Dreams in the Post-war West End of London' in Breward, C. and Gilbert, D. (eds), *Fashion's World Cities* (pp. 159–173) (Oxford: Berg).

Eikhof, D. R. and Haunschild, A. (2006), 'Lifestyle Meets Market: Bohemian Entrepreneurs in Creative Industries', *Creativity and Innovation Management* 15: 3, 234–241.

Elliot, S. (1998), 'The Media Business: Advertising; Starwood is Treating the Opening of a Boutique Business Hotel like a Broadway Premiere', *New York Times* 1 December , www.nytimes.com, accessed 26 August 2004.

Flack U. (2007), 'Hudson Hell: Possibly the Worst Hotel in NYC' blogspot, March 13, http://flacku.blogspot.com/2007/03/hudson-hell-possibly-worst-hotel-in-nyc.html, accessed 12 November 2007.

Gordon, M. (2001), 'The Cool War', *New York Times* 27 May, www.nyt.com, accessed 27 July 2004.

Graham, S. and Thrift, N. (2007), 'Out of Order: Understanding Maintenance and Repair', *Theory, Culture and Society* 24: 1, 1–25.

Gray, C. (1998), 'Streetscapes/The Henry Hudson Hotel, 353 West 57th Street; From Women's Clubhouse to WNET to $75 a Night', *New York Times*, 4 January, accessed via www.nytimes.com, 11 August 2004.

Gross, M. (2004), 'Hotel couture', *Travel + Leisure* June, 166–172, 208–210.

Hamilton, W. and Rich, M. (2003), 'Familiar Faces at the Hard Times Hotel', *New York Times*, 10 August , www.nytimes.com, accessed 15 October 2006.

Hammett, J. and Hammett, K. (eds) (2007), *The Suburbanisation of New York: Is the World's Greatest City Becoming Just Another Town?* (New York: Princeton Architectural Press).

Jim, B. L. (2005), '"Wrecking the Joint": The Razing of City Hotels in the First Half of the Twentieth Century', *The Journal of Decorative and Propaganda Arts* 25, 288–315.

Joie de Vivre (n.d.), *The Secrets to Boutique Success: How Boutique Hotels are Revolutionizing the Hospitality Industry*, http://www.jdvhospitality.com/company, accessed 16 November 2004.

Julier, G. (2000), *The Culture of Design* (London: Sage).

Kopytoff, I. (1986), 'The Cultural Biography of Things: Commoditization as Process' in Appadurai, A. (ed.), *The Social Life of Things: Commodities in Cultural Perspective* (pp. 64–94) (Cambridge: Cambridge University Press).

Levere, J. L. (2005), 'To Bed or to Bar?', *New York Times*, 20 September, accessed via nytimes.com, 9 October 2005.

McKinley, J. (2000), 'It's Hopping in the Lobby as Hotels Party All Night', *New York Times*, 19 November, www.nytimes.com, accessed 28 October 2004.

McNeill, D. (2008), 'The Hotel and the City', *Progress in Human Geography* 32: 3: 383–398.

Miller, D. (1998), *A Theory of Shopping* (Cambridge: Polity).

Muschamp, H. (2000), 'Interior City: Hotel as the New Cosmopolis', *New York Times*, 5 October, www.nytimes.com, accessed 26 July 2004.

Peleggi, M. (2005), 'Consuming Colonial Nostalgia: The Monumentalisation of Historic Hotels in Urban South-East Asia', *Asia Pacific Viewpoint* 46: 3, 255–265.

PKF Consulting. (1996), *Hotel Development* (Washington DC: Urban Land Institute).

Postrel, V. (2004), *The Substance of Style: How the Rise of Aesthetic Value is Remaking Commerce, Culture and Consciousness* (New York: Perennial).

Rantisi, N. (2006), 'How New York Stole Modern Fashion' in Breward, C. and Gilbert, D. (eds), *Fashion's World Cities* (pp. 109–122) (Oxford: Berg).

Reimer, S. and Leslie, D. (2004), 'Identity, Consumption and the Home', *Home Cultures* 1: 2, 187–208.

Riewoldt, O. (1998), *Hotel Design* (London: Laurence King).

Slesin, S. (1990), 'Lobby as Urban Living Room', *New York Times* 9 August, www.nytimes.com, accessed 8 November 2004.

Solomon, C. (2005), 'New York: Royalton Hotel', *New York Times* 18 December, www.nytimes.com, accessed 15 October 2007.

Stein, K. D. (1991), 'Repeat Performance: Paramount Hotel, New York City', *Architectural Record* 179: January, 72–75.

Stern, R. A. M, Fishman, D. and Tilove, J. (2006), *New York 2000: Architecture and Urbanism between the Bicentennial and the Millennium* (New York: Montacelli Press).

Watson, H. (2005), *Hotel Revolution* (Chichester: John Wiley).

Weiss, L. (1997), 'Schrager Outlines Plans for Henry Hudson Hotel', *Real Estate Weekly*, 19 November, accessed via *Highbeam Encyclopedia*, www.encyclopedia.com, 12 November 2007.

Welch, S. J. (2006), 'Itineraries; Let's Meet in the Lobby', *New York Times*, 2 May, www.nytimes.com, accessed 15 October 2006.

Wharton, A. J. (2001), *Building the Cold War: Hilton International Hotels and Modern Architecture* (Chicago: University of Chicago Press).

PART III
Consumption as Connection/ Disconnection/Reconnection

Chapter 8
Manufacturing Meaning along the Chicken Supply Chain: Consumer Anxiety and the Spaces of Production

Peter Jackson, Neil Ward and Polly Russell

Introduction

While consumer trust in food has become an issue throughout Europe (Kjaernes et al. 2007), recent 'food scares' – from BSE and salmonella to the threat of avian flu – have led to a loss of public confidence in the safety of British food. There has been intense debate about who should be charged with protecting the consumer interest in food: government bodies (such as the Food Standards Agency), the corporate sector (through the major supermarket retailers) or lobby groups (see Marsden et al. 2000). Successive reviews have attempted to assess the risks associated with British food production including the Pennington report on the E.coli outbreak in Scotland (Pennington Group 1997) and the Curry Report on the future of food and farming in Britain, following the outbreak of Foot and Mouth Disease in 2001 (Policy Commission on the Future of Food and Farming 2002). The Government's White Paper on *Modern markets, confident consumers* (DTI 1999) spoke of the need for consumers to be able to make informed choices based on reliable information on price, quality and safety, while the Curry Commission described its vision of well-informed consumers exercising choice in a diverse market-place, knowing where their food has come from and how it was produced (Policy Commission on the Future of Food and Farming 2002). These ideas have been further developed in a series of recent reports from the Cabinet Office (2008a, 2008b).

Clearly, food production has become a highly charged issue in contemporary Britain with much at stake in terms of public health and consumer confidence. Questions of food safety and quality need to be balanced with economic demands for a more efficient food and farming sector, able to compete effectively with foreign competition, while responding to calls for greater social and environmental sustainability (Barling and Lang 1993; Goodman 2003). A common feature of recent debates has been the emphasis on *re-connecting* producers and consumers along the food supply chain. Indeed, the central theme of the Curry Commission's report was reconnection, arguing that farming had become detached from the rest of the economy and the environment, and calling for better communication up and down the food supply chain (Policy Commission on the Future of Food and

Farming 2002, 6–10). Our recent work on 'Manufacturing meaning along the food commodity chain' responds to these concerns.[1]

Unlike much previous commodity chain research (cf. Gereffi and Korzeniewicz 1994; Hughes and Reimer 2004), our emphasis is not so much on identifying the points along the chain where value is added and profit extracted (in purely economic terms) but on identifying where and how the distinctive cultural meanings of food are created and negotiated. We argue that this process of 'manufacturing meaning' has direct economic consequences in a commercial climate where food is increasingly 'sold with a story' (Freidberg 2003).[2]

Our methodological approach is also different from most previous commodity chain research, employing a life history approach to record and analyse the personal testimony of key players involved at all points along the supply chain 'from source to salespoint'.[3] The approach allows us to interrogate the interplay between individual accounts of change within the food industry, including personal interests and biographical investments, as well as more public accounts of corporate change within the industry. The life history approach helps to 'humanize' our understanding of commodity chains and to explore the role of subjective ideas, like myth and memory, within contemporary understandings of the food industry. We have focused elsewhere on the popularity of the commodity chain concept in academic and policy discourse, noting that the term is often used in an inconsistent, even chaotic, manner (Jackson et al. 2006). We do not elaborate on that critique here. Instead, we have organized our analysis in terms of three discursive spaces – commodity spaces, commercial spaces and spaces of consumption – which, we argue, enable us to make sense of recent changes within the broiler chicken industry. Our approach also helps us to 'spatialize' the analysis of the chicken supply chain (in the terms suggested by Leslie and Reimer 1999). For our analysis of the chicken industry suggests that the linear logic of a single 'chain' scarcely captures the complexity of the contemporary industry.

Our interview material provides a rich source of information about the commercial changes that have taken place in the food industry within living memory and the various ways that food producers and retailers have responded to changing consumer attitudes. The life histories demonstrate the complex and contradictory meanings of chicken as both a commodity and a living thing, subject

1 The research was funded by the AHRC-ESRC *Cultures of Consumption* programme, award no. RES-143-25-0026.

2 Friedland et al. (1981) used a similar metaphor in their study of 'manufacturing green gold'. But their emphasis was on the political economy of agricultural production, particularly in terms of the impact of mechanization on the labour process, rather than on the cultural meaning of food and its commercial implications.

3 For an introduction to life history research, see Perks and Thomson (2006). In addition to the life history interviews with food producers, we also conducted interviews and focus groups with consumers, and interviews with policy-makers and food campaigners.

to agricultural intensification and commercial exploitation but not without cost in terms of consumer confidence and trust. The commercial space of contemporary chicken production emerges as an anxious and contested space, as producers seek to respond to consumers' fears about food safety, attempting to balance their demand for quality and value. The resolution of these contested meanings can only partly be achieved through increased technocratic control. As important, we argue, is the complex process of 'manufacturing meaning' at various points along the supply chain where subjective notions of myth and memory are as important as the more narrowly-conceived commercial imperatives of technological innovation and product development. Before turning to our interview material, we provide a brief outline of the social and spatial organization of the chicken supply chain.[4]

The Chicken Supply Chain

Our analysis focuses on a single industry – the intensive (broiler) chicken industry – which in many respects epitomizes the recent industrialization of agricultural production (cf. Dixon 2002). Once regarded as a luxury item, the popular consumption of chicken meat as a staple source of protein is a relatively recent phenomenon, dating only from the 1960s (Visser 1999). The development of the refrigerated 'cold chain' and related technological developments in food retailing made fresh rather than frozen chicken widely available. A key trend in the chicken industry has been the steady reduction in the length of time broiler chickens take to grow to their full weight. Today, chickens are reared to their slaughter weight within about 40 days of being hatched while the amount of feed needed to achieve this weight gain has been reduced by almost 40 per cent since 1976 (Compassion in World Farming 2003, 7). These changes resulted in a dramatic intensification of chicken production which now occurs on a truly industrial scale (see Figure 8.1).

In the first half of the twentieth century it was common for farms and small-holdings to keep a limited number of laying chickens to supply the household and local community with eggs (see Figure 8.2).

4 Earlier versions of this chapter were presented at a workshop on 'Managing anxiety: Food scares and the British consumer in Victorian and modern Britain' at Birkbeck College, London (July 2004) and at seminars at Bristol, King's College London and Exeter universities. We would also like to thank the editors for their generous and constructive comments on a previous draft.

Figure 8.1 Contemporary broiler shed
Source: People for the Ethical Treatment of Animals; reproduced with permission.

Figure 8.2 Domestic chicken production, c. 1925
Source: Audrey Kley, private collection; reproduced with permission.

During the Second World War egg rationing saw the expansion of cottage-industry egg production with over one million domestic poultry keepers accounting for 30 per cent of the national laying stock by 1945 (Holroyd 1986, 155). Chickens were predominantly kept for egg production, with poultry meat regarded as an expensive luxury (Burnett 1966, 273). In 1953, following the example of the American broiler industry, the first UK broiler shed was built by Antony Fisher (Holroyd 1986, 160). Fisher's adoption of US poultry production processes was to create the beginnings of the intensive broiler industry.[5] As the number of broiler farms expanded, the price of poultry fell and consumption increased. Between 1960 and 1980 the average person's weekly consumption of poultry meat more than trebled (MAFF 1991, 37). In the space of about fifty years, chicken production in the UK metamorphosed from being a localized cottage industry to being a 'highly concentrated and industrialised sector with production and distribution dominated by a relatively small number of economic actors' (Yakovleva and Flynn 2004a, 229).

There are four key stages in the modern chicken supply chain: breeder farms, hatcheries, growers, and processors. Breeder farms provide the genetic stock for poultry flocks. Only three breeder companies (Aviagen, Cobb-Vantress and Merial) supply the UK with poultry genotypes. The parent stock (mainly Ross and Cobb) produce eggs which are hatched at one of the UK's 12 hatcheries and then sent to grower farms. In May 2007 there were over 24,000 growing premises on the British poultry register, with total holdings of some 250 million birds.[6] Once the chickens have reached the desired culling weight they are sent to a processing plant of which there are over 100 nationwide.

Although the UK broiler population is spread across a large number of holdings, the vast majority of chickens are held in flocks with more than 20,000 birds (Sheppard 2004, iii). It is, then, a highly concentrated as well as a highly intensive industry. The majority of processors have an annual turnover of £1 million or more, with the largest being Grampian Country Foods (£1,206 million), Moy Park (£242 million), Boparan Holdings (£228 million), G.W. Padley Holdings (£222

5 That the intensification of chicken production was shaped by the rise of neo-liberal political ideology is relatively well-established. Less widely recognized is the role of chicken production in the development of neo-liberalism. Fisher's business, Buxted Chickens, grew to command sales of £5million and employed 200 people, handling 17,500 chickens a day by the late 1950s. An enthusiastic free marketeer, Fisher sold the business in 1968 and used some of the proceeds to finance the establishment of the Institute of Economic Affairs (IEA). Milton Friedman later described Fisher as 'the single most important person in the development of Thatcherism' (quoted in Cockett 1995, 122), claiming that 'without the IEA, I doubt very much whether there would have been a Thatcherite revolution' (ibid., 157).

6 Only premises with 50 or more birds are legally bound to register with DEFRA. We are grateful to Matthew Wisbey for researching these data, based on material from the UK Poultry Council (www.poultry.uk.com) combined with other data from DEFRA and the FSA.

million) and Sun Valley Foods (£218 million) (quoted in Yakovleva and Flynn 2004b, 16). The retailing of chicken is also highly concentrated in the UK, with over 70 per cent of fresh chicken sold through the four largest supermarket chains: Tesco, Sainsbury, Asda-Walmart and Safeway (Seth and Randall 1999; Blythman 2004).

Chicken is a remarkable international success story in terms of market share. In 1990, chicken consumption in the US outstripped beef for the first time in what Boyd and Watts call 'a watershed in the history of American dietary culture' (1997, 192). In the UK, the level of domestic production of poultry has increased by almost 20 per cent over the last 10 years, from 1.35 million tonnes to 1.6 million (DEFRA 2005a, 73). In 2004, some 885 million chickens were slaughtered in the UK (DEFRA 2005b). According to the National Food Survey, household expenditure on poultry amounts to 4.8 per cent of the entire domestic expenditure on food and drink (quoted in Yakovleva and Flynn 2004b, 8). Chicken is now the most popular meat in the UK and represents about a third of all meat consumed (Sheppard 2004).

Recognizing the historical role played by Marks & Spencer in shaping the poultry industry in the UK (see Goldenberg 1989; Holroyd 1986; Seth and Randall 1999), we conducted life histories with a number of Marks & Spencer employees working in the poultry department (including technologists, category managers and buyers). We also interviewed some of their suppliers (both hatcheries and growers) though not all of our interviewees were physically involved in the same supply chain (some growers supplied other retailers, for example). In what follows, we focus on the way that one major British retailer (Marks & Spencer) manages the meanings and risks associated with poultry production. As the first retailer to pioneer the sale of fresh (as opposed to frozen) poultry products in the late 1960s, poultry portions in the 1970s and added-value cold chain poultry products in the 1980s, Marks & Spencer has played a key role in determining current poultry production and consumption in the UK. The refrigeration and transportation technology that Marks & Spencer developed in partnership with British Oxygen Company (BOC) guaranteed the microbiological safety of fresh poultry products and led to the widespread expansion of 'cold chain' retail, including the development of the ready-meals market. Today, Marks & Spencer's poultry products have to compete in a market where chicken is a supermarket 'loss leader' and added-value poultry products are common-place. Defining and articulating the difference between Marks & Spencer's products in a saturated market is a key challenge for Marks & Spencer and one which relies upon understanding consumer notions of risk and anxiety.[7]

7 We are pursuing these issues in our current research on 'Consumer culture in an age of anxiety', funded by the European Research Council.

Manufacturing Meaning along the Chicken Supply Chain

Our interview material suggests that chicken producers at various points along the supply chain are responding to consumer anxieties in various ways. These include technological changes and product innovation, improvements in hygiene and animal welfare, and changes in food labelling and marketing strategies. Besides these 'technical' issues, however, our life histories also demonstrate a more subtle process at work which we describe to as 'manufacturing meaning'. By this we refer to the process whereby food producers are not simply manufacturing a product (broiler chicken) but they are also simultaneously attempting to manipulate the meanings which consumers attach to that product. Our interviews with consumers confirm the highly-charged emotions that surround contemporary chicken production. One interviewee claimed that 'The meat that you worry about most is chicken isn't it? Cause it's like you can get so many different things from it' (Caroline, January 2007), while another said: 'You're always warned about salmonella … you have to cook [chicken] properly and if you freeze it, you have to be careful about re-heating and that sort of thing' (Sue, December 2006).[8] Consumers are also concerned about animal welfare and the prospect of chicken being 'squashed together in … some sort of big monstrosity like the Bernard Matthew's thing' (Jane, March 2007).[9]

We organize our discussion of the process of 'manufacturing meaning' in the contemporary British chicken industry in terms of three key discursive spaces: commodity spaces, commercial spaces, and spaces of consumption. The discussion raises questions about the commodification of nature (whether chickens can be reduced to the status of a commodity), about the technocratic control of nature (through the intensification of agricultural production) and about the effects of consumers in shaping the commercial spaces of contemporary chicken production. While we conducted over 30 life history interviews in total, we draw here on extracts from six of the producer interviews, supplemented by extracts from half a dozen of the consumer interviews (see Table 8.1).[10]

---

8 We have used pseudonyms to anonymize the consumer interviewees.

9 Bernard Matthews is the biggest turkey manufacturer in Europe with an annual turnover of £400 million. He was at the centre of a public health crisis in February 2007 when Britain's first case of the H5N1 strain of bird flu broke out at one of his farms in Suffolk. He was subsequently accused of importing poultry from an avian flu exclusion zone in Hungary (*Times Online*, 9 February 2007).

10 The producer interviews lasted from 4 to 12 hours, often conducted over several sessions. They were tape-recorded and transcribed in full. Subject to any restrictions imposed by the interviewees, the tapes, tape summaries and transcripts are publicly available at the British Library Sound Archive and can be accessed via the Library's on-line catalogue (www.bl.uk.cadensa.bl.uk). The consumer interviews were shorter, each lasting around an hour.

Table 8.1 Interviewees

Producers	Occupation	Date of birth
Ray Moore	Hatchery manager	1935
Audrey Kley	Chicken grower	1930
David Gregory	Head of food technology (M&S)	1953
Mark Ranson	Agricultural technologist (M&S)	1967
Andrew Mackenzie	Protein category manager (M&S)	1957
Catherine Lee	Poultry buyer (M&S)	1970
Consumers		
Caroline	PhD student	1982
Deborah	PhD student	1983
Ellie	University administrator	1977
Jane	Clinical scientist (NHS)	1967
Linda	Postdoctoral researcher	1980
Sue	Trainee embryologist	1980
Vicky	School teacher	1977

Commodity spaces

In this section, we consider how our interviewees conceptualize chicken: whether as a commodity (an economic good to be bought and sold) or as a sentient being (a living part of the natural world). We begin with Ray Moore, a chicken hatchery manager, now in his seventies. In the course of the interview, Moore describes chicken in both economic and emotional terms. At some points in the interview he clearly sees chickens as a commodity ('It's all to do with money with chickens really, you shuffle chickens like you shuffle money').[11] Yet he also identifies more closely with the product than many of our other interviewees further down the chain:

> I'm a Ross man …
>
> Chicken talk to you … it don't matter what you say – they talk to you and you know, you'll know whether they're cold, hungry, too hot …
>
> They need loving tender care, you know. Just the same as I look after my wife, just like that. You've got to look after 'em.

11 Ray Moore, interviewed by Polly Russell, November 2003, C821/114/. The code following the date of interview refers to the British Library accession number.

For Moore, chicken are a tradable commodity. His job is to anticipate future demand and to acquire and hatch sufficient chicks to meet that demand (see Figure 8.3).

Figure 8.3 Intensive chicken hatchery, 1970s
Source: Ray Moore, private collection; reproduced with permission.

But he also accounts for them as sentient beings and recognizes the role they have played in shaping his life. Moore's home, situated adjacent to the hatchery, is filled with posters of different poultry breeds, china models of chickens and roosters, and awards celebrating his achievements in the poultry business. It is no exaggeration to say Moore's life has been inextricably bound up with the poultry industry.

Our second interviewee is chicken grower Audrey Kley who takes chicks from Ray Moore's hatchery when they are a few days old and grows them until they reach their slaughter weight and are sent off to the processor.[12] The daughter of a West Country farmer, Audrey Kley was brought up on a farm and has clear memories of her grandmother and mother keeping layer chickens for 'pin-money'. Like Moore, Audrey Kley has spent her life in the poultry industry, starting business as a young woman with a flock of a hundred laying chickens. Asked to reflect upon her feelings towards her flock Audrey Kley articulated a practical, no-nonsense attitude:

12 Chicks from Ray Moore's hatchery can end up as free range or intensively reared broiler chickens. There is no difference between the birds when they leave the hatchery.

> Interviewer: Do you ever think about the fact that your chickens growing down here by you end up [inaudible]?
>
> No, because you … you've got to be hardened to it, haven't you. I mean it's not like your own pet dog when it dies, you get very upset, and you think about it for days … With the chickens, you have, there's no sort of affection or anything towards them, because they all look alike. And, well you just … it's part of your work and you don't think about it. They come in, you know they're going to be killed the next day, but all you're doing is worrying that you've got enough food to see them through, that they're good and they're going to pay you some money, and you don't think about any of the other side of it.[13]

Despite this hard-nosed, practical orientation to her business, Audrey Kley clearly cares for her chickens and describes the process of 'walking the sheds' several times a day, noting from the density and movement of the birds in different parts of the sheds whether there are any problems of heating or ventilation or any shortage of food and water. In Audrey Kley's account, then, chickens are more than a commodity to be bought and sold. She is concerned about their welfare and expends considerable effort to safeguard their well-being.

Our next interviewee is Mark Ranson who works for Marks & Spencer as an agricultural technologist, responsible for meat and poultry products. A key player in defining Marks & Spencer's poultry policies over the last five years, Ranson completed a Masters degree in animal behaviour at Edinburgh University and then worked as an animal welfare officer for the RSPCA before joining Marks & Spencer.[14] Describing himself as 'poacher turned gamekeeper' for moving from the charity to the retail sector, Ranson shared his views with us on the ethics of eating meat and poultry. In reflecting on the differences between species bred for meat he describes chicken as 'much more of a commodity than other protein species'.[15] Ranson comments that one consequence of the cheapness of chicken as a commodity is the high level of waste associated with its production. In particular he describes the waste generated when new products are tasted and whole chickens are disposed of after a couple of slices of meat are tested. Reflecting on this practice, Ranson notes his discomfort:

> When you do a whole bird, you think that, that was a living animal … All right it's been bred for food and there is a reason why we do it, but I think from a moral perspective, if there was a way to, to do the tasting without actually having to kill it … I've got no objection to eating animals and eating food providing

13 Audrey Kley, interviewed by Polly Russell, September 2003, C821/109.

14 We have not discussed our third interviewee, David Gregory, in this section. As head of food technology at Marks & Spencer, he has little direct contact with chickens in his present managerial role.

15 Mark Ranson, interviewed by Polly Russell, January 2004, C821/121.

> that its welfare has been cared for. It's just such, such a waste … I find it a bit objectionable actually, the amount of food which is wasted.

It is the reminder of the 'whole bird' as a 'living animal' that prompts Ranson's concern about levels of waste resulting from the commodification of poultry. He adds, significantly: 'I wouldn't feel so bad if it was just portions, because the rest would've gone to the use it was meant for … it's the waste, it's a reminder of it once being a whole animal'. Chicken, it seems, may be a commodity but the process of commodification – including the routine tasting and wasting – cannot completely erase the memory of chicken as a living animal.

Mark Ranson's colleague, Andrew Mackenzie, a protein category manager at Marks & Spencer, felt that modern customers are indifferent to chicken because they are so cheap and widely available and because the industry has become so clinical and efficient:

> I think people's aspirations and expectations of chicken have lowered in the course of the last number of years. And I describe chicken as a canvas upon which people paint because actually in itself it doesn't offer much. It's a source of protein, but in textural terms unless it's free range or organic, in flavour terms often, it doesn't really deliver. And what delivers is the sauce that you put on it or the way that you cook it. And I think because it's eaten so regularly and we eat such vast quantities of it, and you talked about it down to a unit or a commodity, I think that's a really good analogy because the other thing is that you know, chickens aren't the most appealing of things. They are … people still ask us the question about this, either in a cage as a caged battery egg, or as something they see scratching around on a farmyard, but they would have a very different perception of the way that chickens are actually grown in the modern broiler world. And I think that they don't want to think about it because they're not particularly easy to empathise with.[16]

In the modern broiler world that Mackenzie describes, chickens are removed from the public gaze in large industrial-style sheds. The 'invisibility' of chicken production has direct consequences for the way consumers relate to the product.

Mackenzie went on to reflect upon his own feelings about chickens. He argues that chickens are 'strange things' because they are produced in such large quantities:

> It is a production line, and with chicken even more than anything else, because it works on volume, 9000 an hour or whatever, then that is quite … it's a production, but it's a production where something dies … And I suppose sometimes you do get a bit of a thought around that, but if I'm honest with you, I don't really dwell on that. I don't really dwell on that at all, because I don't think you can. I mean

16 Andrew Mackenzie, interviewed by Polly Russell, February 2004, C821/135.

> if you did, then you'd probably struggle to do the job, I'd certainly struggle to do my job.

In this extract, Mackenzie articulates how the mechanization and acceleration of poultry production creates a distance from any emotional connection with live chickens intended for slaughter. The scale of the production process has the potential to rob the animals of any meaning beyond their economic significance as a commodity. Yet, for Mackenzie at least, the troubling issue remains that this is a production line 'where something dies'.

Commercial spaces

The sense of distance from nature that is created by intensive, highly mechanized forms of production is also relevant to the way meanings are manufactured in the commercial spaces of the contemporary food industry. We start again with hatchery manager, Ray Moore. After completing National Service, aged 19, Moore started working for a Wiltshire poultry entrepreneur, Maurice Millard. Moore described how Millard embraced technological developments and transformed his hatchery from a manual to a predominantly mechanical business between the late 1950s and early 1970s. In 1975, having started work as a farmhand, Moore was given sole responsibility for managing Maurice Millard Hatcheries. In 1987 Moore was made Director of the business, hatching 700,000 chicks each week. Over his working life Moore has experienced the evolution of the poultry industry from a cottage industry to an intensive, integrated business. Given this fact it is perhaps not surprising that Moore's descriptions and reflections on the contemporary poultry industry were framed by his memories of earlier practices.

In recounting his life story Moore situates himself as a witness of change, as a businessman and as a poultry enthusiast. Each of these facets of Moore's life history plays an important part in determining the ambivalence and anxiety he articulates when describing changes in modern commercial poultry production. In a particularly telling anecdote Moore recalls his nervousness when he first encountered an automated hatching machine in the late fifties:

> If we went and picked up ten cases from a breeder in those days, which was 3600 eggs, we used to think we'd picked up a lot of eggs. But now I mean you go out and pick up sort of 300,000 eggs, but it's all sort of grown on you over the years. Well it does grow on you. I think I told you last week. No, maybe I didn't. Had an old boy up the road called Dennis Randall, and he used to be a bloke who was in his day a Black Leghorn breeder, but by the time he sort of got to his age and poor old boy lost his life, and he was just pottering around, but he was a knowledgeable old boy and I went up one day and I said, "Maurice wants me to take over those machines", I said, "But have you seen 'em?" He said, "No", he said. I said, "You must come down and have a look". I said, "They take a hundred thousand eggs". "Ah my boy", he said, "don't worry about that", he

> said. "That'll just seem like the other ones after a couple of weeks, it'll just grow on you." And he was right. It does just grow on you and you get used to it.

Moore's retrospective description highlights his youthful innocence and aligns himself with the experience and insight of the 'knowledgeable old boy'. This story encapsulates an embrace of progress and a simultaneous concern about the scale of change that has characterized Moore's life history and his views of the poultry industry.

Moore's narrative reveals both a pragmatic acceptance of business imperatives and an anxiety about the potential losses incurred through change. Commenting on when he first came into the industry he reflects nostalgically: 'I used to work in the field, we had proper free range chicken, proper free range chicken, not like they try to tell you it is today'. The emphasis on practices in the past being 'proper' and the implication that modern methods fall short of this standard is similarly revealed when Moore was asked to describe how chickens tasted when he was young:

> Interviewer: And can you remember what chickens were like before that? I mean d'you remember what they tasted like before?
>
> Well, yeah, but then we're not in the same league are we? Now, when mother and that would kill a chicken in those days, they would pluck it and then go and hang it up in the old coal shed for two or three days and they wouldn't draw it – and they all say, well you hang beef don't you, and it's better, you hang any meat and it's better. And that is one of the problems I suppose we have got, they haven't got time to hang these chicken … But I mean if you're oldy-fash or traditional or whichever way they want to put it, I don't think you could ever change the thought of hanging 'em in the coal shed and letting 'em mature that way … I mean, I suppose really, when you think about it, when I was coming towards 23, 24 [in the early 1960s], that's when it was all changing. It was all changing then and I mean, still you sort of didn't have chicken like you do now. You never had all these breast fillets and all that jazz, drumsticks and all that, you never had that. Used to just have a chicken.

Moore insists that chicken in the past had a better flavour and that poultry today is 'not in the same league'. Underlying this lament about past practice is a critique of accelerated poultry production that leaves no time for hanging meat and allowing flavour to develop. For Moore, the chicken portions and 'all that jazz' that characterize contemporary poultry products are unfavourably compared to the straightforward appeal of 'just … chicken'. Moore's account conveys a sense of regret and loss about the acceleration of modern broiler production and its consequences.

Moore's regret about the acceleration of production extends beyond a lament for the taste of food in the past. He is concerned about the implications of modern broiler production in terms of animal welfare and increased consumer anxiety. He talked about the need to slow the production cycle down 'to give the bird a bit more chance'. Asked to clarify why the accelerated rate of poultry growth makes him uncomfortable, he replied:

> Well they just, they just grow on fast. I mean you're killing the bird now, within well, less than six weeks. For a 4lb bird. So I mean really, you got to do something somewhere.
>
> Interviewer: And you feel that that has a negative impact on the image of the industry, is that what you …?
>
> It could do, to someone like you yourself, young. They wouldn't understand what's happened – well, they would only want to understand what they see, not what's happened over the years. And yeah, I don't think it gives us a good image at all.

For Moore, the risks associated with accelerated production hinge on his sense that poultry production needs to be 'slowed down' for the benefit of the chickens and because of a related concern about how the industry is viewed from the outside.

Similar concerns are expressed by Audrey Kley whose broiler farm has three sheds each containing about 30,000 birds grown on a six-week cycle. Acting as a contractual supplier within an integrated industry for a processor who supplies all the large multiples means that Audrey Kley has limited control over the growing cycle, numbers of birds grown and the value of her stock. The processor determines where she may purchase chicks, the quantity of birds grown each cycle, the amount she is paid, feed specifications, slaughter weights and shed cleaning rotas. Reflecting on how the intensification of the poultry industry has impacted on her work, Audrey Kley explained that: 'it is a tie, it's a very tedious, and it's a monotonous thing. Because there's nothing, you know, interesting in it, it's just, routine. Only when there's something that goes wrong is there any drama'.

Although she describes contemporary broiler farming as 'tedious' and 'routine', her life history conveyed little nostalgia for poultry farming prior to intensification and mechanization. Instead, the risks and anxieties that Audrey Kley articulated with regard to poultry production were situated in the context of operating as a broiler producer prior to industrial integration and increased government regulation. In other words, for Audrey Kley, regret pertaining to change was not related to a nostalgic memory of pre-mechanized poultry production, but to the late seventies and early eighties when poultry production operated on an open market providing efficient growers like Audrey Kley with opportunities to maximize profits.

Like many farmers, Audrey Kley is hostile to the increasing regulation of the industry, describing those who enforce the regulations as 'failed farmers'. She regards schemes like Farm Assured as pointless and expensive.[17] Such schemes, she feels, are abused by foreign suppliers. She tells a story about going to Tesco to enquire whether chickens described as 'processed in the UK' were actually *produced* in the UK. From her point of view, the retail industry exerts too much power along the entire supply chain: 'Within the whole of the farming industry we have this gut feeling that we are being ripped off by the supermarkets'. She feels that British farmers face unfair competition from abroad because the French 'ignore all the rules' while in Brazil and China chickens 'have been fed on all the antibiotics in the world'. The anxiety that Audrey Kley articulates in relation to poultry growing derives from her feeling that the industry is overly and unfairly regulated, significantly limiting the autonomy of individual actors along the chain.

Within this context David Gregory, head of food technology at Marks & Spencer, provides a more complex account of the industry's response to changing consumer attitudes as well as the business context in which members of the Marks & Spencer poultry team work. He describes an alternative to the stark contrast between safe and unsafe food, arguing that the industry now emphasizes risk management rather than being guided by an absolute notion of food safety: 'No longer do people try and aspire to make safe food, they manage risk to ensure food is safe'.[18] Gregory had considered changing his title to Head of Quality, Trust and Innovation, reflecting the changes taking place within the industry and the fact that: 'today my job is about risk management systems'.

> You really struggle to say that anything is safe but you can think about what are all the potential hazards in a situation and what are all the steps you can take to mitigate these happening. This has really emerged over the UK in the last twenty years.

While his job was previously about food safety, today it is mainly about food quality and delivering new products. Expanding on this argument, Gregory suggests that:

> Safety, which in the previous days was if you like, the word that was used rather than trust, was a narrower area. It was hugely important. It was about making sure that the food that we eat is safe. But by and large the customers of the past really just wanted to know, was the food safe and they wanted to know less about where that food came from, how it was produced, the people who were involved in the production processes and the like. So the job in the past was about food safety, today the job is far broader.

17 Over 78,000 UK farmers and growers are 'farm assured' under the Red Tractor scheme operated by the independent Assured Food Standards organization (www.redtractor.org.uk).

18 David Gregory, interviewed by Polly Russell, February-March 2004, C821/130.

Safety is still an issue, of course, especially with chicken and micro-organisms like campylobacter which are a serious hazard to public health:

> Thirty years ago my predecessors would not even have heard of campylobacter … But of course the food we eat today is actually safer probably than it was twenty-five, thirty years ago, or indeed prior to that. But the big issue and big change isn't about safety and the management of that, it's about trust.[19]

According to Gregory, consumer trust can be understood at three levels: trust for the individual self, trust for others (manifested in concerns for animal welfare and Fair Trade issues), and trust for the future (including concerns about excessive 'food miles' and the impact of pesticides on the environment).[20] Gregory explained how many of these issues involve balancing different, sometimes competing, interests. So, for example, reducing 'food miles' may have impacts on employment levels in developing countries, while reducing the salt content of food products will have benefits for consumer health but may run counter to arguments about promoting consumer choice.

Spaces of consumption

The final discursive space we wish to explore in terms of 'manufacturing meaning' along the chicken supply chain is the way food producers imagine the consumer and how their attitudes in turn shape the process of food production. We start, again, with Ray Moore who expresses a widely-felt anxiety about contemporary consumer knowledge. As part of a quest to bridge the gap in consumer knowledge about poultry production, Moore and his wife have pioneered 'chicken and egg' stands at country shows. These stands include descriptions of the life-cycle of the chicken and provide an opportunity for consumers to watch chicks hatch and handle chickens. In describing the impetus for the 'chicken and egg' displays, Moore reveals a sense of regret about the perceived lack of consumer knowledge and a concern about the distancing of food production from consumers:

> Well, I just think they should, I mean some people might not, little kids might not even know milk comes from a cow. It comes out of a bottle according to them. They don't know how it gets to the bottle, so they more likely don't know where an egg comes from, it comes out of a cardboard case. But there's a chicken at the end of that line. And the same when they're eating a nice piece of chicken or chicken nuggets as these kids will eat, they more likely think that comes out of

19 Recent estimates suggest that as much as half of all the chicken on sale in UK supermarkets is contaminated with campylobacter, the main source of food poisoning in humans, though rarely at levels that pose a threat to human health (Lawrence 2004, 2).

20 The validity of 'food miles' as a measure of the social and environmental costs of lengthening supply chains has been the subject of intense debate (see, for example, DEFRA 2005c).

> a cardboard packet. But there is a live thing at the end, at the beginning rather, where it comes from.

Moore's practical response to the perception that consumers lack understanding about chicken and egg production involves a delicate balance. On the one hand the 'chicken and egg' displays seek to make chicken production more transparent but on the other hand Moore understands that aspects of the intensive poultry industry such as battery cages should remain hidden:

> Interviewer: Isn't there a risk, though, if you tell people about where chicken comes from, particularly children, that if they know that a chicken McNugget was once a living feathered thing, they're less likely to want to eat …
>
> We don't know, but we still think we've got to try and educate 'em to know where it does come from. And to show 'em what we're doing, show 'em what we're doing is not wrong really. We've never put a battery cage at the "chicken and egg" [show], I will say we're wary of that, that's for layers.[21]

Moore's positive memories of the poultry industry before mechanization and his ambivalence towards intensification are related to his recognition that aspects of modern poultry production are likely to be unpopular with consumers. This does not mean, however, that Moore embraces free-range and organic poultry production. He speaks about the dangers of cross-contamination between free-range and organic chickens, a view that recent analyses of the prospect of an avian flu pandemic support (Davis 2005). Moreover, in addition to the risks posed to biological security, Moore believes free-range production raises questions about animal welfare because chickens do not like cold and wet outdoor conditions.

Audrey Kley also has some harsh words to say about contemporary consumers and their alleged ignorance of contemporary food production. She expresses regret about customers' changing attitudes towards poultry, explaining that chicken 'used to be special but now it's everyday'. She does not make a connection between the growth of intensive broiler production and the decline in the economic and social value of chicken. Instead, Audrey Kley holds the customer (and particularly 'the housewife') responsible. She is concerned that housewives are driven purely by price without due regard for quality or provenance:

> Well I'm very worried about the housewife, because she doesn't … she'll go and buy the cheapest that she can, won't she … And she doesn't think to look … where it's come from, or really how it's produced. She doesn't know where the egg comes from, she doesn't even know where the milk comes from. And

21 Indeed, our research has shown that many consumers are confused about the difference between caged battery chickens (kept for eggs) and broiler chickens (kept in sheds and grown for meat).

> really, I think they ought to be taught in the schools, because, I understand that they have quite a lot of people going into the schools teaching the children, you know, from [a] vegetarian point of view, not to eat meat, pushing this vegetarian thing, and we haven't done anything about going in to promote the chicken or our side of, how we do things, and I think that's very wrong.

In her view, housewives are mainly responsible for diseases like salmonella, while 'the poor ruddy grower gets all the blame'. The same is true with cooking chicken:

> I mean they don't, they just think that you take the thing out of its bag, you fling it in the microwave, and that's not the right way to cook a chicken … Or they fling it in the oven, half frozen you see, they don't look in, wash it or anything, do they? … So, I mean, half of the problem with the housewife is, she hasn't been taught to cook good, plain food, because, I understand that the domestic science courses that the children do at school now, they're not taught to make a good stew or a good apple tart or all those sort of things; they make sort of party food, don't they, which you're not going to eat every day. But, show them how to make a good plain meal, they've no idea. And beside that, they don't want to bother with that, do they, they want something quick.

She concludes that the British housewife 'ought to think a little bit more what she's buying'.

Mark Ranson raises some different issues about consumer attitudes, particularly concerning the distinction between whole birds, chicken portions and processed food. Ranson explains how retailers have responded to customers' reluctance to handle fresh meat:

> From a customer perspective I think once an animal has been, has been cut up it loses so much of its … in the customer's mind it's lost so much of its … what it was. I think the customer's questions about or consumer's questioning about chicken in their mind as long as it's a whole bird or a portion then that's fine but once it's become an ingredient in a recipe dish it's kind of lost all its … The same concerns don't exist, as they do for a whole piece of meat … So you know we are developing things like, oven-able trays so the product comes in a tray which can go straight into the oven for cooking rather than them having to lift the product from a piece of cellophane or a plastic wrapper into a tray for them to use themselves. That's a direct request from customers saying they don't like touching raw meat.

In this extract Mark Ranson reflects on the direct consequence of consumer attitudes to chicken in terms of its effect on the process of product development and technical innovation. His inability to express the actual properties of chicken in this extract ('it loses so much of its …', 'it's lost so much of its …', 'it's kind of lost all of its …') also underlines the power of the life history method in revealing our (sometimes inexpressible) anxieties and ambivalences about food, even among those who spend their working lives in the food industry.

Our interviews with consumers clearly support Ranson's argument. There is widespread evidence of consumer squeamishness and anxiety, particularly in relation to whole birds:

> I don't like whole chickens. I don't like cutting them up – the gunge and the fat and everything … all the fat and mess … it's just a mess really [and] you're left [with] your bread board and your kitchen bench and everything all covered in this slimy horrible gunge! (Ellie, January 2007)

Other interviewees spoke about their distaste for the 'bone and gristle' associated with chicken and all the 'veins and things', while some took extreme measures to avoid touching raw meat:

> I hate handling raw chicken. I tend to do it with rubber gloves because I keep thinking salmonella urgh or whatever … and I'm quite careful about how I handle it because I think probably all chicken's got salmonella … I don't like touching it either so I do it with rubber gloves and then wash my hands really well. (Vicky, February 2007)

> I bleach all my knives and boards and wash them down after cooking to make sure it's free of any bugs and then make sure that it's obviously cooked really well … so I don't poison me and my housemates! (Deborah, December 2006)

Consumer attitudes are also relevant to issues of provenance. So, for example, Marks & Spencer source all of their fresh chicken (whole birds and portions) from the UK but import chicken for ready-meals and other processed products from Europe. Mark Ranson suggests that consumers are less interested in the provenance of added-value chicken as opposed to fresh raw chicken:

> When something's being cut up and is not visibly a piece of raw meat, people then tend to lose their [interest in understanding] where it's come from or even think about it. But I think, certainly where there has been issues with where it's come from, it's either because customers think that the standards there are really bad or they're perceived as bad because they're reported that way by the media, or you get things like … the media reporting that, you know, meat from the Netherlands or Denmark or Brazil is either full of hormones or antibiotics or it's been injected with water or whatever … But again, from a retail perspective, retailers would put their standards in place whichever country they would trade in.[22]

22 The use of additives in Dutch chicken was reported by the BBC's Panomara programme (The Chicken Run, 23 May 2003). Other instances of adulterated chicken are documented in Lawrence (2004).

Ranson's explanation indicates how retailer specifications for poultry products – whether fresh or added-value – are developed in anticipation of a range of customer anxieties. These anxieties, about provenance and quality, are not applied equally to all poultry products but depend upon the level of 'interest' or emotional investment customers associate with a given product. A whole fresh chicken is a reminder of a living animal and therefore its provenance (and the imagined life it has lived) is salient. Poultry in added-value products is disassociated from the 'live-ness' of the bird, so provenance and freshness are less important. Like Moore, Ranson feels strongly that consumers should understand the process of meat and poultry production. For Ranson this is a question of consumer ethics and relates once again to his concern about waste:

> Ethically people, people should know where it's come from and how it's being produced, because I think … if there was an understanding of the lengths that had gone into producing that animal, then [pause] … I think there would be a greater appreciation for it … and again not wasting product and wasting food.

Ranson feels that most British consumers are quite distanced from where food is produced, both because of the mediating role of the retail industry and because of most consumers' lack of direct contact with agriculture.

Our final interviewee, brings many of these issues together, linking the commodification of chicken and the intensification of production with the ignorance about farming that has allegedly arisen as consumers have become more distant from producers. Catherine Lee, a poultry buyer at Marks & Spencer, is adamant that customers do not want to be educated about farming practices:

> Customers are interested from when they see it in the pack, to what it delivers once they've eaten it. And they're interested in the bit in between that. They don't want to know that that's a dead body sitting in front of them.[23]

The reason for this lack of interest, she feels, is because consumers have become so far removed from the place of production:

> We've moved so away from, so away from a rural environment … that the majority of the population live in a town, you know, they don't really see a live chicken on a day to day basis any more and therefore they've become squeamish about dealing with the consequences of that. They've become disassociated with it you know.

23 Catherine Lee, interviewed by Polly Russell, February 2004, C821/129. Our consumer interviews certainly support this view: 'I don't think about the production of chicken; it's not something I want to think about really' (Ellie, January 2007); 'I tend not to think that much about erm the birds' (Jane, March 2007).

Again, our consumer interviews would support Catherine Lee's concerns about the effects of customers becoming increasingly distanced from the site of production. As one interviewee suggested: 'I think we're probably, in this kind of Western culture, far too removed from what we're actually eating ... It's not because I'm squeamish about what I'm eating ... I think there's this whole kind of silly thing, that's almost making what we're eating as less like an animal as possible which is daft because at the end of the day, it is an animal' (Linda, December 2006).

This sense of increasing distance from agricultural production and the 'sqeamishness' to which it can give rise poses particular problems for firms like Marks & Spencer that seek to differentiate their product in quality terms from rivals like Safeway and Asda who compete mainly in terms of price.[24] Like Andrew Mackenzie, Catherine Lee argues that it is hard to influence the texture and flavour of chicken. While customers will debate the texture of a steak, she argues, they don't want to reflect on what the chicken ate because 'then it's a live thing'. This is particularly problematic for retailers at the top end of the market, who cannot compete simply on price:

> If I was Asda, it would make my life a lot easier, but when you're at the other end of the marketplace where you're trying to get customers to understand quality and what you're doing and a whole proposition of agriculture when they don't really want to know, is extremely hard. Because you want to, you want the customer to understand why it costs ten pence more in M&S than it does in Tesco's or Asda, and it's very difficult to actually tell them why without making them squeamish.

Catherine Lee describes the commercial challenge posed by consumer anxieties (and customers who 'don't want to know') for firms like Marks & Spencer who seek to emphasize the 'whole proposition of agriculture' involved in improved animal welfare, lower stocking densities, better diet and longer growing cycles. While Marks & Spencer have introduced these kinds of improvement across their entire range with the introduction of their Oakham brand, other supermarkets are beginning to follow suit with premium brands like Sainsbury's Devonshire Red and Tesco's Willow Farm chicken. As we have argued elsewhere (Jackson et al. 2007), the development of Oakham chicken was a direct result of consumer concerns about the quality and safety of intensively-reared chicken and their nostalgic desire for chicken 'as it used to be'.

---

24 According to Seth and Randall, Marks & Spencer have a reputation as relentless innovators, particularly in the field of chilled ready-meals. Their development into a national institution and leading British retailer was based on the principle of 'quality first', developing long-term partnerships with suppliers, exacting standards of supply chain management and the application of science and technology to product development (1999, 116–20).

Conclusion

To summarize, we have suggested that the chicken supply chain is a particularly good example of the process of 'manufacturing meaning' within the contemporary food industry as producers have been forced to respond to increasing consumer anxieties about food safety and quality. As well as making 'technical' changes in product design and delivery, they have also been required to respond to the changing values that customers bring to the table. This is particularly true for 'industrialized' products like broiler chicken where the connection between consumers and the conditions of production has been severed. For chicken is arguably the paradigmatic case of agricultural intensification in Britain, with the growing cycle having been halved since the 1960s through improvements in feed-weight ratios. These changes have been driven by the retail industry through the development of the 'cold chain' and related technological innovations. While these changes have been a great success in commercial terms, they have been accompanied by growing consumer anxieties such that risk management and the restoration of consumer trust are now key concerns for the British food industry.

Our interview material suggests that there are inherent limits to the industrialization of a natural product such as chicken where (as Ray Moore recalls), there is 'a live thing' at the end of the supply chain. Rather than seeing its goal as the elimination of risk or a search for absolute food safety, the food industry is now engaged in a process of risk management through a variety of socio-technical strategies. Chicken's organic properties – including its susceptibility to damp, cold and disease – constantly threaten to undermine the imposition of technocratic control. In managing the risks associated with chicken production, we suggest, the food industry is faced with a number of tensions most readily apparent in the desire to justify premium prices through revealing more about animal welfare, quality and provenance without making consumers 'squeamish' by providing too much information. These tensions are illustrative of the process we describe in terms of 'manufacturing meaning' along the food supply chain and are intrinsic to the process of risk management where the maintenance of consumer trust must always be balanced against the commercial need for constant innovation. Consumer anxiety is, we argue, an inevitable consequence of the intensification of the industry, a direct result of the commodification of nature and the increasing distance separating consumers from producers.

We also suggest that there are limitations to thinking of chicken production according to the linear logic of a simple 'supply chain'. Rather, we suggest, that research should focus on some of the complexities and inter-relationships that characterize the contemporary chicken industry. Here, we have argued, a life history approach may be helpful in 'humanizing' our understanding of the industry and in analysing the complex process of manufacturing meaning that occurs as chickens move 'from farm to fork'. Our analysis of this life history material suggests that it is possible to make sense of recent changes in the chicken industry by focusing on three discursive spaces associated with commodities, commerce and consumers. The key distinctions within these spaces concern whether chickens are conceptualized as live animals or as commodities (and how this may depend on whether consumers are thinking about whole birds, chicken portions or processed food); the degree to which the commercial success of the industry has relied on the relentless intensification of production; and the extent to which consumer trust has been jeopardized in this process of intensification as consumers have become increasingly distanced from producers.

References

Barling, D. and Lang, T. (2003), 'A Reluctant Food Policy? The First Five Years of Food Policy under Labour', *Political Quarterly* 74, 8–18.

Blythman, J. (2004), *Shopped: The Shocking Power of British Supermarkets* (London: Fourth Estate).

Boyd, W. and Watts, M. (1997), 'Agro-industrial Just-in-Time: The Chicken Industry and Post War American Capitalism' in Goodman, D. and Watts, M. J. (eds), *Globalizing Food* (pp. 192–225) (London: Routledge).

Burnett, J. (1989 [1966]), *Plenty and Want: A Social History of Diet in England from 1815 to the Present Day* (London: Routledge).

Cabinet Office (2008a), *Food: An Analysis of the Issues* (London: Cabinet Office Strategy Unit).

Cabinet Office (2008b), *Food Matters: Towards a Strategy for the 21st Century* (London: Cabinet Office Strategy Unit).

Cockett, R. (1995), *Thinking the Unthinkable: Think Tanks and the Economic Counter-revolution, 1931–1983* (London: Fontana).

Compassion in World Farming. (2003), *The Welfare of Broiler Chickens in the European Union* (Petersfield: Compassion in World Farming Trust).

Davis, M. (2005), *The Monster at our Door: The Global Threat of Avian Flu* (New York: The New Press).

Department for Environment, Food and Rural Affairs (DEFRA). (2005a), *Poultry and Poultry Meat Statistics Notice*, 29 September 2005. Available at: http://statistics.defra.gov.uk/esg/statnot/ppntc.pdf.

Department for Environment, Food and Rural Affairs (DEFRA). (2005b), *Agriculture in the UK* (London: Stationary Office).

Department for Environment, Food and Rural Affairs (DEFRA). (2005c), 'The Validity of Food Miles as an Indicator of Sustainable Development', Final Report No. ED50254.

Department of Trade and Industry (DTI). (1999), *Modern Markets, Confident Consumers* (London: The Stationery Office).

Dixon, J. (2002), *The Changing Chicken* (Sydney: University of New South Wales Press).

Freidberg, S. (2003), 'Editorial: Not All Sweetness and Light: New Cultural Geographies of Food', *Social and Cultural Geography* 4, 3–6.

Friedland, W. H., Barton, A. E. and Thomas, R. J. (1981), *Manufacturing Green Gold: Capital, Labor, and Technology in the Lettuce Industry* (New York: Cambridge University Press).

Gereffi, G. and Korzeniewicz, M. (eds) (1994). *Commodity Chains and Global Capitalism* (Westport: Greenwood Press).

Goldenberg, N. (1989), *Thought for Food: A Study of the Development of the Food Division, Marks & Spencer* (London: Food Trade Press).

Goodman, D. (2003), 'The Quality "Turn" and Alternative Food Practices: Reflections and Agenda', *Journal of Rural Studies* 19, 1–7.

Holroyd, P. (1986), *History of the Institute of Poultry Husbandry* (Telford: Harper Adams Agricultural College).

Hughes, A. and Reimer, S. (eds) (2004), *Geographies of Commodity Chains* (London: Routledge).

Jackson, P., Russell, P. and Ward, N. (2007). 'The Appropriation of "Alternative" Discourses by "Mainstream" Food Retailers' in Maye, D., Holloway, L. and Kneafsey, M. (eds), *Alternative Food Geographies: Representation and Practice* (pp. 309–330) (Amsterdam: Elsevier).

Jackson, P., Ward, N. and Russell, P. (2006), 'Mobilising the Commodity Chain Concept in the Politics of Food and Farming', *Journal of Rural Studies* 22, 129–141.

Kjaernes, U., Harvey, M. and Warde, A. (2007), *Trust in Food: A Comparative and Institutional Analysis* (Basingstoke: Palgrave Macmillan).

Lawrence, F. (2004), *Not on the Label: What Really Goes into the Food on Your Plate* (London: Penguin Books).

Leslie, D. and Reimer, S. (1999), 'Spatializing Commodity Chains', *Progress in Human Geography* 23, 401–20.

Ministry of Agriculture, Fisheries and Food (MAFF). (1991), *Fifty Years of the National Food Survey* (London: Ministry of Agriculture, Fisheries and Food).

Marsden, T., Flynn, A. and Harrison, M. (2000), *Consuming Interests: The Social Provision of Foods* (London: UCL Press).

Pennington Group. (1997), *Report on the Circumstances Leading to the 1996 Outbreak of Infection with E.Coli 0157 in Central Scotland, the Implication for Food Safety and the Lessons to be Learned* (Edinburgh: The Stationery Office).

Perks, R. and Thomson, A. (eds) (2006), *The Oral History Reader*, 2nd Edition (London: Routledge).

Policy Commission on the Future of Food and Farming. (2002), *Farming and Food: A Sustainable Future* (London: Cabinet Office).

Seth, A. and Randall, G. (1999), *The Grocers: The Rise and Rise of the Supermarket Chains* (London: Kogan Page).

Sheppard, A. (2004), *The Structure and Economics of Broiler Production in England* (Exeter: Centre for Rural Research, University of Exeter).

Visser, M. (1999), *Much Depends on Dinner* (New York: Grove Press).

Yakovleva, N. and Flynn, A. (2004a), 'Innovation and Sustainability in the Food System: A Case of Chicken Production and Consumption in the UK', *Journal for Environmental Policy and Planning* 6, 227–250.

Yakovleva, N. and Flynn, A. (2004b), 'Innovation and the Food Supply Chain: A Case Study of Chicken', Centre for Business Relationships, Accountability, Sustainability and Society, Working Paper Series No. 20, University of Cardiff.

Chapter 9
Place and Space in Alternative Food Networks: Connecting Production and Consumption

David Goodman

Introduction

This chapter examines the 'alternative food economy' constituted by the provisioning of 'quality' food products to what at present is a relatively narrow segment of consumers. Typically, though not exclusively, these are rich in economic and cultural capital, the 'worried well', as the phrase goes. In short, there is a strong class dimension to the social relations of consumption of the 'organic', the 'local', the 'regional', and the 'alternative' (Tregear 2005). The social positioning of these foods, seen through the prism of class, race and inequality, and the corollary of how to democratize access, should be kept in mind throughout the following discussion.

The analysis focuses on the 'geographical imaginations' (Harvey 1990; Gregory 1994) and conventions of quality underlying the construction of the spaces of alternative production and consumption, the socio-ecological relations holding these networks together, and how these spaces are being redefined by the countervailing forces of dissolution and assimilation into conventional corporate supply channels. The main theoretical issues concern the conceptualization of the geographies—the places and spaces—of alternative food production and consumption and the material and discursive trafficking between them. Are these linkages to be found in unconventional, contested knowledges and shared ethical values or, more instrumentally, in the marketing opportunities afforded by the anxieties of the 'risk society' and efforts by rich, self-interested and predominantly white consumers to quarantine themselves from food risks? (cf. Szasz 2007).

Alternative food networks clearly are reconfiguring an expanding subset of production-consumption relations and nature is being commodified in different, hopefully more sustainable, ways. However, it is well to remember that these networks and new economic forms are embedded in capitalist societies rather than inhabiting a more benign, parallel universe. Accordingly, it is important to bring critical social analysis to bear on the recent literature on 'alternative', local and 'quality' food networks and examine the distribution of gains arising from these new economic activities between local and extra-local actors, and locally, along the lines of class, gender, race, and ethnicity. In this respect, the present chapter seeks to redress the limited attention given to the unequal relations of consumption and the uneven retail geographies of quality food in the literature on the alternative food economy.

Context

The 'turn' to quality foods and the emergence of alternative food provisioning networks reflects the confluence of longer-term developments and more conjunctural factors. The following are among the most prominent of these developments, although this discussion is avowedly schematic rather than exhaustive.

As many commentators have suggested, the unnerving frequency of 'food scares' in Western Europe, especially the BSE pandemic and recent outbreaks of Foot and Mouth disease in the UK, have provoked a 'crisis of trust' (Murdoch and Miele 2004b, 170) among consumers and prompted closer interrogation of the practices of large-scale corporate agriculture. In effect, the sanitarian-Fordist pact between the corporate food industry, with its standardized, highly processed product lines, and middle-class consumers has been undermined, provoking a flight to 'quality' and encouraging the growth of alternatives to supermarket provisioning. This crisis of confidence in mass-produced 'placeless and faceless' foods is articulated mainly, though not exclusively, by higher income consumers with the means to opt out from mainstream food provisioning.

These 'fugitives' from the industrial food system, although still restricted by time-space constraints and open to the appeal of supermarket convenience, have reinforced parallel cultural changes in contemporary eating habits. For the more affluent, these changes include the attractions of culinary diversity, quality foods, and gastronomic distinction, to which the media panders with the cult of celebrity chefs and food writers, elite restaurants and food shops, and the titillation of food pornography liberally featured in up-market newspapers, magazines, and lifestyle publications. The August, 2004 cover of *The Observer Food Monthly* bearing the headline 'Dish of the day' and featuring the buxom, bikini-clad wife of Salmon Rushdie reclining on a table of lobsters surrounded by male chefs epitomizes this culture of celebrity and privileged excess.

Local food initiatives have been among the beneficiaries of the continuing, if halting, process of reform of the Common Agricultural Policy (CAP). For these initiatives, milestones in the 1990s included the closer targeting of the European Union (EU) Structural Funds on local rural development in depressed agricultural regions and support for endogenous, 'bottom up' socio-economic development under LEADER: Liasons Entre Actions de Developpement de l'Economie Rural (Ward and McNicholas 1998; Ray 2000). At the macro-level, reforms of the CAP responded to the trade liberalization mandate of the World Trade Organization (WTO) by prioritizing non-production related payments, such as environmental stewardship schemes, which are considered not to distort international trade and so are classified as 'green box' programmes under WTO criteria. These so-called Second Pillar programmes, by de-coupling farm payments from production, are gradually re-orientating the CAP away from its narrow, sectoral focus towards a more decentralized model in which a multi-functional agriculture is the centre-piece of a more pluralistic, integrated and endogenous approach to rural development.

However, this shift is glacial and strongly contested (Potter and Tilzey 2005), with significantly different approaches to its implementation among EU member states (Lowe et al. 2002), and increasingly is dominated by a rearguard struggle led by France to defend current levels of EU agricultural support, which are due to be renegotiated by 2013. These reform dynamics have been further complicated by concerns for food security. These were well and truly aroused in 2007/2008 by record commodity prices, food price inflation and declarations of a world food crisis at the G-8 Summit in July, 2008. Commodity price inflation has since receded with the global 'credit crunch' but food security is now firmly on the policy radar and whether or not local food networks will be given greater priority over conventional agriculture remains very much an open question at this point.

Since 2003, myriad direct EU production subsidies have been consolidated into a 'single farm payment' system, which gives greater scope to market forces in commodity agriculture and concomitantly highlights the social and environmental rationale of continued farm support, strengthening prospects for the possible re-nationalization of farm policy. Localized quality food networks and 'Short Food Supply Chains' (SFSCs) have an integral role in the market-based Second Pillar strategy to restructure the productivist model of the post-WWII agricultural settlement. The institutional space and budgetary resources created by CAP reform have encouraged some scholars to envision alternative food networks (AFNs) as forerunners of the 'new' model of rural development, as advocated in the 1996 Cork Declaration, the Agenda 2000 reforms, and more recent EU policy statements (Ploeg et al. 2000).

Alternative food provisioning networks, where premium prices or economic rents still can be earned, provide opportunities for more diversified farm livelihoods. Such networks promise relief from the oppressive cost-price pressures on farmers exerted by the combination of diminishing direct, production-related subsidies and the market power of oligopolistic retail multiples, with their harsh, exploitative techniques of supply chain management (Marsden, Flynn and Harrison 2000; Marsden 2004; Sustainable Development Commission 2008).

With this broad, contextual overview of the 'turn' to quality in local food networks, the discussion moves to the spaces of localized production and spaces of consumption in these networks, and examines several neglected questions found in the interstices between these categories. The final section briefly considers the uneven retail and moral geographies of quality food provisioning, with some illustrative examples from London, and argues for a renewed emphasis on social justice in food production-consumption practices as the foundation of a re-invigorated food policy.

Spaces of localized production

Attempts to theorize the emergence of AFNs have engaged an eclectic range of meso-level perspectives, including commodity chain analysis, systems of provision, convention theory, actant-network theory, culture economy approaches, and discussion of the aesthetics of local and regional gastronomic networks. A second focus has provided 'thick' ethnographic descriptions of individual local networks and detailed analyses of the plethora of food quality accreditation and promotional schemes devised to provide new bases for rural development (Morris and Young 2000; Ilbery and Kneafsey 1999; Tregear 2003; Parrott et al. 2002)

Taken broadly, these perspectives are primarily production-centred, concerned mainly with different modalities of economic coordination, with consumers and consumption seen as appendages of the production process. Among the meso-level approaches, the more avowedly sectoral or industry focus offered by commodity chain analysis has enjoyed a recent resurgence, as in the work of Brian Ilbery and Damian Maye (2005, 2006) and Peter Jackson and his colleagues (2006).[1]

However, convention theory has been easily the most influential in AFN research since it speaks directly to the 'economy of quality' (Murdoch and Miele 1999; Morgan et al. 2006). This approach offers a general typology to distinguish product quality in terms of 'orders of worth' that specify the different logics orchestrating their production and governance. It thus identifies the norms, qualifications, and organizational forms involved in network coordination and which uphold the different conventions of quality (Wilkinson 1997).

In this literature on the quality 'turn' to alternative food provisioning, the concepts of 'quality' and the 'local' are frequently used inter-changeably, although, oddly enough, only quality is theorized as a socially constructed category. Quality is defined in terms of various elements drawn from Granovetterian sociological analyses of regional industrial networks, notably interpersonal relations, trust, embeddeness, localized tacit knowledge, and other 'untraded dependencies' (Storpor 1999). Local provenance has become virtually synonymous with 'quality', and this is forcefully conveyed by the concept of re-embeddedness in local social and ecological relations developed by Murdoch et al. (2000).

In policy terms, SFSCs are formulated primarily as farm-based initiatives that are regarded collectively as the catalyst of an alternative rural development, promising a different trajectory from the productivist model of the post-WWII agricultural settlement, whose dynamics have decapitalized farms, concentrated resource access and land ownership, impoverished rural communities, accelerated out-migration, devalued localized tacit knowledge, and devastated local ecologies. According to protagonists of this new trajectory, the 'earlier modernization

1 These recent contributions draw on archetypal commodity chain methodologies, bypassing earlier re-workings of this tradition, including the cultural economy approach of Jane Dixon (1999) and the systems of provision perspective developed by Ben Fine and his colleagues (Fine et al. 1996; Fine 2002).

paradigm ... is increasingly at odds with society's expectations of agriculture' (392) and the 'reconstruction of agriculture and countryside and their realignment with European society and culture is imperative' (Ploeg et al. 2000, 396).

It is revealing to analyse the emergence of alternative or local quality food networks in terms of several intertwined or overlapping geographical imaginaries, again centred mainly on production, which are articulated in the literature and debates on the reconstruction of European rural development policy. These imaginaries cast AFNs as vectors of resistance, cultural identity, and rural regeneration.

Resistance

In this imaginary, local food provisioning is depicted as a site of 'resistance' to the anomic forces of a globalizing corporate food system and against further incursions of placeless, homogenized foods and standardized gastronomic practices (cf. Miele and Murdoch 2002; Holloway and Kneafsey 2004; DeLind 2003; Kloppenburg et al. 1996). Here, the local is seen as the normative realm of resistance to the global corporate food regime, characterized by the time-space compression of production-consumption, and 'the corporate principles of distance and durability' (Friedmann 1994, 30). This 'revival of the local' also finds resonance in the wider context of international trade negotiations, where the notion of multifunctionality, as the essence of the '*exception europeanne*', is advanced by the EU to legitimize continued agricultural support (Buller 2001; Potter and Burney 2002), and thereby maintain a bulwark against the homogenizing pressures of trade liberalization (Potter and Tilzey 2005, 2007). The intersections between agri-environmental public goods, rural cultures, local foodways and traditional inhabited landscapes here are woven into a narrative to underpin the model of European agricultural exceptionalism.

Cultural identity

The imaginary of resistance against 'placeless and faceless' foods is closely allied to narratives of cultural identity, as reflected in President Francois Mitterand's reference to a 'certain kind of rural civilisation' in Europe and the importance of preserving it for future generations (*The Times*, 7 February, 1987), a sentiment echoed more recently by Christine Lagarde, then France's trade minister, who considers farming as 'fundamental to our identity' (*The Economist*, 10 December, 2006, 28). This cultural narrative embraces the defence and conservation of local agro-food networks, historical landscapes and *terroirs*, tacit knowledges and craft skills, and regionalized culinary networks, epitomized by the agenda set out by the Slow Food Movement (Murdoch and Miele 1999, 2004a, 2004b; Miele and Murdoch 2002). Here, neo-populist yearnings can be detected in the characterization of the 'new rural development paradigm' in terms of processes of 're-peasantisation' (Ploeg et al. 2000). This same narrative is the central thread running through earlier advocacy of endogenous rural development and operationalized by Jan van der Ploeg's notion of regionally differentiated 'styles of farming' (Ploeg, 1990, 1993). In this imaginary, rural development grounded in farm-based quality food enterprises promises to revitalize rural society and redress the more egregious social and ecological consequences of post-WWII productivism.

Territorial valorization

These geographical imaginaries of resistance and cultural identity are complemented and reinforced by a market-oriented developmental imaginary of territorial valorization. This is neatly illustrated by the tenor of the 2002 Curry Commission report, *The Future of Food and Farming*, issued in the wake of the severe UK outbreak of Foot and Mouth disease. Farmers are advised to develop new livelihood opportunities and exploit untapped sources of value added by making a 'reconnection' with consumers through new markets for quality local produce.

These new livelihood opportunities and sources of value added are open to those producers who can successfully adopt quality conventions that demonstrate territorial provenance or embeddedness in localized socio-ecological processes. In this economic discourse of market-based 're-connection' through reconfigured producer-consumer relations, the focus is on the local capture of value added, which is envisaged as a means to arrest, or even reverse, the historical decline of the share of farm activities in the value stream of the agro-food system. In short, territorial valorization is seen as an entrepreneurial opportunity, a farm livelihoods strategy and the cornerstone of a revitalized rural economy (cf. Marsden and Smith 2005).

This shift towards the production of quality local foods, as opposed to the generic 'placeless' commodities of productivist agriculture, which often are sold into the closed 'internal markets' of conventional supply chains and contract production relations, is variously conceptualized as the re-embedding, re-socializing, and re-localizing of food systems. SCFCs are a major institutional expression of these reconfigured production-consumption relations, whether in the form of direct, face-to-face contact at farmers' markets, for example, or narrated to distant consumers by symbols, logos and labels of quality and 'qualification' of place, process, and product, or the 'three Ps', according to Ilbery et al. (2005).

Thus farmers are encouraged to 'short-circuit' industrial supply chains and to reconstruct the producer-consumer interface by engaging with different conventions and constructions of quality 'that evoke locality/region or speciality and nature' (Marsden, Banks and Bristow 2000, 425). With 'their capacity to re-socialise and re-spatialise food', SFSCs are in a position to 'redefine the producer-consumer relation by giving clear signals as to the origin of the food product' (425), since these new relations play a key role in 'constructing value and meaning "and enhance" the potential of products "to command a premium price"' (425).

In a nutshell, SFSCs are seen as the vehicle for capturing the economic rents arising from the commodification of the 'local', with these returns accruing, at least in theory, to owners of local material and intellectual property rights (cf. Moran 1993; Guthman 2004a). However, as we observe below, these property holders must contend with extra-local actors, who have the power to capture shares of this value stream at other sites as these quality products navigate the spaces of consumption.

Interstitial Questions

Several interstitial questions arise between the spaces of localized production and the 'translation' of territorial resource endowments and their semiotics into value in spaces of consumption. Although their unfolding bears directly on the relations of power in the production-consumption spaces of the new 'economy of quality', these issues are linked by their relative neglect in the literature on quality food networks.

Rent-seeking and the competitive control of quality

The premium prices on which the territorial valorization model of rural development is founded encourage rent-seeking behaviour by off-farm, downstream actors and are vulnerable to competitive erosion. This is particularly the case where these excess profits reflect product differentiation strategies, such as the adoption of 'territorial identity labels and 'traditional', 'local' and 'organic' designations that are broadly *generic* in character' (Goodman, D. 2004, 9, original emphasis). Key issues here concern 'the durability and magnitude of these income flows and the location of the actors who capture them' (ibid., 8–9). Other authors, notably Valceschini et al. (2002), Buller (2000) and Ray (2000), also recognize that the logic of territorial valorization is susceptible to the proliferation of competing quality schemes, labels and logos as public agencies utilize territorial identity to promote regional development (Mollard et al. 2005). Corporate food interests, notably supermarket chains, also have responded to the new constructions of quality, and particularly the marketing focus on provenance and traceability, by developing own-label, locally-sourced product lines and quality food brands, such as the 'SO: Sainsbury's Organic' range, launched in 2005. 'This combination of imitative expansion and strategic convergence (is) accentuating downward pressures on price margins and threatening to shift economic rents away from the farm and local level' (Goodman, D. 2004, 9).

Remarkably, however, apart from Terry Marsden's (2004) recent contribution, the rent-seeking behaviour of downstream actors, notably retail multiples, has received little attention. In addressing this neglect of food chain relationships, Marsden analyses the production of new quality food 'spaces' and the concomitant struggle to dominate the material and discursive construction of quality. 'Competitive control of quality', he argues, confers power to delineate '"competitive spaces", boundaries and markets' between retailer-led commodity chains and AFNs (147). In short, control of the cultural meanings of quality can be translated into excess profits or economic rent. Marsden's prognosis for AFNs in the UK is not optimistic since the large retailers, backed by 'a supportive state', hold an almost impregnable advantage over other actors in this struggle due to 'the continued institutional and regulatory dominance of retailer-led food governance' (Marsden 2004, 144). This analysis of quality struggles and power relations across the spaces of food provisioning hopefully will stimulate further research on this critical theme.

This raises the question of how AFNs/SFSCs will be 'folded' into the conventional food system. Will there be co-habitation and complementary yet relatively autonomous growth or will these networks be selectively co-opted and 'mainstreamed' and increasingly subjected to the practices and downward cost-price pressures typical of the conventional system? If the latter is the case, AFNs will have brought only temporary respite from exploitative supply chain management and the income 'squeeze' on farmers.

This increasingly seems to be the fate of organic agriculture in the US (Guthman 2004a), where supermarket chains, such as Wal-Mart, Safeway, Kroger's, Target and Trader Joe's, in a fine demonstration of Tim Lang's simile of oligopolistic competition as 'synchronized swimming', are rapidly expanding their own-label lines of fresh and processed organic products. With annual rates of growth of 15–20 per cent in 1998–2005 and organic food sales estimated at US$13.8 billion in 2005, this 'space of consumption' clearly has captured the attention of the leading US food retailers. The rising intensity of competition is reflected in the US$565 million acquisition of rival Wild Oats Marketplace by Whole Foods Market, which pioneered organic retailing in the early 1980s, increasing its number of stores from 193 to 303 in the US, Canada, and Britain.[2]

This mainstream future also is being consolidated in the UK, where the leading multiples hold a 75 per cent share of sales of organic products. The increasingly corporate face of UK organics was again revealed in October, 2007, when Abel and Cole, already a giant in the organic box delivery sector with some 50,000 customers and sales of $56 million, announced that it had sold an interest in the business to the private equity firm Phoenix. The case of Rachel's organic yogurt, the brand name originally of a single dairy farm in Aberystwyth, the first certified organic farm in Britain, and subsequently used to designate its supply network of organic Welsh dairy farms, provides a cautionary tale of the appetite of the mainstream food system. This SFSC subsequently scaled up to supply Sainsbury's before being acquired by Horizon, the dominant organic milk supplier in the US, whose large-scale, 'factory farm' production methods were challenged legally in 2007 by US organic consumer groups as usurping organic regulatory standards.

2 The US Federal Trade Commission approved this merger in 2008 after initially filing a legal complaint on 6 June, 2007 to block the proposed takeover.

Horizon, in turn, has been swallowed up by the multinational conglomerate, Dean Foods, leaving behind only the Rachel's Organic brand name in memory of local 'resistance' and eco-cultural identity.

These illustrative cases warn of the potency of 'mainstreaming' and brand imitation in diluting and disempowering the counter-narratives and imaginaries of 'local', 'organic' and 'quality' foods.

Social relations of consumption

While the accelerating incorporation of organic and local products into the mainstream noted above holds out a promise of more democratic access, at present, the social relations of consumption underlying these new forms of food provisioning are highly unequal. Their markedly higher prices, the time-space commitments needed to acquire and prepare these alternative and local foods, and the associated food knowledges involved strongly suggest that significant levels of economic and cultural capital are required to gain access to these provisioning systems. Moreover, within organics, the Soil Association's 2006 market report noted that a premium sector is now emerging.[3]

Despite higher prices, this report reveals that the annual growth in sales of organic produce has averaged 27 per cent over the past decade, although they still account for less than 1.6 per cent of total UK food sales. As we have seen, AFN/SFSC and other actors, notably supermarkets, clearly are responding to this increasingly differentiated market, whose emergence raises the prospect of a new, multi-tiered food system stratified by income and other class markers. In the main, only highly privileged consumers are in a position to join this 'flight to quality', leaving others as the 'missing guests at the table' (Goodman, D. 2004, 12–13). This is a critical issue of social policy, as recent work on the poverty-food-health nexus (Dowler and Turner 2001) and the stubborn persistence of child poverty under New Labour, with currently 2.8 million children below the poverty line (*Financial Times*, 28 March 2007), make abundantly clear.[4]

Noteworthy efforts to break down these 'socially exclusive niches' include the growth of community food co-operatives and particularly the Public Sector Food Procurement Initiative introduced by DEFRA in 2003, with its emphasis on sustainable food, better access for minority groups and improved working conditions for public sector catering workers (Morgan 2007). Media coverage of

---

3 The onset of the 'credit crunch' since September, 2008 and a return to 'value for money' in food shopping suggest that continued rapid growth in the organic and quality food sector is in serious jeopardy. Whether these developments will widen or narrow relative price differences between organic and conventional foods remains to be seen.

4 In the US, two legislative mechanisms directly enhance the access of food-aid recipients and low-income families to farmers' markets: the Seniors Farmers' Market Program and the Farmers' Market Nutrition Program, a component of food-aid provision ear-marked for Women, Infants and Children (WIC).

Jamie Oliver's damning indictment of school meals in 2005 gave further momentum to this process and school meals have become 'a litmus test of New Labour's avowed commitment to public health, social justice and sustainable development' (ibid., 8). Public sector food procurement should be seen as only the foundation stone of a new, well-funded national food policy if the question of inequitable access and the public health consequences are to be tackled comprehensively (Morgan and Sonnino 2008). The critical issue still is how to create large-scale provisioning systems that democratize access and make nutritious food affordable. In addition, this must be done in ways that avoid the re-imposition on farmers of the exploitative practices and cost-price squeeze currently associated with retailer-led supply chain management (Marsden et al. 2000; Marsden 2004). In the absence of such new systems, the mainstreaming of alternative quality foods represents a Faustian bargain for producers.

In this respect, a recent *Time* magazine (12 March 2007, 11) notes that the advice to 'Forget Organic, Eat Local' offers cold comfort to organic producers whose premium price margins have been eroded by the multiples under competitive pressure to 'get big or get out'. European experience also suggests that the economic rents implicit in the slogan 'Eat Local' are vulnerable to imitation and appropriation. This vulnerability is revealed in a survey undertaken in 2007 by the Guild of Fine Foods, whose members account for some 25 per cent of the deli and farm shop sector in the UK. According to press reports, the director of the guild, Bob Farrand, said 'Waitrose actively mimics delis and farm shops with speciality and locally sourced foods, but often at lower prices because of its buying power' (*The Guardian*, 3 July 2007, 11). Guild members expressed their apprehension at Waitrose's plans to expand nationally from its base in southern England, initially by targeting smaller market towns. A possible way to square this particular circle is to find mechanisms to reduce supermarkets' oligopolistic power over supply chains, such as the proposal by the Sustainable Development Commission (2008) to introduce domestic 'fair trade' principles to re-structure relationships between producers, retailers and consumers.

Without a more interventionist state and root-and-branch changes in UK competition policy, the continued dominance of the oligopolistic supermarkets is unlikely to be disturbed, as Marsden (2004) contends. The Competition Commission's report into the grocery sector published on 30 April 2008, indicates that this scepticism continues to be well-founded. Narrowly equating the public interest with consumer choice in local retail markets, the report broadly accepts the market power of the Big Four and Tesco's pre-eminence and offers only nominal relief to their beleaguered suppliers. That is, it acknowledges the ineffectiveness of the previously voluntary code of conduct in addressing the current gross imbalances in supermarket-supplier relationships and recommends the creation of an ombudsman charged with its enforcement. This is little more than a mild annoyance to supermarkets since it leaves the oligopolistic market structures that account for the powerlessness of growers and suppliers firmly in place.

Spaces of Consumption: The Elusive Consumer

By and large, consumers in agro-food studies have been described in the instrumental, abstract terms of neo-classical demand theory or, at best, been assigned to the stratified categories beloved of market research profiles. That is, to paraphrase Arjun Appadurai (1986, 31), consumers emerge as private, atomistic and passive rather than being 'eminently social, relational and active'. In a very real sense, the theoretical challenge facing agro-food studies in attempting to bring consumption 'back in' is to explore each of these dimensions fully (Goodman and DuPuis 2002). Rather than using consumption as a means to talk mainly about quality and its production, the everyday social practices of food consumption and the knowledges that inform these routine, habitual practices should share analytical priority.

In analysing the consumer politics of alternative food networks, Goodman and DuPuis (2002) invoke the figure of the 'reflexive consumer' in order to access forms of agency that are not encompassed by emancipatory mass politics, the Marxian terrain of class struggle in production or formally organized, collective social movements.[5] Taking this post-Foucauldian approach, the political can then be re-defined as a more de-centred 'capacity to act' (Baker 1990), which extends into the 'private' sphere of everyday life. Several recent papers have broadly incorporated this perspective in analyses of the quality 'turn' to alternative modes of food provisioning.

Contested knowledge practices: 'Growing food and knowing food'

This emphasis on forms of agency and politics embedded in everyday social practices opens up the issue of consumption in AFNs to analysis in terms of competing knowledge systems. This formulation only recently has been extended to the consumer and 'knowing food' that we eat, although it has been quite widely used to characterize sustainable agricultural production or 'growing food' as a struggle between formal and tacit knowledge systems. (Kloppenburg 1991; Kloppenburg and Hassanein 1995; Ploeg 2003). That is, 'By linking these struggles over knowledges, we begin to see the politics of the food system as involving alternative "modes of ordering" in which food is an arena of contestation rather than a veil over reality' (Goodman and DuPuis 2002, 16). On this approach, food is no longer polarized conceptually between Durkheimian totem—a symbol which represents social relationships—and Marxian fetish—a symbol that conceals social relationships—but emerges as an arena of struggle, as well as a realm of relationality (ibid., 15). An analytical focus on the contested knowledges of 'growing food' and 'knowing food' provides a bridge between the spaces of

5 Goodman and Dupuis (2002) essentialize the figure of 'the reflexive consumer' by failing to consider the various understandings of 'reflexivity' and the different ways in which these contribute to social change. I grasped this point after reading Littler (2005).

production and consumption, private and public spaces, and highlights alliances across these spaces in which consumers engage as active and relational participants. This amounts to the simple, yet long ignored, recognition that food, its producers and consumers, are entangled in a politically contested, ever-changing discourse of competing knowledge claims.

Although not examined at length here (see Goodman 2008), the nature of the 'politics' in a reflexive and ethical politics of consumption is attracting increasing controversy and debate. Scepticism remains deeply rooted (Bernstein and Campling 2006; Guthman 2004b) but several recent contributions regard consumption as a potentially important arena of citizenship and of a participatory, civic 'republican' politics (cf. Needham 2003), and one where consumer advocacy organizations can articulate forms of 'individualised collective action' (Micheletti 2003; Clarke et al. 2007; see also Miller 1995). In this respect, Soper's (2004) concept of an 'alternative hedonist' ethics and politics offers an innovative, nuanced analysis of the shift to ethical and 'green' lifestyles in affluent societies and social groups (see also Soper and Thomas 2006).

The issue of who 'knows food' and has the resources to act upon this knowledge raises serious questions about the class and racial character of food provisioning cultures (cf. Slocum 2006, 2007). Thus in an earlier paper, Guthman (2002, 299) discerns an 'elite sense of reflexivity' in recent work on consumption politics, arguing that '... implicit to notions of reflexivity is that mass-taste is pre-determined, unreflective, and based on a cultural economy of food to which the reflexive eater objects'. While food knowledges are not the exclusive property of the affluent elite, the power to opt out of conventional provisioning clearly is constrained by income and unequally distributed. In this context, Jaffee et al's (2004) call for the recognition of 'fair provisioning' and social justice in alternative production-consumption relations exposes a neglected and critical dimension of the politics of alternative food provision. As they succinctly observe, 'Analyses of the social embeddedness associated with alternative and direct marketing of food have not foregrounded considerations of equity' (175).

Contested aesthetics of food

Questions of knowledges, social justice and equity in spaces of consumption arise in a rather different guise in recent work by Jonathan Murdoch and Mara Miele, who explore the changing and increasingly contested aesthetics of food in an anxious post-BSE world. These aesthetics and their contingent socio-cultural construction offer an explanation of the reproduction of differentiated spaces of production-consumption. In an initial paper on this theme, they contrast an 'aesthetic of entertainment' developed by standardized, mass food chains, such as McDonalds and Burger King, where the quality of food is secondary and disguised by the 'restaurant experience', and the 'gastronomic aesthetic', the hallmark of the Slow Food Movement (SFM), whose reference points are freshness, seasonality, and typicality, and whose dominant rhetoric is *terroir* (Murdoch and Miele 1999).

A subsequent paper argues that food scares have made consumers more reflexive about food and its origins and led them 'to rediscover the product "behind the sign"' (Murdoch and Miele 2004b, 160). This cognitive shift is described in terms of a 'relational aesthetic', which is conceptualized as a 'double movement' involving disconnection to establish critical distance for reflection and new forms of (re)connection. They go on to analyse quality conventions articulated in terms of cultural heritage, ecology and social justice by the SFM, Soil Association and Fair Trade organizations, respectively, and the role these social movements play in orchestrating this relational aesthetic to promote and commodify certain notions of quality, and so construct 'alternative' spaces of food production-consumption.

As Murdoch and Miele (2004b, 170, original emphasis) observe, these new social movements

> ... aim to regulate and monitor alternative food chains while simultaneously attracting consumers to their products ... [C]harter marks and logos are employed to draw consumers into a new relationship with the environments of production. In short, the marks and logos *fold* complex sets of relations into food products in ways that permit easy consumer appreciation. They therefore promote a new aesthetic of food, one based on *connectedness* to those natural and social conditions that are thought to ultimately determine the quality of food.

In this respect, these NSMs are acting like trade organizations, setting trade standards, regulating a market, and using their control of quality conventions to exercise economic co-ordination and governance. Such organizations exemplify Micheletti's (2003) notion of 'individualized collective action' and 'political consumerism' and provide one response to the question of how 'ordinary' ethical values performed routinely by individual consumer-actors can be 'translated' or mobilized in ways that can have wider systemic impacts (Barnett et al. 2005; see also Popke 2006).

Murdoch and Miele's work is a significant and coherent attempt to theorize the linkages between the spaces of production and consumption in alternative food networks. Nevertheless, the sense of these reflexive consumers is that they are still 'led' and 'reactive', and the emphasis falls strongly on the activities of the civil society organizations that regulate and orchestrate supply chains and their markets. Membership of this undifferentiated category of reflexive consumers is achieved by making choices that are said to involve critical judgement. This requires that they engage and identify with the marketing of selected quality conventions and their respective logos, which seek to convey, and so purvey, distinctive forms of 'connectedness' with producers and spaces of production.

As such distinctiveness implies, however, only certain, selected socio-ecological relations are being revealed or made transparent by these logos and the auditing practices on which they are based, while others remain obscured. These different aesthetics are fetishized and alienated forms of relationality, reflecting what Andries Du Toit (2002, 371) calls the 'technology of ethics', where labels induce a kind of placebo effect; in this case, what he designates as an 'ethics effect' is experienced simply by engaging in the act of consumption. As some authors have stressed, such aesthetic moves can be described in terms of a very different 'double movement' of de-fetishization and re-fetishization. A common example is organic agriculture, where the technical or ecological relations of production are exposed to view but the social relations, including the wage-relation, working conditions and civil rights, remain hidden 'behind the sign' of the organic, as both Allen (2004) and Guthman (2004a) have emphasized in the case of California.

In the case of Fair Trade, Bryant and Goodman (2004, 359) similarly analyse the de-fetishization moves whereby knowledge of eco-social production relations is revealed 'to allow consumers, it is hoped, to make moral and economic connections to the producers of the food they ingest'. They go on to suggest, however, that in this very act, 'the effect is to commodify, in turn, the ethical relationship at the heart of fair trade—that is, small-scale farmers, producer cooperatives and 'sustainably' managed second nature. Fair trade knowledge flows thus act to *re-work* the fetish surrounding fair trade commodities into a new kind of 'alternative' spectacle for Northern consumers' (original emphasis).

The failure to acknowledge the limited access of the vast majority of consumers to quality 'slow' foods is a further example of selective de-fetishization. The relational aesthetics 'branded' in the logos of alternative NSMs re-enact and re-inscribe the fetish in their respective markets, subsuming these inequities in food distribution and consumption. This theme is extended in the following section, which looks briefly at the 'alternative' retail geographies of quality foods and the 'moral charge' of these political geographies of aestheticised consumerism.

Retail Food Geographies, Moral Geographies?

It is easy to suggest that quality foods and their associated knowledges can be sources of cultural and economic 'consumer surplus' and a form of social capital, fitting neatly into Bourdieu's analysis of consumption and social distinction. Such distributional inequalities also can be mapped on to the uneven retail geographies of 'quality food spaces' described in *Time Out*'s 'London Eating and Drinking, 2005'. Charting their rapid growth and spatial concentration, *Time Out* (2005, 8) suggests that

> If there has been one major trend among food lovers (it is that) … more and more (they) are shopping in local street markets instead of lamenting their decline. Farmers' markets are springing up so fast … that it's hard to keep track. Borough Food Market—the jewel in the crown among London's food markets, but, incredibly, only a few years old—now has more visitors than Madame Tussaud's … In addition, there has also been a rise in the number of quality food shops with a significant diversity of stock, from "ethnic" stores to high quality delis and specialists.

The resurgence of food shops has occurred, it is argued, as a reaction against the 'dreary shopping environments 'created by supermarkets and because, while the multiples have given almost everyone else access to high-quality, inexpensive food ... better-off people can afford to be more discerning' (8). In a paean to affluent discernment and its food spaces, Time Out selects two 'foodie neighbourhoods' as emblematic of these unequal geographies and social relations of quality food consumption: Marylebone High Street and Northcote Road in Battersea, noting that 'both have seen a vast transformation in their food shops, but for very different reasons' (8; see Figure 9.1).

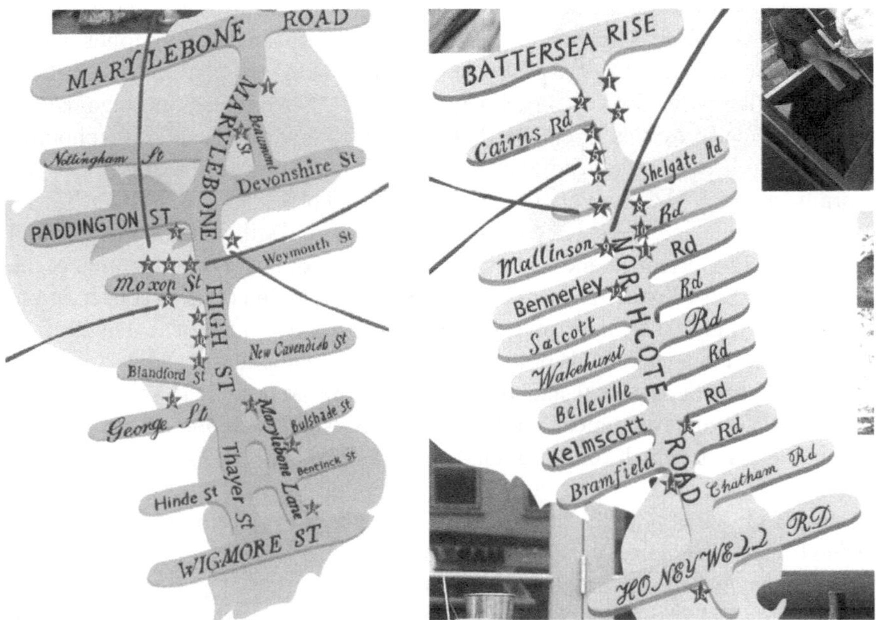

Figure 9.1 Maps of two London 'foodie neighbourhoods'
Source: *Time Out*, London Eating and Drinking, 2005; reproduced with permission.

In the case of Marylebone 'Village', 'the glut of fine food shops' is attributed to recent pro-active efforts of the Howard de Walden Estate, which owns Marylebone High Street and the surrounding 92 acres, and its 'promise to its high-rent residents to provide a high level of services' (9). A partial list of 'desirable' shops attracted to the area by the Estate includes La Fromagerie, Rococo Chocolates, Total Organics, Marylebone Farmers' Market (Sundays), Patisserie Valerie, Ginger Pig and Divertimenti, 'arguably the best tableware and cookware shop in London' (9). Marylebone High Street also is home to The Natural Kitchen, with a shop front advertising 'Organic, Wild and Artisan Food', whose statement of 'Who We Are' reassures its customers that 'We do not do mass-produced food '. The culinary landscape of Northcote Road has been transformed by the 1990s gentrification of the Victorian terraced houses in the area. Soaring property prices and the influx of those able to afford them, mainly white professional classes, have revived the traditional street market and established food shops and attracted newcomers. These range from a cheese specialist, with 'an appealing array of accoutrements, including several types of oatcake', high-quality vintners, and master bakers to a new Italian deli, I Sapori, whose 'stock is resolutely top-class stuff' (12). No food deserts here!

These two cases of 'foodie gentrification' emphatically underline, even to the point of caricature, the huge social and spatial inequalities that generally are typical of the consumption relations of organic and artisanal quality foods. Close 'connections among property, privilege and paler skin' are constitutive of these spaces of alternative food praxis (Slocum 2007, 7). Nevertheless, protagonists of alternative food provisioning contend that quality foods and their cultural economies represent embedded moral geographies and that, like fair trade products, they are morally charged. The more discerning 'better-off people', as Time Out has it, can eat their organic and exclusive, 'handcrafted' foods and still claim the moral high ground. On this argument, the eco-social imaginaries of quality food—think of the discourse of Slow Food—connect consumer lifestyles to an ethic of conservation, embracing traditional landscapes, rural communities, historical legacies of *savoir faire*, humane methods of animal husbandry, and cultural and gastronomic difference. Insofar as local and quality foods are 'represented as embodying "alternative rural development", producers and consumers are linked together in a discursive field in which consumption "makes a difference"' (Goodman, M. 2004, 902).

In effect, a Leopoldian land ethic of stewardship is performed *corporeally* through a food aesthetic, a performance to which higher income, and predominantly white, consumers have privileged access. This 'politics of perfection' around 'right living' and 'right eating' 'universalizes and elevates particular ways of eating as ideal' (DuPuis and Goodman 2005, 362). A middle-class, white food aesthetic is the 'unmarked category' in these politics, which embrace naturalized, unacknowledged possessive investments in class, race, gender and place/space (Lipsitz 1998). As DuPuis and Goodman (2005, 368) observe, 'it is important to pay more attention to the ways in which our possessive investments in our own racial privilege influence how we define problems and solutions'. To emphasize again, Leopoldian stewardship enacted in the privileged white spaces of alternative food provisioning is not about social justice, whether in labour relations from 'field to shelf' or equal access.[6]

As Barnett et al. (2005, 8) observe, drawing on Bourdieu (1984), 'Ethical consumption often works by presenting consumers with what are essentially

6 On the raced dimensions of AFNs in the US, see Allen (2004), Guthman (2004a) and Slocum (2006, 2007).

positional goods', which reinforce social and cultural differentiation. Nevertheless, if alternative food practice currently is ordered by economic and cultural privilege and possessive investment in whiteness, critique should not involve outright rejection and blanket condemnation. To the contrary, this ordering is contingent and the possibility that other, more socially just, ethical engagements may emerge, as evidenced by the recent growth of public procurement programmes and food cooperatives in the UK, needs to be recognized and explored (Slocum 2007). Also, as Bryant and Goodman (2004) imply, alternative food practice can represent a form of counter-consumerism and such disenchantment with mainstream collective modes of provision may be an important precursor of a wider 'alternative hedonist' politics (Soper 2004; Soper and Thomas 2006). This brief commentary reveals the complex analytics of politicized consumption, with its ambivalences of reflexive, disaffected privilege, and it is high time that such questions moved centre-stage in the literature and praxis of alternative food provisioning.

Concluding Observations

This chapter has explored the recent research literature on the places and spaces of production-consumption opened up by the growth of alternative food provisioning networks. Particular attention has been given to the construction of quality and its contested meanings. The new quality conventions that inspire alternative food practices have created emergent markets, new sources of economic value, and unsettled power relations across the spaces of food provisioning. Further analysis of the evolving interrelationships between conventional and alternative food provisioning, captured by the notion of 'mainstreaming', and of how power asymmetries operate to influence rent distribution between actors in these value chains should be high on the research agenda.

Socially just food provisioning remains prominent in the imaginary of alternative food systems (cf. Slocum 2007) but material progress on this front, with the exception of current efforts in the public procurement realm (Morgan 2007; Morgan and Sonnino 2008), so far has involved mainly isolated, micro-level initiatives. Apart from bemoaning its neglect, the research literature has tended to put this issue to one side. This chapter has suggested that recent critical analyses of moral food geographies and ethical consumption offer potentially fruitful ways of reviving debates on alternative food provisioning and social justice.

More pragmatically, however, social justice needs to be the centre-piece of a reinvigorated food policy to promote equal access to nutritious quality foods and in ways which sustain 'fair' livelihoods for farmers, farm workers and other actors in food provisioning. In turn, this requires parallel, root-and-branch measures to eradicate oligopolistic practices used by supermarkets to secure economic rents and dominate the value chain. Without deep-seated political and institutional change, alternative food networks are likely to be confined to provisioning 'better-off people' in narrowly circumscribed spaces of consumption whose boundaries are patrolled by rent-seeking retailers hungry for new opportunities to differentiate their product lines.

References

Allen, P. (2004), *Together at the Table: Sustainability and Sustenance in the American Agri-Food System* (University Park: Pennsylvania State University Press).

Appadurai, A. (1986), 'Introduction: Commodities and the Politics of Value' in Appadurai, A. (ed.), *The Social Life of Things* (pp. 3–63) (Cambridge: Cambridge University Press).

Baker, P. (1990), 'The Domestication of Politics: Women and American Political Society, 1780–1920' in Gordon, L. (ed.), *Women, the State and Welfare* (Madison: University of Wisconsin Press).

Barnett, C., Cloke, P., Clarke, N. and Malpass, A. (2005), 'Consuming Ethics: Articulating the Subjects and Spaces of Ethical Consumption', *Antipode* 37: 1, 23–45.

Bernstein, H. and Campling, L. (2006), 'Commodity Studies and Commodity Fetishism II: "Profits with Principles"?', *Journal of Agrarian Change* 6: 3, 414–447.

Bourdieu, P. (1984), *Distinction: A Social Critique of the Judgement of Taste* (London: Routledge).

Bryant, R. and Goodman, M. (2004), 'Consuming Narratives: The Political Ecology of "Alternative" Consumption', *Transactions of the Institute of British Geographers NS* 29, 344–366.

Buller, H. (2000), 'Re-creating Rural Territories: LEADER in France', *Sociologia Ruralis* 40: 2, 190–199.

Buller, H. (2001), 'Is this the European Model?' in Buller, H. and Hoggart, K. (eds), *Agricultural Transformation, Food and Environment* (pp. 1–8) (Aldershot: Ashgate).

Clarke, N., Barnett, C., Cloke, P. and Malpass, A. (2007), 'Globalizing the Consumer: Doing Politics in an Ethical Register', *Political Geography* 26, 231–249.

Curry Report. (2002), 'Farming and Food: A Sustainable Future', Report of the Policy Commission on the Future of Food and Farming (London: UK Cabinet Office).

DeLind, L. (2003), 'Considerably More than Vegetables, a Lot Less than Community: The Dilemma of Community-supported Agriculture' in Adams, J. (ed.), *Fighting for the Farm: Rural America Transformed* (pp. 192–206) (Philadelphia: University of Pennsylvania Press).

Dixon, J. (1999), 'A Cultural Economy Model for Studying Food Systems', *Agriculture and Human Values* 16, 151–160.

Dowler, S. and Turner, S. (2001), *Poverty Bites: Food, Health and Poor Families* (London: Child Poverty Action Group).

DuPuis, M. and Goodman, D. (2005), 'Should We Go "Home" to Eat?: Toward a Reflexive Politics of Localism', *Journal of Rural Studies* 21, 359–371.

Du Toit, A. (2002), 'Globalizing Ethics: Social Technologies of Private Regulation and the South African Wine Industry', *Journal of Agrarian Change* 2: 3, 356–380.

Fine, B. (2002), *The World of Consumption: The Material and Cultural Revisited*, 2nd Edition (London: Routledge).

Fine, B., Heasman, M. and Wright, J. (1996), *Consumption in the Age of Affluence: The World of Food* (London: Routledge).

Friedmann, H. (1994), 'Food Politics: New Dangers, New Possibilities' in McMichael, P. (ed.), *Food and Agrarian Orders in the World Economy* (pp. 15–33) (Westport: Praeger).

Goodman, D. (2004), 'Rural Europe Redux? Reflections on alternative agro-food networks and paradigm change', *Sociologia Ruralis* 44: 1, 3–16.

Goodman, D. (2008), 'Pragmatists, "Satisficers" and Radical Structuralists: Reflections on Ethical Consumption and the (New) Politics of Food', Draft paper.

Goodman, D. and DuPuis, E. M. (2002), 'Knowing Food and Growing Food: Beyond the Production-Consumption Debate in Agro-Food Studies', *Sociologia Ruralis* 42: 1, 6–23.

Goodman, M. (2004), 'Reading Fair Trade: Political Ecological Imaginary and the Moral Economy of Fair Trade Foods', *Political Geography* 23: 7, 891–915.

Gregory, D. (1994), *Geographical Imaginations* (Oxford: Blackwell).

Guthman, J. (2002), 'Commodified Meanings, Meaningful Commodities: Re-thinking Production-Consumption Linkages in the Organic System of Provision', *Sociologia Ruralis* 42: 4, 295–311.

Guthman, J. (2004a), *Agrarian Dreams: The Paradox of Organic Farming in California*.(Berkeley: University of California Press).

Guthman J. (2004b), 'The "Organic" Commodity and Other Anomalies in the Politics of Consumption' in Hughes, A. and Reimer, S. (eds), *Geographies of Commodity Chains* (pp. 233–249) (London: Routledge).

Harvey, D. (1990), 'Between Space and Time: Reflections on the Geographical Imagination', *Annals of the Association of American Geographers* 80: 3, 418–34.

Holloway, L. and Kneafsey, M. (2004), 'Producing-Consuming Food: Closeness, Connectedness and Rurality' in Holloway, L. and Kneafsey, M. (eds), *Geographies of Rural Cultures and Society* (pp. 262–282) (London: Ashgate).

Ilbery B. and Kneafsey, M. (1999), 'Niche Markets and Regional Speciality Food Products in Europe: Towards a Research Agenda', *Environment and Planning A* 31, 2207–2222.

Ilbery, B. and Maye, D. (2005), 'Food Supply Chains and Sustainability: Evidence from Specialist Food Producers in the Scottish/English Borders', *Land Use Policy* 22: 4, 331–344.

Ilbery, B. and Maye. D. (2006), 'Retailing Local Food in the Scottish/English Borders; A Supply Chain Perspective', *Geoforum* 37: 3, 352–367.

Ilbery, B., Morris, C., Buller, H., Maye, D. and Kneafsey, M. (2005), 'Product, Process and Place: An Examination of Food Marketing and Labelling Schemes in Europe and North America', *European Urban and Regional Studies* 12: 2, 116–132.

Jackson, P., Ward, N. and Russell, P. (2006), 'Mobilising the Commodity Chain Concept in the Politics of Food and Farming', *Journal of Rural Studies* 22: 2, 129–141.

Jaffee, D., Kloppenburg, J. Jr. and Monroy, M. (2004), 'Bringing the "Moral Charge" Home: Fair Trade within the North and within the South', *Rural Sociology* 69: 2, 169–196.

Kloppenburg, J. Jr. (1991), 'Social Theory and the De/reconstruction of Agricultural Science; Local Knowledge for an Alternative Agriculture', *Rural Sociology* 56: 4, 516–548.

Kloppenburg, J. Jr. and Hassanein, N. (1995), 'Where the Green Grass Grows Again: Knowledge Exchange in the Sustainable Agriculture Movement', *Rural Sociology* 60: 4, 721–740.

Kloppenburg, J. Jr., Hendrickson, J. and Stevenson, G. W. (1996), 'Coming into the Foodshed', *Agriculture and Human Values* 13: 3, 33–42.

Lipsitz, G. (1998), *The Possessive Investment in Whiteness: How White People Benefit from Identity Politics* (Philadelphia: Temple University Press).

Littler, J. (2005), 'Beyond the Boycott: Anti-consumerism, Cultural Change and the Limits of Reflexivity', *Cultural Studies* 19: 2, 227–252.

Lowe, P., Buller, H. and Ward, N. (2002), 'Setting the Next Agenda? British and French Approaches to the Second Pillar of the Common Agricultural Policy', *Journal of Rural Studies* 18: 1, 1–17.

Marsden, T. (2004), 'Theorizing Food Quality: Some Key Issues in Understanding its Competitive Production Regulation' in Harvey, M., McKeekin, A. and Warde, A. (eds), *Qualities of Food* (pp. 129–155) (Manchester: Manchester University Press).

Marsden, T., Banks, J. and Bristow, G. (2000), 'Food Supply Chain Approaches: Exploring their Role in Rural Development', *Sociologia Ruralis* 40: 4, 424–438.

Marsden, T., Flynn, A. and Harrison, M. (2000), *Consuming Interests: The Social Provision of Foods* (London: UCL Press).

Marsden, T. and Smith, E. (2005), 'Ecological Entrepreneurship: Sustainable Development in Local Communities through Quality Food Production and Local Branding', *Geoforum* 36: 4, 440–451.

Micheletti, M. (2003), *Political Virtue and Shopping: Individuals, Consumerism and Collective Action* (London: Palgrave).

Miele, M. and Murdoch, J. (2002), 'The Practical Aesthetics of Traditional Cuisines: Slow Food in Tuscany', *Sociologia Ruralis* 42: 2, 312–328.

Miller, D. (1995), 'Consumption as the Vanguard of History: A Polemic by Way of an Introduction', in Miller, D. (ed.), *Acknowledging Consumption: A Review of New Studies* (pp. 1–57) (London: Routledge).

Mollard, A., Hirczak, M., Moalla, M., Pecqueur, B., Rambonilaza, M. and Vollet, D. (2005), 'From the Basket of Goods to a More General Model of Territorialized Complex Goods: Concepts, Analysis Grid and Questions', draft ms.

Moran, W. (1993), 'The Wine Appellation as Territory in France and California', *Annals of the Association of American Geographers* 83: 4, 694–717.

Morgan, K. (2007), 'Greening the Realm: Sustainable Food Chains and the Public Plate', unpublished ms.

Morgan, K., Marsden T. and Murdoch, J. (2006), *Worlds of Food: Place, Power and Provenance in the Food Chain* (Oxford: Oxford University Press).

Morgan, K. and Sonnino, R. (2008), *The School Food Revolution: Public Food and the Challenge of Sustainable Development* (London: Earthscan).

Morris, C. and Young, C. (2000), '"Seed to Shelf", "Teat to Table", "Barley to Beer" and "Womb to Tomb": Discourses of Food Quality and Quality Assurance Schemes in the UK', *Journal of Rural Studies* 16, 103–115.

Murdoch, J., Marsden, T. and Banks, J. (2000), 'Quality, Nature and Embeddedness: Some Theoretical Considerations in the Context of the Food Sector', *Economic Geography* 76, 107–125.

Murdoch, J. and Miele, M. (1999), '"Back to Nature": Changing "Worlds of Production" in the Food Sector', *Sociologia Ruralis* 39: 4, 465–483.

Murdoch, J. and Miele, M. (2004a), 'A New Aesthetic of Food? Relational Reflexivity in the Alternative Food Movement' in Harvey, M., McMeekin A. and Warde, A. (eds), *Qualities of Food* (pp. 156–175) (Manchester: University of Manchester Press).

Murdoch, J. and Miele, M. (2004b), 'Culinary Networks and Cultural Connections; A Conventions Perspective' in Hughes, A. and Reimer, S. (eds), *Geographies of Commodity Chains* (pp. 102–119) (London: Routledge).

Needham, C. (2003), *Citizen-Consumers: New Labour's Marketplace Democracy* (London: Catalyst).

Parrott, N., Wilson, N. and Murdoch, J. (2002), 'Spatializing Quality: Regional Protection and the Alternative Geography of Food', *European Urban and Regional Studies* 9: 3, 241–261.

Ploeg, J., van der (1990), *Labor, Markets and Agricultural Production* (Boulder: Westview Press).

Ploeg, J., van der (1993), 'Rural Sociology and the New Agrarian Question: A Perspective from the Netherlands', *Sociologia Ruralis* 32: 2, 240–260.

Ploeg, J., van der (2003), *The Virtual Farmer: Past, Present and Future of the Dutch Peasantry* (Assen: Royal Van Gorcum).

Ploeg, J., van der, Renting, H., Brunori, G., Knickel, K., Mannion, J., Marsden, T., de Roest, K., Sevilla-Guzman, E. and Ventura, F. (2000), 'Rural Development: From Practices and Policies towards Theory', *Sociologia Ruralis* 40: 4, 391–408.

Popke, J. (2006), 'Geography and Ethics: Everyday Mediations Through Care and Consumption', *Progress in Human Geography* 30: 4, 504–512.

Potter, C. and Burney, J. (2002), 'Agricultural Multifunctionality in the WTO: Legitimate Non-trade Concern or Disguised Protection?', *Journal of Rural Studies* 18, 35–47.

Potter, C. and Tilzey, M. (2005), 'Agricultural Policy Discourses in the European Post-Fordist Transition: Neoliberalism, Neomercantilism and Multifunctionality', *Progress in Human Geography* 29: 5, 581–600.

Potter, C. and Tilzey, M. (2007), 'Agricultural Multifunctionality, Environmental Sustainability and the WTO: Resistance or Accommodation to the Neo-liberal Project for Agriculture?', *Geoforum* 38, 1290–1303.

Ray, C. (2000), 'The EU LEADER Programme: Rural Development "Laboratory"', *Sociologia Ruralis* 40: 2, 163–171.

Slocum, R. (2006), 'Anti-racist Practice and the Work of Community Food Organizations', *Antipode* 38, 327–349.

Slocum, R. (2007), 'Whiteness, Space and Alternative Food Practice', *Geoforum*, 38: 3, 520–533.

Soper, K. (2004), 'Rethinking the "Good Life": The Consumer as Citizen', *Capitalism, Nature, Socialism* 15: 3, 111–116.

Soper, K. and Thomas, L. (2006), '"Alternative Hedonism" and the Critique of "Consumerism"', ESRC/AHRC Cultures of Consumption Research Programme, Working Paper No. 31.

Storpor, M. (1999), *The Regional World: Territorial Development in a Global Economy* (New York and London: Guilford Press).

Sustainable Development Commission. (2008), 'Green, Healthy and Fair: A Review of Government's Role in Supporting Sustainable Supermarket Food', Sustainable Development Commission, UK.

Szasz, A. (2007), *Shopping Our Way to Safety. How We Changed from Protecting the Environment to Protecting Ourselves* (Minneapolis: University of Minnesota Press).

Time Out. (2004), *Eating and Drinking, 2005* (London: Time Out Group Limited).

Tregear, A. (2003), 'From Stilton to Vimto: Using Food to Re-think Typical Products in Rural Development', *Sociologia Ruralis* 43: 2, 91–100.

Tregear, A. (2005), 'Origins of Taste: Marketing and Consumption of Regional Foods in the UK', paper presented at the ESRC Seminar on Global Consumption in a Global Context, University of Cardiff, 29 November.

Valceshini, E. et al. (2002), 'Agriculture and Quality in 2015: The Outlook Based on Four Case-studies', unpublished paper, Working Group on Alternative food Networks, COST A12, Rural Innovation, Brussels.

Ward, N. and McNicholas, K. (1998), 'Reconfiguring Rural Development in the UK: Objective 5b and the New Rural Governance', *Journal of Rural Studies*, 14: 1, 27–39.

Wilkinson, J. (1997), 'A New Paradigm for Economic Analysis? Recent Convergences in French Social Science and an Exploration of the Convention Theory Approach, with a Consideration of its Applicability to the Analysis of the Agro-food System', *Economy and Society* 26: 3, 305–339.

PART IV
Consumption as Production and Production as Consumption

Chapter 10
Creating Palate Geographies: Chilean Wine and UK Consumption Spaces

Robert N. Gwynne

Introduction

This chapter will attempt to make connections between worlds of production and worlds of consumption. More specifically the chapter will introduce the concept of palate geographies and argue that the concept provides a useful lens through which to refract our understandings of food/wine networks at the global scale and their economic and cultural underpinnings. At the general level, I argue that palate geographies reflect how food and wine networks are intimately linked to the changing tastes of consumers and how key global actors (such as supermarket buyers) interact with shifts in consumer taste in the markets in which they operate. Meanwhile, at the production end of these global networks, palate geographies reflect the need for wine producers to make efforts to change their production mixes and 'create' new tastes given their perception of the changing palates of consumers in distant and often culturally very different consuming spaces.

The vehicle used to explore the concept of palate geographies in this chapter is that of the wine trade between the worlds of wine production (in Chile) and the worlds of wine consumption in the most important of Chile's global markets, that of the UK. Why am I using the lens of the Chile/UK wine trade to explore the concept of palate geographies? One key point is that the act of consuming wine in the UK is both a growing and a relatively 'open' phenomenon.

UK consumers are increasing their per capita consumption of wine. As with other North European consumers, the drinking of alcohol in the UK has historically been dominated by beer consumption and, although this is still the case, consumption of wine has been growing much faster than that of beer in recent years. This is in contrast to South European countries (including France) where per capita wine consumption has been seriously declining for many years – admittedly from much higher levels than those in the UK. Per capita wine consumption in France has fallen from over 100 litres a year in 1977 to 50 litres in the early twenty-first century (Robinson 2006, 279).

Secondly, the UK wine consumer tends to be relatively 'open' to new types and sources of wine – even more so than other North European consumers such as those from Norway (Sanchex Hernandez 2004). This is unlike the French or Spanish consumer for example whose consumption patterns are more distinctly rooted in French and Spanish wine producing spaces; in France imports command less than 5 percent of the market and there is a refusal to drink foreign wines (Rachman 1999, 107). In the UK, partly because the UK produces very little wine itself, there are no such (political and cultural) restrictions on the importing of non-national wine. This introduces the idea of the UK market as being one of the few permitted 'battlegrounds' where the rival attractions of Old World (Western Europe) and New World (Americas, Australasia, South Africa) wines can be played out through the decisions and 'palate geographies' of large numbers of consumers. Indeed, since 2004, the UK market has received more wine in volume terms from Australia than from France (IWSR 2007, 19), the latter being historically the major source of wine consumed in the UK.

The UK consumer is also very conscious of price when buying wine, generally searching for a good quality-price ratio when purchasing a bottle of wine. In November 2006, the average price of a bottle of wine sold in the UK was £3.99 with £3.82 being the average price of a bottle of wine sold in the supermarket chain (Tesco) that is responsible for 33 percent of all supermarket wine sales in the UK (Davis 2006). The actual cost of the wine in a £3.82 bottle in the UK market (taking account of freight charges, dry goods costs, duty, taxes, distributor margins and supermarket margins) would only be between 50 pence and 75 pence depending on the level of promotion that the various actors (wine producer, distributor, supermarket) decided on.

How then do Chilean wine producers fit into this highly competitive, price-sensitive but expanding UK wine market? In June 2007 the average price of a Chilean wine bottle sold in the off-trade in the UK was £3.82 (Wines of Chile, 2007), indicating that Chilean wine producers have been able to achieve an appropriate quality-price ratio for a cost-conscious market: 'The grocery multiples have had a lot of trade in the lower end and the consumer has got to think of Chile as being a comparatively cheap place for bargain wine' (Room 2006). Hence the UK market provided a key target for Chilean wine producers as they became engaged in the search for export growth since the early 1990s.

The organizing principle of this paper is to develop the concept of palate geographies. Palate geographies can combine theoretical themes from material culture, value chain analysis and even convention theory. The key theme here is how the issue of taste can be transmitted along global value chains. There are many methodological difficulties in tracing the value (or commodity) chain all the way from the producer to the consumer. Most researchers that have focused on the production networks behind commodity chains (Gereffi 1994; Humphrey and Schmitz 2002) have not attempted this. Meanwhile those that have traced commodity chains from the perspective of the consumer have often done this in relation to food as with the papaya (Cook 2004). Cook (2006) recognizes this lack of dialogue between those researchers focusing on the networks of production in global commodity chains and those examining these chains through the prism of cultures of consumption.

> A key question that food geographers are asking is how the widely-acknowledged and longstanding division between an … agri-food studies literature dominated by political economy … and a cultural studies of food literature dominated by post-structuralism and qualitative research can be bridged. (Cook 2006, 4)

Cook (2006) goes on to argue that there is a need for more multidisciplinary approaches to food research and that the organizing principles for this research could focus on specific foods and ingredients.

This chapter will begin by exploring the concept of palate geographies before examining four themes in relation to the links between British wine consumers and Chilean wine producers: UK consumers and the power of the supermarket; knowledge transfers and cooperation along the wine value chain between Chilean producers and UK supermarket buyers; the creation of producing spaces in Chile; and wine producers trading up through taste.

Much of the analysis comes from a two-year British Academy research project (2005–2007) which examined the impacts of globalization on export-oriented wine firms in Chile's Colchagua Valley and the nature of collaboration and knowledge transfer between these firms and key purchasing companies and distributors within the UK market. Twenty one wine firms were interviewed in Chile and five buyers of Chilean wine from leading UK supermarkets and specialist retail chains.

Creating Palate Geographies

To paraphrase Cook et al. (1998) can one taste places? Wine as a commodity is distinguished by having a huge range of tastes – probably more than any other commodity. Historically, the wines of a producing space would largely be consumed within or near to that space (Friedland 2005). Now, however, there are more and more producing spaces (particularly in the New World) and more and more wine producers that are gaining access to global distribution networks. The wide range of different tastes is partly linked to an increasing number of 'global' grape varieties, the numerous blends of these grape varieties and the great variety of producing spaces from which they come. Wines have lives before and after they appear on the supermarket shelves. Palate geographies are concerned with the many different places, people and social institutions that wines travel through as they move from vineyard to glass.

The geographies of the winemaking process (the technology of manufacture) can create very different tastes for the consumer's palate from the same grape – such as the winemaking differences between the Rhône and Australia in terms of the Syrah/Shiraz grape.

> Viticulturally Shiraz is identical with Syrah but the resulting wines taste very different, with Australian versions tasting much sweeter and riper, more

> suggestive of chocolate than the pepper and spices often associated with Syrah in the Rhône. (Robinson 2006, 627)

Chilean wineries have observed these contrasts with this particular grape variety and have attempted to develop wines 'in between' the French and Australian versions. Understandably they use the term Shiraz on their bottles when they are verging towards the Australian style (as with the wines of Luis Felipe Edwards – see below) and favour the Syrah label when the wine is more restrained (as with Montes in the Colchagua Valley). Thus consumers can now recognize a wide-ranging palate geography of the Syrah/Shiraz grape at the global scale as wineries in France, Chile and Australia emphasize different components of taste in their final product.

In this initial exploration of the concept of palate geographies, I will examine three processes that link worlds of production with worlds of consumption: production, regional branding and *terroir*; knowledge, taste and the role of supermarket buyers; and the relationship between the consumer and the supermarket in terms of relative risk and purchase effort.

Production, regional branding and terroir

From a production perspective, palate geographies can be differentiated through the contrasting winemaking of a number of 'global' grape varieties. Stevenson (2005, 36–43) identifies eleven white and ten red 'global' grape varieties. Most originate from France, but others come from Germany (Riesling), Austria (Sylvaner) and Italy (Nebbiolo). However, sometimes the grape origin can be more obscure as with the case of California's Zinfandel:

> Once thought to be the only indigenous American Vitis vinifera grape, Zinfandel has now been positively identified by Isozyme "finger printing" as the Primitivo grape of southern Italy. However, the origins of this grape are Croatian, where it is known as the Crljenak Kastelanski. (Stevenson 2005, 43)

So the taste from California's signature red wine is based on a grape variety with complex origins. More generally, the travels of wine grape varieties from Old World to New World have often created new 'palate' geographies.

Wine-producing countries or regions have been particularly successful if they have developed a distinctive 'palate' or regional branding for their wines that is recognized by a wide range of consumers and not just by wine specialists. New Zealand Sauvignon Blanc (Hayward and Lewis 2008), Californian Zinfandel and Chilean Carmenère (see below) or Cabernet Sauvignon could be seen as examples of New World producing spaces developing examples of a distinctive 'palate'. The most successful regional example at the global scale is that of Champagne.

Palate geographies also introduce the highly-contested concept of *terroir*. Gade (2004) argued that the special quality of an agricultural product such as a particular wine is determined by the character of the place from which it comes. Some (Fanet 2004; Wilson 1998) see the place factor as the synergistic effect of soil, bedrock, landforms, climate (both macro and micro) and exposure in creating growing conditions for quality production. However, Gade (2004) pointed out that the concept of *terroir* could include unique human factors acquired from the past, such as the skills or practices passed on from one generation to the next. This required cooperation and knowledge transfer through history between winemakers in the *terroir,* as with the example of small-scale producers in Burgundy. Barham (2003) maintained that the *terroir* concept, although perceived as legacy, actually centres on an active and conscious social construction of the past, which shapes present place-based identity. However, Moran (2001, 1) uses a (translated) quote from Louis Latour, the Burgundy wine trader, to add a critical note to the use of *terroir* in the Old World: 'when you try to analyse in detail the diverse elements of *terroir* you find yourself with such uncertainties that it's better not to stick your nose in too far. That said, *terroir* is an excellent marketing tool'.

There is nevertheless a critical distinction between the Old World and New World perceptions of wine and *terroir*. In the Old World, the concept of *terroir* looks to the past. Place identity for quality wine production looks to rural roots and attempts to combine local physical and human geographies. Meanwhile, in New World wine regions, the concept of *terroir* is being constructed. Firms look to the present and future to create place identity for quality wine production. *Terroir* becomes more a combination of physical geography and modern technology. For this reason Bisson et al. (2002, 697) argue that the complex interplay of physiological, genetic and environmental factors that underpin consumer choice need to be more clearly understood.

Furthermore, the concept of *terroir* in Chile and the New World is being created within a market-oriented system in which the decisions of enterprises, both individually and collectively in terms of their support of place-based identity, are very significant in 'creating' global recognition for the products of a particular place. As Guthey (2004, 17) argues 'the only way to show that one's wine has *terroir* is to make it and then tell people about it'. Wine enterprises and winemakers thus create and transmit local knowledges about their wines through the invitation of wine writers to taste the wines of their *terroir* (either physically or virtually), through special tastings for wine enthusiasts, through wine tourism and varied forms of place-based advertising of their products in retail and other outlets.

Guthey (2004) sees the concept of *terroir* as making connections between producers and consumers. He sees the claim to *terroir* in California as a convention and argues that one needs to see it in terms of the social, political and economic practices of producers. Wine enterprises try to make connections between quality and place. This can be partly about the connections between nature (climate, soils and aspect) and wine in a particular locality. But it is often the case that a range of wine qualities is produced from a particular *terroir*, such as from the Colchagua Valley in Chile.

Knowledge transfer, taste and the role of the supermarket buyers

Thus, in New World wine regions one could argue that geographical indication has to be created – particularly by wine enterprises but also by local and national business associations linked to the wine sector. This could be seen as an example of what the conventions literature (Ponte and Gibbon 2005) calls 'qualified markets' – markets in which the definition of quality relies upon crucial communication between producers and consumers about what constitutes quality. Ponte and Gibbon (2005, 2–3) argue that the management of quality may be seen as a question of competition and/or cooperation between actors in the same value chain. The wine value chain incorporates such different actors as the grape producer, winery, exporting firms in the producing country, importing firms and distributors in the consuming country, wine retailers and, of course, the consumer. Each actor has only partial access to information in this value chain. Producers have access to information on the product and related production and process methods – but they do not have the information on how different products engage with the consumers of crucial global markets – distributors and wine retailers have more control of this information. Hence in order to ensure the management of quality in any particular chain cooperation and knowledge transfer between actors is essential and can potentially benefit all members of the chain.

Palate geographies can thus highlight the negotiated interface between aspects of production (physical, cultural, economic) and those of consumption (cultural, economic, geographic). It can specifically involve some interpretative analysis of taste and its varied meanings across space. Who are the crucial actors here? It could be argued that taste in consuming countries is increasingly being managed by a small number of very large supermarket firms and the wine buyers within them.

Such is the power of the supermarkets that some analysts (Hayward and Lewis 2008) have talked of the global wine value chain as being buyer-driven (Gereffi 1994). Certainly supermarkets must be seen as one of the key agencies in the global wine value chain. One could refer to Gereffi's (1999) 'requirements thesis' of key actors in this context. Although supermarkets and other end-clients do not have direct access to the technology of their suppliers, they draw up requirements – in this case for the wine-exporting companies. Wine producers need to find the relevant technologies and knowledges from other sources if they do not have it 'in house'. This has been linked to the growing significance of the flying winemaker (Lagendijk 2004). However supermarket requirements often revolve around the key theme of taste, such as when supermarket buyers request wineries to develop a new taste or grape variety combination in order to see how the development of the new taste coincides with the palates (and pockets) of 'their' consumers.

This links palate geographies with themes of retail concentration. Humphrey (2006, 574) argues that concentration in the retailing of fresh and processed food has led to a substantial reorganization of agribusiness value chains: 'Large buyers have transformed themselves from resellers of products made by others into firms that go out to find suppliers for the products that they want for their customers'. This would also appear to be very much the case in the global wine value chain. Supermarkets in such core country markets as the UK see sales of wine as critical for attracting higher-spending customers and have developed competitive strategies based on increasing the range and quality of wine.

The consumer and the supermarket: Relative risk and purchase effort

One element of supermarket strategy has been to develop brand image and increase the significance of own-label products. This has meant that supermarkets have taken an active role in product innovation and supply chain management. Humphrey (2005, 3) shows that own-label penetration of retailing in the UK rose from around 22 percent in 1980 to around 43 percent in 2001. In focusing on the UK consumer, one could adapt Spawton's (1991) model of hierarchical ranging in relation to relative risk and purchase effort (see Figure 10.1).

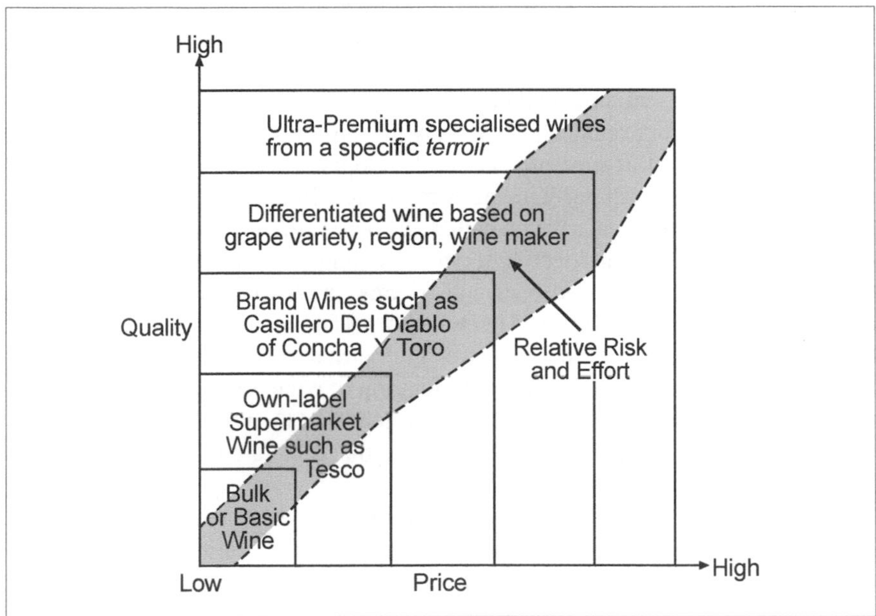

Figure 10.1 Hierarchical ranking and the price, quality, effort and risk classification for consumers of Chilean wine products in the UK market

Note: 1) The relative risk and purchase effort of the consumer is represented by the relative width of the shaded area; 2) The quality/price axes and the area of the classifications of the wine product types are not representative of market proportions.

Source: Adapted from Spanton, 1991.

The risk and effort to the UK consumer is seen as basically low in terms of purchasing cheap basic wine and own-label supermarket brands. It should be emphasized that in Figure 10.1 the quality/price axes and the area of the classifications of the wine product types are not representative of market proportions. In terms of the UK's largest supermarket, Tesco, about 50 percent of wine sales are accounted for by own-label products (Davis 2006). Furthermore, supermarket chains have succeeded in establishing credibility in their own brands so that consumers do not just perceive 'the own-label brand as a "cheap" alternative but as a worthy competitor to the manufactured brand' (Chaney 2004, 5).

Relative risk and purchase effort somewhat increase when the consumer is faced with branded wines from distant wine producing districts. However, the highest risk and effort is involved in the purchase of wine differentiated by grape variety, estate, region or wine-maker as the price (along with the presumed quality) begins to rise sharply. As the very high prices (and quality) of ultra-premium wines are reached, the risk and effort of the consumer lessens as the importance of local *terroir* identities (such as Champagne) play a part. Spawton (1991, 27) argues that these involve little risk 'as the number of buyers is limited to a small band of specialists who are wine-knowledgeable and are prepared to pay for the exclusivity of the product'.

It can be argued that supermarkets reflect the changing lifestyles of consumers. In the UK market, for example, per capita wine consumption is increasing – hence the particular interest of supermarkets in developing own-label brands for those consumers who see drinking wine as part of their lifestyle. Analysts of regular wine consumers focus on what is called the *wine drinking moment* and see it as involving at least four lifestyle themes: feelings of escape and indulgence; connecting with family, loved ones and friends; unwinding and relaxation; and sophistication (Halstead 2006). The wine drinking moment is when the UK consumer's palate receives the impact of the particular taste that the winemaker had wished to impart when the final combination of wine was selected and bottled months (or years) earlier. Halstead (2006) argues that the wine drinking moment can be both a 'self' or 'sociable' occasion and that there are considerable gender differences within these occasions.

Discussions of 'palate geographies' could involve discussions of the consumer awareness of grape variety and wine region. Certain grape varieties, most notably Chardonnay, have high recognition – 80 percent of women in Halstead's (2006) survey associated Chardonnay with a white grape variety. However, consumer awareness of wine regions is much lower. Only 30 percent of men and 16 percent of women correctly identified 'Stellenbosch' as a wine region of South Africa. Such a response to a 'difficult' question tempted Halstead to estimate that only around one third of male consumers and one sixth of female consumers are interested in (or knowledgeable about) wine regions and the characteristics of wine that these regions produce.

Palate geographies therefore allow for the interaction of the consumer with wine producing spaces. What knowledge consumers have can potentially be a significant factor in wine choices. They can be concerned with the actual production process of wine – for example, whether it is fair-trade or organic. It can also reflect wider associations between places and wine quality, such that geographical names or imagery help to construct consumer perception of product quality (Cook et al. 1998). One issue concerns the institutions and sites in which consumers can place responsibility for knowledge of wine provision. This leads us to examine the issues that link UK consumers with wine producing districts in Chile.

UK Consumers and the Power of the Supermarket

The UK consumer has provided a key barometer for many Chilean wine producers as they developed export-oriented strategies. According to one of Chile's leading exporters, Luis Felipe Edwards, UK consumers are 'five years ahead of the rest of the world in terms of taste, price and requirements' (Edwards 2005). What exactly does this mean? From the point of view of palate geographies, it means that the UK consumer is open to new tastes from new producing spaces – particularly if the wines are at the right price.

Chile's two wine business associations (*Asociación de Viñas de Chile* and *Chilevid*) recognized the strategic nature of the UK wine market and created 'Wines of Chile' in 2002 as a non-profit-organization, funded by fees from both associations and with a limited amount of assistance from the Chilean government's export promotion body (Cox 2005). The initial strategy was to set up only one office outside Chile and that was to be in the main market for Chilean wine, the UK. A disproportionate amount of financial resources was then directed to this office, which was used not only to increase the sales of Chilean wine in the 'battleground' market of the UK but also to be a template for future offices in other world markets. 'Wines of Chile' started up in June 2003 and recorded significant success in boosting Chilean wine sales in the UK. The share of Chilean wine rose from 5.9 percent in 2003 to 6.9 percent in mid-2007 in the UK's off-trade (supermarkets and retailers) market and from 6.5 percent in 2004 to 8 percent in 2007 in the on-trade (restaurants, pubs and hotels) market (Wines of Chile 2007). Press releases from 'Wines of Chile' illustrate the competitive nature of the UK market in terms of rival international producers: 'Chile now has its highest ever share of the UK retail market – 6.9 percent and leapfrogs Spain to lie in 6th position in the country league table – with South Africa (5th) now firmly in its sights' (Wines of Chile 2007).

Supermarkets see sales of wine as expanding and as critical for attracting higher-spending customers. They have been very flexible in developing supply contracts with new producers from the New World in order to increase the range and quality of their wines. The adaptation of Spawton's model (see Figure 10.1) is particularly appropriate in terms of how supermarkets see UK consumer behaviour. They have attempted to reduce the risk and effort of UK consumers by developing a wide range of own-label supermarket brands.

> The role of own-label brands for Chilean wine is fundamental. In terms of possible consumer confusion about price and quality there is always more consumer confidence in buying a Tesco branded wine. As a result Tesco branded wine accounts for nearly 50 percent of Chilean wine sales in Tesco branches. (Davis 2006)

This must be one of the key figures of how the UK consumer relates to Chilean wine within a retail environment of extreme concentration. Supermarkets control 72 percent of off-trade sales and Tesco alone controls 33 percent of this. Of the 22 Chilean red wine products sold in Tesco at end-2006, only seven were own-label. However, they were responsible for nearly 50 percent of Chilean red wine sales. When the UK consumer buys the own-label Chilean red from Tesco, the purchase requires less risk and effort given the price-quality imagery surrounding the Tesco brand. Meanwhile, Tesco's own-label quality Chilean brand (Tesco *Finest*) has greater than half of total Chilean wine sales in the higher price range, indicating that Tesco consumers 'believe' in the supermarket brand image even more as the quality increases.

Beyond the purchase of own-label brands, the relative risk and purchase effort of the consumer increases with brand wines (see Figure 10.1), such as Casillero del Diablo, the key brand of Chile's largest wine company, Concha y Toro. However, then the relative risk and purchase effort expands considerably as higher quality and higher-priced wines are reached in the hierarchy; these would be wines differentiated by producer, grape variety, region or sub-region of production, estate, winemaker and other factors. High risk and purchase effort can continue into the very expensive range of highly specialized wines, where wines can come from a distinctive *terroir* (as with Casa Lapostolle's Clos Apalta).

A brief mention of the downstream value chain for Chilean wine provides useful context to the theme of British consumption. Most wine market share analysis only deals with the off-trade. However, in Figure 10.2, and using the database of the International Wine and Spirit Record (IWSR) I have estimated the consumption of Chilean wine in terms of both the on-trade and off-trade.

The on-trade is highly differentiated and has high mark-ups on the sale of wine. However, drinkers of Chilean wine in restaurants, pubs and hotels are responsible for about 21 percent of consumption in the UK market. Hence 79 percent is consumed through the off-trade where supermarkets dominate – they are responsible for as much as 72 percent of off-trade Chilean wine sales, with the top four supermarkets (Tesco, Asda, Sainsburys, Morrisons) being responsible for around 70 percent of supermarket sales. This leaves 28 percent of off-trade Chilean wine sales to specialist retail chains (such as Majestic, Oddbins, Threshers) and independent wine merchants (see Figure 10.2). What perception does the UK consumer have of Chile and Chilean wine? One theme to discuss is how the British consumer views Chilean wine versus wine from other countries, such as France. First of all it should be noted that the British consumer of wine is faced with a highly differentiated product on the supermarket or retailer shelf. In some retail outlets (as in the specialist retail chain, Majestic) bottles of wine are differentiated by country, and then, in the case of France, by region. Consumer perceptions of global spaces in relation to wine production (in terms of both countries and regions) can become important criteria for selection or non-selection. This is measured and recognized by the retailers. In Majestic, the full range of wines is reviewed every six months in terms of sales growth or decline and shelf space for a country's wine is expanded or reduced accordingly; since 2004 Chile has been taking shelf space off wines from Australia and other countries in Majestic outlets (Pym 2006).

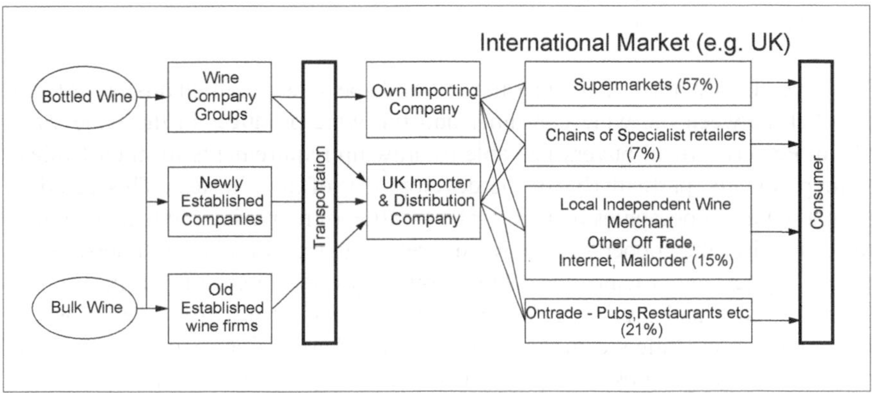

Figure 10.2 Chilean wine: Value downstream to UK market
Note: Percentages relate to IWSR estimate of UK market in volume terms.

In the British wine market, brand 'Chile' is associated with 'good value for money' at the lower end of the market (Eguiguren 2006). Most British wine consumers have little knowledge of Chile and have little possibility of visiting it on holiday (unlike Spain) and changing their views on the product *in situ*. Some supermarkets have been concerned to 'educate' their consumers into the tastes of Chilean wine 'both by direct promotion and through shaping the wider provision environment' (Cook et al. 2000). Waitrose has been one supermarket to have recognized this, producing an eight-page free leaflet during Chilean wine promotions. The leaflet, 'Chile Uncorked', contains eight photos, one map and an introduction to the nine wine regions which source the Chilean wine that Waitrose sells. This effort to provide place-based information by an up-market supermarket chain reflects the fact that an important proportion of Waitrose customers are interested in provenance (Room 2006). Providing such information allows Waitrose consumers not only to taste a variety of Chilean wines but also to taste the complex geographies and wine-producing places of Chile.

Apart from spatial categories, wine bottles are also differentiated by grape variety, brand and category (varietal, premium, super-premium, ultra-premium).

For the consumer these relate to different combinations of price and quality. As already noted, price can often be the determining factor. This has benefited the insertion of Chilean wine into the market given that the 'average' British consumer is aware of the relationship between low price and high quality in most Chilean wine (Cox 2005). Wine Intelligence (2004) saw Chilean wine as a good value proposition for UK consumers, particularly for big red wines 'and the spending profile is creeping upwards from a low base'.

Knowledge Transfers and Cooperation along the Wine Value Chain: New Palates in Action

We have already noted that supermarkets and specialist retail chains provide the key link between the wine consumer and the wine producer in the wine value chain. Powerful retail buyers feel able to draw up requirements for their Chilean wine producers, particularly for the development of new 'tastes'. This could be seen as palate geographies in action. The creation of new wine products with new tastes requires the cooperation between a retail buyer, with his/her understanding of UK palates, and a winemaker with knowledges of winemaking from different grape varieties and the 'expected' characteristics of grapes from new vineyard locations within Chile's complex geography (Richards 2006). Collaborating and cooperating along a global value chain can be a long and involved process.

One example of a high level of collaboration and cooperation between UK retail buyer and Chilean winemaker is the relationship between Matt Pym, the Chilean buyer for the UK specialist retail chain, Majestic, and the chief winemaker, Marcelo Papa, for Concha y Toro, Chile's largest wine company. Matt Pym collaborates with Marcelo Papa in jointly creating Winemaker Lot 'parcels' – new types of wines (new vineyard locations and wine tastes) with small volumes requested in the first instance (Pym 2006).

> With the Winemaker Lot wines, Marcelo normally offers me four or five possibilities from various regions for each style. For example, with the 2006 Merlot, I tasted five different wines back in August, from various valleys, before deciding on one that fits my requirements best. I then look at the time I want it to spend on oak, when I want it bottled and released. I then taste twice more over the next six months to check on the progress and fine-tune when to bottle. Clearly I'm not a winemaker, so I am guided by Marcelo. Likewise though, I do know what the market wants, so he is guided by me … it's a very collaborative process. (Pym 2006)

This was an example of key actors making choices and cooperating within the context of a global value chain in order to produce a quality product – and very much geared to a specialist's perception of the UK palate.

In terms of actors linking production and consumption in the wine value chain, an important role can be allocated to the distributor. Chilean wine companies have been characterized by seeking exclusive contracts with one UK distributor – in other words the UK distributor is not permitted to represent more than one Chilean wine company. Pym (2006) argues that the choice of UK distributor is one of the key factors in determining the success of an export-oriented Chilean wine company in the UK market. Certain companies, such as Montes, chose well as the Montes distributor, HwCg, 'understands the dynamics of the UK market better than the Montes management in Chile and have performed well for Montes in the quality restaurant and hotel sector' (Pym 2006). In contrast, some Chilean wine companies, such as Undurraga, have recorded serious decline in UK market share due to problems with their choice of UK distributor (Salvestrini 2005).

Distributors can be seen as a critical actor between the Chilean wine producer and a potentially wide range of off-trade (supermarkets and specialist retailers) and on-trade (restaurants, pubs and hotels) in the UK market (see Figure 10.2). In terms of the Colchagua winery, Luis Felipe Edwards (LFE), the main impetus to improve its international marketing in general and its sales to the UK market in particular came from its UK distributor, D & D wines from Cheshire (Eusabiaga 2005). On a visit to Chile in 1998, D & D suggested that LFE should stop following the French school in terms of Cabernet Sauvignon and other red wine production as it was too astringent and gave too little fruit for the British palate. As a result, LFE decided to change course and to make wine more in the Australian tradition and hired an Australian 'flying' winemaker, Mike Farmilo, in December 1998. Farmilo made three or four visits a year and managed to produce more fruit-driven reds for LFE. Subsequently contracts with Tesco and other key retailers were negotiated in the UK market and rapid expansion of UK market share has followed.

Thus knowledge in the wine value chain is not necessarily a one-way flow of knowledge from producer to consumer. This may be the main element of the knowledge transfer as the wine producer is trying to market wine based on a greater understanding of local *terroir*, geographical indication and grape variety. However, knowledge transfer can also be interactive with winemakers listening to distributors and retail buyers in order to adapt their future planting and winemaking strategies to what they perceive to be the tastes of distant consuming spaces.

Important knowledges can also be created outside the wine value chain but have a significant impact on it. The increasingly varied products of Chilean winemakers are assessed not only by individual consumers but also through tests, trials, and codified measurements backed by recognized legal or social institutions. One interesting theme here is the role of wine writers codifying the quality of wine. The most influential wine writer in the US is Robert Parker who assesses individual wines out of 100 through the *Wine Advocate*. Meanwhile in the UK Jancis Robinson evaluates wine out of a maximum of 20 and the best-selling magazine for wine buffs, *Decanter*, gives a more simple five star rating – five stars (outstanding), four stars (very good to excellent), three stars (good), two stars (fair) and one star (poor). Wine writers undoubtedly have developed the function of making core economy consumers aware (or not) of wines from 'new' producers in New World countries such as Chile. Colchagua Valley reds (see below) in general and Casa Lapostolle (advised by Michel Rolland) in particular do particularly well in the US market because of high ratings from Robert Parker.

The Creation of Producing Spaces in Chile

How has Chile reacted in defining its wine producing spaces for international consumers and their palate geographies? Wine producing districts can be defined at least at two scales – the geographical region as delimited by government legislation or by *terroir*, the local combination of characteristics of physical and human geographies (Moran 1993). The main legislation defining the nature of Chile's wine regions was passed in 1995 and was more flexible than the legislation imposed on producing districts in France and other Old World countries. For example, no region was restricted by the limited choice of grape variety. The 1995 Wine Law focused on giving a spatial framework to Chilean production that could be 'easily' understood in foreign markets – particularly by European Union Council decision-makers who needed to approve the legislation in order for Chilean exporters to maintain access to EU markets (which receive 55 percent of Chilean wine exports). The framework was based on internationally recognized spatial categories – region, sub-region, zone and area (place). All regions, sub-regions and zones became permitted EU regional descriptors.

The 1995 legislation introduced a spatial framework that reflected the power of key actors in the Chilean wine industry – the large wine companies and business associations. The spatial framework provided a distinctly uneven distribution of wine production. Five regions were created (see Figure 10.3) but just one region produced over 80 percent of all wine – what was termed the Central Valley.

The large Chilean companies (such as Concha y Toro) had lobbied for such flexibility as they owned land in the five east-west valleys crossing the 'longitudinal' Central Valley. The key reason behind such a regional classification in 1995 was that it allowed the large wine companies not only to blend wine from these five main producing valleys of Chile – Maipo, Rapel (divided into Cachapoal and Colchagua), Curico and Maule – but also to label it with a regional descriptor that could be readily understood in European and North American markets. The valley connotation was seen as important for international consumers – even though Dudley Moore represented the Central Valley as akin to a mountainous Andean valley rather than a long lowland basin in Tesco advertisements of the late 1990s.

Other parts of the legislation closely followed that which had been in place in Australia. First, if a grape variety were to be mentioned on the label, at least 85 percent of the grape input must be from that grape variety. Secondly, if a region (or sub-region, zone, area) is mentioned on the label, at least 85 percent of the grape must come from the relevant geographical indication.

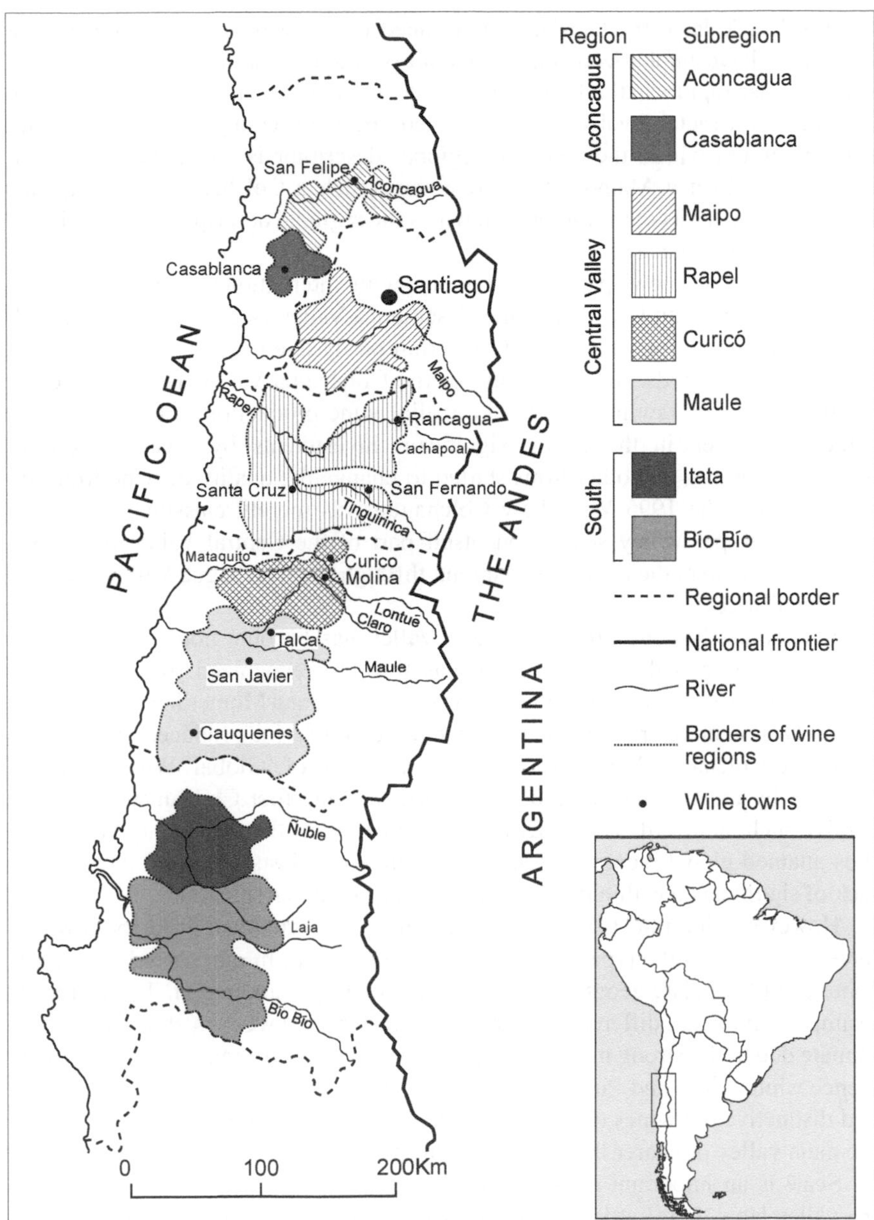

Figure 10.3 Chile's main wine regions

Over a decade later, the legislation is in need of renewal as most Chilean wine companies have emphasized enhancing quality and innovation since wine exports started to rise rapidly in the 1990s. The 'Central Valley' descriptor is now normally reserved for varietal-level wines where blending is necessary. For premium and super-premium wines, the favoured regional descriptor is either the sub-region (such as Casablanca, Maipo or Curico) or zone (such as Colchagua or Cachapoal). Meanwhile, in terms of ultra-premium wines the regional descriptor is often linked to the concept of place or *terroir*.

A number of 'new' producing regions with international recognition have appeared. One example would be the Casablanca Valley (see Figure 10.3), which has become noted for white wine production partly due to the impact of coastal fogs on the vineyards. Another would be the Colchagua Valley. Before the early 1990s, there were few investments in quality wine production in the Colchagua Valley. However, in the 1990s, significant investments by Casa Lapostolle, Montes, Montgras and other firms started to change the quality of wine from the valley. Under the 1995 Wine Law Colchagua was merely classified as a zone within the Rapel Valley sub-region, itself part of the Central Valley region (to complicate matters the main river running through the Colchagua Valley is called the Tinguiririca).

Since then, however, the Colchagua Valley has become noted as a Chilean producing space with high quality levels that are recognized by international retailers, wine writers and consumers. Casa Lapostolle and Montes have established themselves on price grounds as two of the top quality wine producers in Chile. In 2006, the Colchagua Valley was awarded the prize of 'Global' Wine Valley of the Year by the US magazine *Wine Enthusiast* – the first Chilean wine area to receive such an award. The geographical indication of the Colchagua Valley has thus attained global recognition amongst wine enthusiasts – despite its apparent lack of significance within the government's regional classification.

However, within the Colchagua Valley, there are significant differences which have an impact on the type and taste of wine produced. In other words, there are some complex palate geographies within this quality wine region. There are, for example, significant differences between the eastern and western areas in terms of climate due to the strong maritime influence on the western margins of the valley. Hence white wines and Pinot Noir are being produced nearer to the Pacific coast and distinctive red wines (Carmenère, Cabernet Sauvignon, Syrah and Merlot) in the main valley or nearer the Andean piedmont (Richards 2006).

Scale is an important consideration here. Most export-oriented producers in the valley label their bottles as coming from the Colchagua Valley – they see the value of the increasing global recognition for this geographical indication amongst a significant number of global wine consumers. However, given the flexibility of the Chilean Wine Law, there is no agency assessing and monitoring the quality of wine which has a Colchagua Valley geographical indication.

Meanwhile, within the valley some notable examples of *terroir* have been found, such as at the south-facing semi-circular escarpment and basin of Apalta

(Casa Lapostolle, Montes). Even at this scale, however, the link between *terroir*, grape variety and quality can be varied. Whereas the Syrah grape produces quality wine from vineyards recently established on the escarpment, ultra-premium Cabernet Sauvignon comes from long-established vineyards on the flatter slopes near the escarpment edge.

Thus the palate geographies of Chilean wine producing districts, such as the Colchagua Valley, can become very complex indeed, particularly when the issue of scale is actively addressed. The system of geographical indication introduced by the 1995 Wine Law needs to be drastically reformed in order to reflect the growth of major new regions – and at the local scale the creation of new *terroir* which can produce wines with distinctive taste. Nevertheless Chilean wine producing districts are still in the process of being created – as key actors such as vineyard managers, winemakers, distributors and supermarket buyers collaborate over the negotiated interface between the geographies of production and the geographies of consumption.

Trading Up Through Taste

The palate geographies of Chile's wine producers have evolved substantially since the growth of export-oriented production. A number of related processes have been at work. First of all, wine producers have changed from their former concentration on red wines for the domestic market. Whereas in the 1980s the balance of Chilean production was 80 percent red and 20 percent white, it has subsequently changed to a 60/40 percent split between red and white wine production (Cox 2005). This has largely been the result of the demands of international supermarket chains that require a more even 50/50 split between the supply of red and white wine, particularly in the lower price categories.

Secondly, Chilean producers have focused on improving both the quality and diversity of taste for overseas consumers. In terms of the former, there has been a notable process of increased vertical integration with most leading wineries buying up more and more land for new vineyard plantings over the past decade. This is allowing the wineries to source more and more of their wine grapes from vineyards over which they have complete quality control – reducing yields per vine, increasing density of planting, developing new forms of canopy management. Winemakers and vineyard managers thus liaise on the tastes required for overseas consumers. Hence, the wines produced from these new vineyards are normally destined for the higher categories of wines – premium, super-premium and ultra-premium.

As regards increasing diversity of taste, most producers have introduced a significant range of new grape varieties that did not exist when the industry was oriented to the domestic market in the 1980s. Most of Chile's historic varieties (such as Cabernet Sauvignon, Merlot, Sauvignon Blanc) came from the Bordeaux area. But Chilean producers have found success with 'new' grape varieties from

Burgundy (Chardonnay, Pinot Noir), the Rhone Valley (Syrah, Viognier), Germany (Riesling, Gewurztraminer) and neighbouring Argentina (Malbec) – and many others.

Chilean producers are also developing a distinctive 'palate' from a grape variety that (so far) cannot be produced at a similar quality and price anywhere else in the world. This is the 'resurrected' grape variety known as Carmenère. Carmenère was originally one of the mainstays of Bordeaux. But after the phylloxera disease caused massive destruction of vineyards in the Bordeaux area in the 1870s, the vineyard owners that survived the subsequent economic chaos did not replant with the Carmenère variety. It had developed a reputation of unreliable yields, late ripening and susceptibility to mildew and other diseases. Carmenère rootstock had, however, traveled to Chile in the 1850s along with other Bordeaux varieties. As Chile did not suffer from phylloxera, Carmenère vineyards continued to be planted – although it was called *Merlot Nacional* or *Merlot Chileno* – given some similarities to the Merlot variety. It was not until 1994 that the DNA of Carmenère was identified as being quite distinct from Merlot (Duijker 1999) by a French oenologist and Chile had found a 'resurrected' grape variety.

Carmenère is a vigorous vine that is also very late ripening. It is normally harvested in Chile in May (the equivalent of November in the northern hemisphere) – and two months later than Merlot. If the grape is harvested too early, it can give virulently green flavours – hence its lack of success when planted alongside Merlot in Chile before the discovery of its DNA. However, since 1994, vineyard management of the Carmenère grape has progressed remarkably with better harvest timing, low yields, good canopy management and the selection of appropriate (warm and sheltered) sites for planting. The better Carmenère wines now show 'ripe, black fruits with savoury notes of grilled red pepper, roasted herbs, tar and paprika, with a broad, smooth, and round palate' (Richards 2006, 23). Warm and sheltered sites within such valleys as Colchagua have thus produced a combination of tastes that was not previously known – a completely new 'palate' geography in other words as Carmenère is tasted by wine consumers.

Through investments in new vineyards and new winemaking technologies, and through greater awareness of the changing tastes of overseas consumers, the majority of Chilean wine producers have been successful in 'trading up' or steadily improving the quality, price and taste of their final product through time. Trading up is a process by which wine producers start producing for buyers at the low end of the market and then strategically upgrade the quality of their products and target more sophisticated market segments (with similar or different buyers).

The Chilean producer, Montgras, provides a clear case study of trading-up and involved three distinctive stages, at least in terms of UK consumers and supermarkets (Middleton 2005). Between 1992 and 1998 the emphasis was on supplying wine for own-label brands for such supermarkets as Tesco, Sainsburys and Waitrose. Between 1998 and end-2004, the strategy was to improve quality and concentrate on massive own brand (Montgras) promotions in order to establish the brand in the market place. It developed an innovative and potentially high-risk strategy with one UK Supermarket in particular, Sainsbury's. The strategy was to become involved in Sainsbury's 'killer' Christmas promotions in which prices are reported to be halved on the supermarket shelf. Montgras engaged in four such promotions from 2001 to 2004. During 2004 this Sainsbury's promotion was alone responsible for nearly 50 percent of sales – 250,000 cases (divided equally between Merlot and Chardonnay) out of a total production figure of 600,000 cases.

By early 2005 the decision was taken to improve quality even more by mainly using grapes from Montgras vineyards. Own-brand promotions were reduced and sales to supermarkets (now including Waitrose) focused more on premium level wines. Within a decade, Montgras had moved from bulk wine producer to a winemaker increasingly noted for premium wines.

The record of trading up in Chile may reflect a special case in terms of agricultural and agro-industrial producers from developing countries as they forge links with supermarkets from core country economies and start to export. In some ways it is in distinct contrast to many agricultural producers in Africa who have had to 'trade down' in the sense that the contracted prices from the buying supermarkets are continually squeezed and stricter conditions placed on them (Gibbon and Ponte 2005). Why might Chilean wine be a special case? One point is that wine can be a product with considerable value-added and hence more of a manufactured than an agricultural product. The knowledge transfer between producers and consumers effectively operates to expand the range of palates and differentiate between them. Tastes are being produced which are valued more highly by distant consumers – and who are willing to pay the higher prices for them. It is fitting to point out that the record of Chilean table grape production since the early 1990s equates more with the 'trading down' principle (Gwynne 2003).

Conclusions: Palate Geographies and the Chilean Case

This chapter has provided an early exposition of the theme of palate geographies. Through the medium of taste it has tried to make links between methodologies of global value chains and agri-food research based on material cultures. More specifically it has linked spaces of production and consumption with wine which has such an important component of *terroir* in constructing the taste, quality, cost and 'knowledge' of wine as commodity. Through using the notion of the palate, I have attempted to analyse the negotiated interface between wine producers in the New World (as they actively interpret the evolving palates of distant markets and consumers) and consumers (as their tastes and wine drinking efforts are catered for or managed by the increasingly powerful retail chains which have evolved in most markets).

Another aim of the chapter has been to make links between Chilean wine producing districts and UK consuming spaces by reference to the actors involved in the value chain that links these spaces together. Chilean wine firms that have been successful (such as LFE or Montes) have depended on knowledge transfer in each chain being interactive – with winemakers listening to distributors and retail buyers (who themselves are interacting with the tastes and palate geographies of consumers) in order to adapt their future planting and winemaking strategies to what they perceive to be the tastes of distant consuming spaces. The wine buyers of the large supermarkets and specialist retail chains may be assigned the role of key actors in the chain. In some ways they have become the arbiters and mediators of taste between the UK consumer and the Chilean wine producer.

Retail concentration is important to note here; the wine buyers of the ten largest supermarkets and specialist retail chains are responsible for around 80 percent of all Chilean wine entering the UK off-trade. In the Chile/UK wine value chain there is also a diffusion of power between producers and buyers. Buyers do not control the relevant technologies and have neither the local knowledges of local *terroir* and vineyard managers nor the more global knowledges of winemaking techniques through flying winemakers (Lagendijk 2004). They therefore rely on Chilean wine producers for these knowledges, and, as a result, there can be significant collaboration and cooperation through the value chain.

There are some interesting policy implications which emerge from the Chilean case study. We have already noted the contrast between the 'trading up' scenario of Chilean wine producers in terms of how their relationships with supermarkets have evolved and the more common pattern of 'trading down'. However, it would be interesting to investigate which actors appropriate the gains from trading up.

The Chilean case shows that export-oriented wine producers must be concerned about and be able to adapt to the knowledges about taste transmitted to them by such key value chain actors as supermarket buyers. This could be of particular relevance to emerging wine countries in south-eastern Europe, such as Bulgaria. In 1996, Bulgaria had more of the UK wine market than Chile. However, by 2005, Chile's share of the UK wine market had risen to 7.2 percent and Bulgaria's had plummeted from 4.5 to a mere 0.3 percent (IWSR 2007). The emphasis in Bulgaria had been on continuing to produce cheap wines. In contrast, Chilean producers were active in creating wines with distinctive tastes for UK (and other) consumers. As Balkan countries begin to focus again on the potential of their wine industries for export growth, the Chilean model is instructive. The challenge of developing knowledges of consumer tastes in overseas markets and creating wines that appeal to those palates should be a key concern for export-oriented wine producers.

This may be one reason why the wine value chain between Chile and the UK could be seen as a benign escalator unlike the footwear sector (Schmitz and Knorringa 2000). There is considerable evidence that UK supermarkets and specialist retail chains not only act as the lead firms in the chain but also that they facilitate knowledge transfer rather than obstruct it up and down through the chain. Hence by Chilean wine firms becoming increasingly aware of the palate geographies of the UK consumer, Chilean wine has become more and more successful within the UK – despite the significant physical (and cultural) differences involved. Nevertheless in the end the critical relationship of taste and palate must be between the winemaker at one end of the world and the consumer at the other. As Rafael Urrejola the winemaker of Viña Leyda states: 'I'm trying to make wines that invite you to have another glass' (Richards 2006, 32).

Acknowledgements

I would like to thank the British Academy for funding for this research, my former colleague Ian Cook for early discussions about palate geographies, Michael Cox for his generous help in providing data and contacts in the Chilean wine sector and the editors for their incisive comments on an earlier draft.

References

Barham, E. (2003), 'Translating Terroir: The Global Challenge of French AOC Labeling', *Journal of Rural Studies* 19: 1, 127–138.

Bisson, L. F., Waterhouse, A. L., Ebeler, S. E., Walker, M. A. and Lapsley, J. T. (2002), 'The Present and Future of the International Wine Industry', *Nature* 418, 696–99.

Chaney, I. M. (2004). 'Own-Label in the UK Grocery Market', *International Journal of Wine Marketing* 16: 3, 5–9.

Cook, I. (2004), 'Follow the Thing: Papaya', *Antipode*, 36: 4, 642–64.

Cook, I., Crang, P. and Thorpe, M. (1998), 'Biographies and Geographies: Consumer Understandings of the Origins of Foods', *British Food Journal* 100: 3, 162–167.

Cook, I., Crang P. and Thorpe, M. (2000), '"Have you got the Customer's Permission?" Category Management and Circuits of Knowledge in the UK Food Business' in Bryson, J., Daniels, P., Henry, N. and Pollard, J. (eds), *Knowledge, Space, Economy* (pp. 242–260) (London: Routledge).

Cook, I. et al. (2006), 'Geographies of Food: Following', *Progress in Human Geography* 30: 5, 655–66.

Cox, M. (2005), Interview with UK Director, Wines of Chile, 31 March.

Davis, J. (2006), Interview with Chile Wine Buying Manager for Tesco, 28 November.

Duijker, H. (1999), *The Wines of Chile* (Utrecht: Spectrum).

Edwards, L. F. (2005), Interview with Managing Director, Luis Felipe Edwards, Santiago, 1 September.

Eguiguren, P. (2006), Interview with Managing Director of Casa Lapostolle, Colchagua Valley, 4 August.

Eusabiaga, J. (2005), Interview with Winemaker of Luis Felipe Edwards, Nancagua, Colchagua Valley, 8 September.

Fanet, J. (2004), *Great Wine Terroirs* (Berkeley: University of California Press).

Friedland, W. H. (2005), 'Commodity Systems: Forward to Comparative Analysis' in Fold, N. and Pritchard, B. (eds), *Cross-continental Agro-food Chains* (pp. 25–38) (London: Taylor and Francis).
Gade, D. W. (2004), 'Tradition, Territory and Terroir in French Viticulture: Cassis, France, and Appelation Controlee', *Annals of the Association of American Geographers* 94: 4, 848–67.
Gereffi, G. (1994), 'The Organization of Buyer-driven Global Commodity Chains: How US Retailers Shape Overseas Production Networks' in Gereffi, G. and Korzeniewicz, M. (eds), *Commodity Chains and Global Capitalism* (pp. 67–92) (Westport: Praeger).
Gereffi, G. (1999), 'International Trade and Industrial Upgrading in the Apparel Commodity Chain', *Journal of International Economics* 48: 37–76.
Gibbon, P. and Ponte, S. (2005), *Trading Down: Africa, Value Chains and the Global Economy* (Philadelphia: Temple University Press).
Guthey, G. (2004), 'Wine, *Terroir* and Agro-industry in Napa and Sonoma Counties', paper presented at the Annual Conference of the Association of American Geographers, Philadelphia, 17 March.
Gwynne, R. N. (2003), 'Transnational Capitalism and Local Transformation in Chile', *Tijdschrift voor Economische en Sociale Geografie* 94: 310–21.
Halstead, R. (2006), 'Understanding the Wine Drinking Moment', paper presented to the XXIX World Congress of Wine and Vine, Logroño, Spain, 27 June.
Hayward, D. and Lewis, N. (2008), 'Regional Dynamics in the Globalizing Wine Industry: The Case of Marlborough, New Zealand', *The Geographical Journal* 2: 124–137.
Humphrey, J. (2005), 'Shaping Value Chains for Development: Global Value Chains for Development', Report commissioned by Germany's Federal Ministry for Economic Cooperation and Development.
Humphrey, J. (2006), 'Policy Implications of Trends in Agribusiness Value Chains', *The European Journal of Development Research* 18: 4, 572–592.
Humphrey, J. and Schmitz, H. (2002), 'How Does Insertion in Global Value Chains Affect Upgrading in Industrial Clusters?', *Regional Studies* 36: 9, 1017–1027.
International Wine and Spirit Record (IWSR). (2007), 'The UK Wine Market Turns into a Share Game', *Drinks Record* May, 19–20.
Lagendijk, A. (2004), 'Global "Lifeworlds" Versus Local "Systemworlds": How Flying Winemakers produce Global Wines in Interconnected Locales', *Tijdschrift voor Economische en Sociale Geografie* 95: 5, 511–526.
Middleton, P. (2005), Interview with Managing Director, Montgras, Colchagua Valley, 25 August.
Moran, W. (1993), 'The Wine Appellation as Territory in France and California', *Annals of the Association of American Geographers* 83: 4, 694–717.
Moran, W. (2001), '*Terroir* – The Human Factor', paper presented at the Pinot Noir New Zealand Conference, 25–28 January.
Pym, M. (2006), Interview with Chile Wine Buyer for Majestic, 23 May.

Rachman, G. (1999), 'The Globe in a Glass: Survey of Wine', *Economist* 18 December, 107–123.
Richards, P. (2006), *The Wines of Chile* (London: Mitchell Beazley).
Robinson, J. (ed.) (2006), *The Oxford Companion to Wine*, 3rd Edition (Oxford: Oxford University Press).
Room, N. (2006), Interview with Chile Wine Buying Manager for Waitrose, 28 November.
Salvestrini, H. (2005), Interview with Market Manager of Undurraga, Santiago, 11 August.
Sanchez Hernandez, J. L. (2004), 'La Liberalización Controlada del Mercado del Vino en Noruega (1996–2002)', *Investigaciones Geograficas* 35, 85–102.
Schmitz, H. and Knorringa, P. (2000), 'Learning from Global Buyers', *Journal of Development Studies* 37: 2, 177–205.
Spawton, T. (1991), 'Wine and the Marketing Mix', *European Journal of Marketing* 25: 3, 19–32.
Stevenson, T. (2005), *The Sotheby's Wine Encyclopedia*, 4th Edition (London: Dorling Kindersley).
Wilson, J. E. (1998), *Terroir* (Berkeley: University of California Press).
Wines of Chile. (2007), 'Market Statistics Summary 2003–2007 and Press Release' (Ascot).
Wine Intelligence. (2004), 'Wines of Chile: Measuring Purchasing Behaviour and Attitudes to Chilean Wines', 8 December.

Chapter 11

Consuming Burmese Teak: Anatomy of a Violent Luxury Resource

Raymond L. Bryant

The history of the development of the teak industry is a record of the creation of a luxury resource steeped in chronic and unspeakable violence. Notably centred on Burma (re-labelled Myanmar by the country's military rulers) where most of the world's teak supplies are located, teak production has long been a Jekyll and Hyde tale of murder and marketing, of distinction and extinction, all masked behind a carefully crafted image of the mystery and romance of 'green gold'. It is a tale where the production of consumption has been inextricably linked to the consumption of production – or, at least, a mythical version of that production.

This chapter examines selected themes in this process focusing on a Burmese teak trail littered over time with policies and practices involving brutal production, coercive conservation and seductive consumption. It is only through such an intertwined approach to the history of this commodity that the anatomy of a violent luxury good can be properly understood. Indeed, to understand fully the social life of teak (or other commodities for that matter), there is a need to explore spaces of production and consumption as inseparably tied together through often deeply exploitative histories (colonial and post-colonial) that notably include violence, as well as contested perceptions of use and exchange value, luxury status plus aesthetic beauty. In the process, teak is linked to romanticised place – all the better to *simultaneously* create a mystique surrounding the consumption of this increasingly rare 'king of the timbers' even as harsh production conditions are thereby swept from view. The aim in this chapter therefore is to provide a 'rich thick description' of a distinctive and historically situated empirical case – that of Burmese teak – as part of this book's wider effort to probe some of the ambiguities and tensions that characterize spaces of consumption and production in an era of acute political, economic, cultural and ecological trade-offs.

The intersection of spaces of consumption and production over teak has indeed been a process ripe with ambiguity and tension in Burma – one of Southeast Asia's less well-known yet historically important countries. In fact, it is no exaggeration to say that Burmese history has long been closely associated with the teak tree. Identification of the country as 'the land of teak' was reinforced in numerous fiction and non-fiction accounts. The elaboration of a large-scale export-oriented timber industry in the nineteenth and twentieth centuries demonstrated the economic importance of this resource. As such, teak was popularized and romanticized even as its commercial value was realized (Brandis 1888; Kelly Talbot 1912; Stebbing 1922–1926).

Political interest centered on controlling teak forests and their inhabitants. Burma's rulers – pre-colonial monarchs, British officials, post-colonial civilian and military elites – have all grappled with this problem, even as forest residents – shifting cultivators, villagers, timber traders – have sought to evade central control. In short, teak has been a perennial focus of struggle (Bryant 1997; Global Witness 2003).

Indeed, as often happens where valuable natural resources are involved, teak exploitation in Burma has been associated with the disruption if not devastation of the livelihoods of the rural poor living in or near by the teak forests. True, some forest residents earned an income in the timber industry as they felled, processed and transported teak to market. Others were involved in shipbuilding as well as the 'sustainable' management of teak forest – Karen shifting cultivators being a case in point. And yet, it would be wrong to exaggerate the importance of these benefits given the overall impact of teak exploitation on resident populations. As powerful Burmese and non-Burmese elites came to prize teakwood, a series of political, military and economic measures were taken that served to violently disrupt pre-existing local 'life-worlds' – comprising not only material welfare but also cultural and spiritual attachments to place. What for some was 'green gold' became for others 'blood timber' in a process that exposed the inescapably contested nature of this natural resource.

Anatomy of Production

To describe Burmese teak as blood timber – that is, as a resource whose record of exploitation can be viewed as a *bad* thing – is clearly to go against numerous published historical accounts extolling the virtues of its exploitation. These accounts – written by and/or for elites – present one particular version of history that is 'factual' in tone, partial in scope, and de-politicized in presentation. In all their diversity, they present a narrative of teak production and the context of production that laid the basis for the transformation of teak into a premiere world luxury wood and its consumption.

Whether in the dry bureaucratic form of government officials or the florid style of travel writers, 'progress' in Burma's forests was often about mapping, extracting, planting or simply experiencing the majestic teak forests (e.g. Kelly Talbot 1912; Nisbet 1901; Stebbing 1922–1926). Thus, the colonial travel writer, R. Kelly Talbot (1912, 101), gushed:

> Surrounded by scenes of the supremest beauty, each day's ride seemed more beautiful than the last, a gradual crescendo of loveliness which only increased as familiarity aided appreciation, and of which no words of mine could ever give an

> adequate impression. I had never anticipated anything so completely fascinating as these Burmese forests proved to be.

Yet, this discourse of progress served to romanticize forest intervention that, when viewed from a different angle, was all about the bloody production of a timber and its subsequent ethical cleansing. The production of Burmese teak has thus been inescapably associated with the production of a particular moral geography that has been deeply politicized.

It is not difficult to see why both Burmese and British elites were enamored with the teak tree. As a durable, attractive and long-lasting hardwood resistant to the predation of many wood-boring insects, teak (*Tectona grandis*) was used in the construction industry, notably as the preferred source for the homes of pre-colonial Burmese royalty and nobility. Significantly, teak was also the preferred timber of a pre-colonial shipbuilding industry based in southern Burma from at least the seventeenth century. This was so because it held up well under prolonged exposure to the corrosive effects of seawater as well as containing an essential oil which prevented metal corrosion (Brandis 1888; Nisbet 1901).

As the reputation of teak grew, pre-colonial Burma became the focus of a thriving export trade in the eighteenth and nineteenth centuries. Burmese shipbuilders constructed a variety of ships from small coastal boats to larger square-rigged vessels based on the European model. Between 1786 and 1824 alone, 111 European-style vessels were built in Rangoon with an aggregate tonnage of 35,000 (Pearn 1939, 71; Lieberman 1984, 119–20). Labourers flocked south from elsewhere in the country to become part of this economic boom attracted by the prospect of relatively well-paid semi-skilled work (Htin Aung 1968, 10–11). A complex if fluid regulatory system was in place that was designed to tax teak production and control labour in the teak forests, along the major transport routes and at the site of shipbuilding itself (Koenig 1990; Taylor 1987). As a valued timber, there was a royal monopoly on teak from at least the eighteenth century that was enforced in the forests by specially appointed guards empowered to fine or arrest anyone involved in its illegal extraction. Such extraction persisted given both the high value of teak and the relatively weak administrative capacity of the pre-colonial Burmese State (Bryant 1997, 39–41). And yet, however small scale, it was in late pre-colonial Burma that a political economy of teak was crafted that proved remarkably durable and which was based on great commercial value of the product, export-oriented production and political coercion.

Teak was thus influencing Burmese history even before the nineteenth century intervention of the British in that country. Indeed, that intervention was itself shaped by a growing international awareness of 'the land of teak'. By the early nineteenth century, the international identification of Burma with teak was ever more pronounced thereby setting the stage for a fierce geopolitics of teak. There were diverse aspects to this process. Thus, teak was becoming a sought after resource by the European powers on military grounds. Imperial shipbuilders had discovered what their Burmese counterparts already knew – namely, that teak was an ideal timber for the construction of warships. Further, at a time of heightened Anglo-French conflict, the British were desperately searching for new timber supplies for their navy as oak forests were depleted in Europe and limited teak forests were felled in southern India (Ribbentrop 1900, 64–66). Finally, eyewitness accounts by foreigners reported that Burma was home to the largest teak forests in the world. The actual extent and quality of those forests would only be known following the British conquest of Burma (between 1824 and 1886) once the Burma forest department was able to complete a systematic survey (Bryant 1997). Yet, if anything, feverish expectations of 'limitless' Burmese teak forests in the early to mid-nineteenth century only hastened the process whereby Burma became identified as the home of teak.

From the mid-nineteenth century to the Japanese invasion of Burma in early 1942, the consolidation of British control went hand in hand with the elaboration of the world's leading export-oriented teak industry. As such, teak became a prime *imperial* resource with a political-economic context defined by four elements. First, a specially created forest department facilitated a rise in teak production as forests across British-ruled Burma were systematically mapped and exploited. Whereas 20,462 tons were produced in 1859 (albeit, from British-ruled Lower Burma only) by 1900 the figure was 205,000 tons while in the early 1920s it exceeded 513,000 tons. While production levels fluctuated (notably due to shifting market conditions), the basic situation was nonetheless a considerable increase in teak production as more and more of the national teak forest was incorporated into the realm of market activity. From a British viewpoint, teak was the main forest crop with the corollary of that being that other timber and non-timber species were 'minor forest products'.

Second, virtually all teak extracted was destined for export. A classic case of export-oriented resource production, teak was shipped to markets in British-ruled India as well as in Europe where it was converted into everything from railway sleepers to park benches, warships to office paneling (Brandis 1888, 104; see below). Burmese teak dominated the world market. Just prior to the Second World War, for example, Burma provided 85 per cent of world teak exports (Gallant 1957, 2).

Third, foreign firms dominated the Burma teak industry especially after 1900 as forests were 'privatized'. Led by the mighty Bombay Burmah Trading Corporation Limited (BBTCL), five European firms controlled the key teak forests while much smaller Burmese firms were left to fight over the residual tracts. Indeed, between 1904 and 1924, outturn by Burmese firms fell from 23 per cent of total production to under 5 per cent whereas outturn by European firms climbed from just under 44 per cent of total production to more than 74 per cent (Bryant 1997, 103). In a pattern familiar in many parts of the South today, foreign firms dominated this key natural resource sector at the expense of smaller local rivals – thereby generating local resentment. The substantial profits generated by these firms were repatriated to Europe and when global market conditions turned sour, as they did in the early

1930s, the colonial state cut their royalty rates by 30 per cent – a move not extended to Burmese firms in the industry at that time.

Finally, official revenue earned from the lucrative teak industry was used to sustain the British Indian colonial administration (of which Burma was a part) with a substantial proportion of that revenue being siphoned off to support the overall running of that administration. Indeed, in the early 1890s, the annual net revenue from Burma's forests (virtually all teak-related) amounted to 45 per cent of the total for all of British India, up from 39 per cent in 1870 (Bryant 1997, 57). Burma's teak forests thus helped to sustain not only the colonial administration in Burma itself but also imperial endeavors in British India as a whole. If colonial rule needed to be a 'paying proposition' then teak played a notable role in the propagation of the British Empire. Relatively little of this revenue was devoted to the improvement of the lives of the conquered Burmese. Indeed, even where benefits occurred – for example, as a result of the construction of a railway network that facilitated the movement of people and goods – there is nonetheless a need to relate benefits to costs. To take the railway example, it is important to recognize its role in enabling a far more effective and systematic suppression of internal dissent than was hitherto possible.

Independence did not end Burma's close association with teak. To the contrary, it marked the start of an increasingly bitter chapter in that history. Teak has remained a sought after commodity destined for export that political and economic elites struggle to control at the expense of many Burmese. As in pre-colonial and colonial times, post-colonial teak production generates enormous profits that are not reinvested in the economic improvement of the country but are rather used to improve the lives of those who control the Burmese State as well as their political and economic allies.

From *Raison d'etat* to Reason to Shop

So much for how teak was linked to shifting political economies of timber production in Burma – but what of the consumption patterns that spurred on this process? Far from being simply a matter of 'natural' market activity, the consumption of teak was profoundly shaped by military and political concerns that notably reflected the vicissitudes of empire. Burmese teak played a role indeed in the pomp and (bloody) circumstance of British imperialism.

As noted, strong British interest in Burma in the nineteenth century – and certainly a key factor in the decision to conquer that country – related to the need to source new supplies of durable wood for the British navy. By the beginning of the nineteenth century, that old mainstay of naval construction, oak, was in clear decline – especially quality oak timber that met the demanding and highly specific standards of the British Admiralty. At the same time, the superiority of teak over oak in naval construction was becoming ever more apparent: teak was more resistant to marine borers such as the dread mollusc (*teredo navalis*) even as its natural oils (unlike tannic acid in oak sap) preserved metals in sea water meaning that ships could be built using iron spikes (on naval shipbuilding in British India, see Lambert 1996).

Throughout the nineteenth century and even into the early twentieth century, Burmese teak played a leading role in the production of quality warships for the British Admiralty even as it also found much use in the commercial arena. Given the demanding requirements of naval shipbuilding, only best quality teak referred to as Europe or English first class quality grade was used – with exports to UK shipyards led by the Clyde and the Tyne and distribution centres led by London varying sharply from year to year, but broadly increasing over time. Thus, for example, teak imports to the UK (mostly from Burma) in 1846 amounted to 8,712 tons, rising to 25,112 tons in 1860, before climbing in 1883 to 45,539 tons (Simmonds 1885, 353–54). In 1900 that figure was 63,598 tons before falling back slightly in the years thereafter (Andrews 1931, 31).

The general upward trend in teak imports to the UK reflected shifts in how this expensive timber was used in naval construction. Thus, on the one hand, a steady deterioration in the quality of teak supplies (as prime trees were soon felled) as well as the substitution of iron for teak sidings meant that the quantity of teak used in any one ship fell greatly during the century (Albion 1926). By the late nineteenth century, gone were the days when a military or merchant vessel was made largely of teak – as was the case with the famous tea clipper the 'Cutty Sark' which was built on the Clyde in 1869 (and which was a popular tourist attraction in London until its devastation by fire in 2007). On the other hand, many more ships were being built for the British navy (and the other major world powers) as an imperial arms race got under way in the decades leading up to the First World War even as more and more teak was being used in the construction of merchant and passenger ships. By the 1920s, modern shipbuilding technologies as well as a smaller peacetime fleet meant that the British Admiralty's teak requirements 'henceforth ceased to be the direct and active factor in the markets in the UK' that it had previously been (Andrews 1931, 24).

During the nineteenth century, Burmese teak of inferior quality was sent to India where it replaced local teak that had been all but exhausted by that time. Here, teak was popular for a wide variety of purposes including use in the general building trade (houses, temples, etc.), public works (official buildings), the railways (carriages, sleepers or railway ties, bridges) and local shipbuilding in Bombay and Calcutta. As time went by and the price of teak climbed, teak was used in India mainly for higher value shipbuilding and public works with substitutes found for other prior uses such as the substitution of non-teak hardwoods for teak in railway sleepers as well as steel for window frames (Hopwood 1935). Still, the size of the expanding Indian market meant that throughout this period, most of the teak produced in Burma was consumed in India.

By the time of the First World War, teak was well established as one of the most valuable timbers that the world had ever known. If its world renown was still closely linked to its supreme utility in the shipbuilding trade, the properties that made teak a wood that was a 'cut above the rest' meant that it was also well placed to expand into new markets in the non-military and non-maritime sector. In the prospering imperial heartland, teak was to find a new role as a marker of social distinction in the homes and bureaus of the affluent.

It is not really surprising that teak moved from being a piece of essential military hardware to a status where it was a much sought after luxury good. Four factors can be briefly noted here. First, there was the 'demonstration effect' as imperial elites became aware of the beauty and durability of teakwood from having seen it in use in colonial offices, railway carriages, and on board passenger vessels. That teak was literally embedded in key symbols of imperial modernity (the railway, the passenger liner) only added to its allure. Both of the world's two largest liners at the start of the Second World War, the Queen Mary and the Queen Elizabeth, for example, contained about 1,000 tons of teak used notably for decking, gangways, handrails and window frames (Gardiner 1942, 749). Second, there was the way in which the prospering British middle classes more and more came to value highly quality possessions as markers of moral living, 'good' taste and social 'respectability'. As Deborah Cohen (2006: xv) notes, middle-class consumerism gathered pace as the British became obsessed with shopping for household goods (including luxury furniture and other wood fittings) as part of a wider process of 'middle-class self-fashioning' in keeping with modern 'taste-full' living. Third, there was the close association of Burmese teak with a sense of the 'exotic' as well as adventure – replayed in journalist accounts (e.g. Geary 1886), imperial popular 'guides' (e.g. Ferrars and Ferrars 1900; Kelly Talbot 1912), and memoirs written by those who had served in Burma (e.g. Colonel Williams in his popular book *Elephant Bill* first published in 1950). To consume teak was thus to 'imbibe' a 'glorious' history of imperial adventure. Finally, there was sheer practicality of teak as a hard wearing and long-lasting timber that also displayed little shrinkage in drying and very little subsequent movement in variable climates. These properties meant that it was ideal, for example, as a piece of garden furniture or even as decking and fittings for private yachts – where its virtue for 'staying put in ship's decking, windows and doors, furniture and fittings exposed to sun and rain' (Scott 1945, 82) was unbeatable.

Teak first made its social mark in the tropics where its appeal was simultaneously one of great luxury and no nonsense practicality. Not only were supplies relatively close at hand (at least in South and Southeast Asia) but the wood was one of the few construction timbers that was immune to the attack of white ants (Scott 1945). Just as teak was used in public buildings and religious temples, so too was it commonly used in the exteriors and interiors of the bungalows that were the home of many a colonial official: window frames, verandas, roof trusses and shingles, doors, panelling, flooring and furniture. Similarly, the social world where colonial and sometimes local elites mingled was often literally framed by this wood: in the case of the Royal Bombay Yacht Club, for instance, the ceiling, flooring, panelling, staircase and tables were all made from teak (Gardiner 1942, 737). Writing about Burma where he had served in the Imperial Police between 1922 and 1928, George

Orwell described in his first novel, *Burmese Days* (written in 1934) the European Club as 'a teak-walled place smelling of earth-oil' – as a ubiquitous 'institution' in British India that was 'the spiritual citadel, the real seat of the British power' (1987, 17–18).

Where the Empire led, the imperial heartland soon followed. Teak panelled and floored offices became a hallmark of quality and success in the business world – designed to impress both clients and partners. Even in America, where there were still supplies of local hardwood trees, the Ford Institute Museum in Detroit Michigan completed just prior to the Second World War was decked out entirely in teak – covering an area of 340,000 square feet or about 8 acres, this was the largest single expanse of teak flooring in the world at that time (Gardiner 1942). The affluent middle classes in Europe (and to a lesser extent in North America) also acquired a taste for teak – with flooring, panelling, stairways, window seats, doors, fire places, as well as garden furniture and green-houses but a few examples of the spread of teak into the household (Morehead 1944). At a more modest level, teak ornaments added an exotic touch to the modern imperial living room – at the 1938 Glasgow world exhibition, 'many thousands of hand-carved teak elephants, carved ash trays, flower bowls, labelled 'genuine Burmese teak', have been eagerly purchased by the public' (Myat Tun 1938, 808). In the process, more and more people came to possess 'a little bit of empire' in their own homes.

The organized bloodbath that was the Second World War interrupted the steady march of teak into the boardrooms and homes of the imperial heartland. Indeed, Burma ('the land of teak') was the focus of ferocious fighting during the war as the country's teak and other forests became widespread killing fields. As the Second World War gave way to a prolonged and bitter civil war, Burmese teak only re-entered the international market in dribs and drabs – and only then under the barrel of a gun (see below). Postcolonial import restrictions in South Asia (where much of the teak had hitherto gone) limited teak exports from Burma in the 1950s, as did the slow pace of recovery in the UK and in Western Europe. Conversely, as traditional markets began to reopen and expand in the 1960s and 1970s, Burmese supplies were limited by ongoing bloodshed in the country as well as the disastrous autarchic economic policies that embodied the 'Burmese Way to Socialism' (Steinberg 1981; Bryant 1997).

By the late 1980s and early 1990s, teak was ready to resume its re-conquest of the luxury timber market. Political and economic conditions were propitious. On the supply side, the Burmese army (or *tatmadaw*) had largely cleared the country of insurgents – at a bloody cost. Closer links with China meanwhile provided ideal business arrangements for the export and processing of teak (and other hardwoods) for the world market, thereby neatly bypassing Northern NGO-backed attempts to boycott Burma's brutal military regime (Global Witness 2003). On the demand side, newly prosperous middle classes in the North (and increasingly in the South too) 'rediscovered' the aesthetic beauty and practicality of teak – above all in the garden and on water. Thus, imports of Burmese teak have climbed rapidly – in recent years, to an estimated $37 million per year to the US alone (and worth $100 million per year in consumer purchases there) (EarthRights International 2005, 2). In a matter of only a few years it seemed, 'outdoor living' was predicated on the teak dining set, garden bench, sun lounger and even teak deck. In the process, teak once more became an essential part of modern elegant living. As one UK-based firm (Gloster n.d., 1) puts it:

> Choose a space under the open sky and make it your own. In a hectic world, Gloster lets you create an environment that is not confined by walls, but defined by a sense of personal space, an oasis of peace, relaxation and freedom. View the outside as an extension of your home, an expression of your individual style, every bit as important as any other room in the house, and then furnish it with beautiful things. Choose from teak, metal, sling or woven furniture in a variety of styles, from the traditional to the contemporary. And remember, outside is a far tougher environment than inside, so accept nothing but the best. (Gloster)

In Singapore, the middle classes were also being exhorted to accept 'nothing but the best' as they were encouraged to buy teak furniture that would enable them to 'enjoy resort living in [their] own home' (OriginAsia n.d).

For those with even more money, as well as a maritime disposition, the yachting world is awash with teak – for those who hanker for stylish onboard living. As in colonial times, yachts are today kitted out with teak fittings, doors, decks and trimming – supplied by specialist suppliers such as Jamestown Distributors and East Teak Trading Group (both in the US) and Teak Decking Limited in the UK. And yet, unlike in those times, the yachting industry is a much bigger phenomenon than before as yachts (and power boats) become as much a part of the lives of the affluent as golf memberships and second (third, etc.) homes. At the same time, the size of yachts keeps getting bigger and bigger – with the rich and famous at the top end of the market seeking to outdo one another in the race to have the biggest super yachts. In the process, teak remains a favourite because, as the website of Teak Decking Ltd (UK) points out, it will 'provide unmistakable lasting beauty, adding value and character to any boat'. In all of this, discussion over the relative merits of using teak for decking, cockpits, trims, and so on is widespread in the yachting industry. And yet this is a wholly 'technical' and aesthetic conversation oblivious to the 'dark side of teak'. Whether based on wilful forgetfulness or blithe obliviousness, this attitude is undoubtedly part-and-parcel of the modern romanticization of teak – all the better to enhance a consumption experience that revolves around feeling and looking good.

A River of Blood Runs Through It

Yet even as teak wood contributed to imperial grandeur and post-imperial fine living by providing a marker of distinction for both the already well-to-do and the socially up-and-coming, it was also the focus of vicious strife in the forests from whence it came. Neither a rip-roaring boy's own adventure nor a romantic 'crescendo of loveliness' (Kelly Talbot 1912, 101), Burmese teak extraction has been a brutal and tawdry tale of state repression, local displacement, popular fear and loathing, and out-and-out murder.

How teak geopolitics has led to battles for control of Burma's teak forests is central to the anatomy of a 'blood timber'. Resource militarization is long associated with the quest for teak through a process that has produced a 'violent environment' in which many lives have been blighted (cf. Peluso and Watts 2001). True, forest violence is not only associated with the quest for teak – other natural resources have been at stake (such as tin and jade). Further, violence has also been motivated by *non*-resource-related objectives, notably the crushing of ethnic opposition to the Burmese State (Smith 1999). Still, the production of violent environments has been part-and-parcel of the extraction of teak in both colonial and post-colonial times.

Teak forests have been notable hotspots in the long-standing production of violent environments. Indeed, they have long been home to individuals and groups opposed to the country's lowland rulers – 'a traditional hiding-place for malcontents' (Foucar 1956: 72). Historical and contemporary accounts abound with descriptions of 'bandits', 'banditry' and pitched battles in the forest as weaker political opponents have retreated to areas into which powerful political groups were at a tactical disadvantage due to the terrain (Mills 1979; Adas 1982). A case in point occurred in the early 1930s – the Hsaya San rebellion was centered on the teak-bearing Pegu Yoma and took the British Indian army several years to quell (Maung Maung 1976, 187–88). In other cases, 'banditry' has been of a more 'home-grown' variety, as residents mobilized against outsiders intent on disrupting local livelihoods – the Burma forest department was often a focal point of such resistance. Clearly, the circumstances and protagonists varied over space and time. However, the capture of teak logs in the forest was a favored practice. As forests were transformed into 'bandit country', though, residents almost inevitably became ensnared in fighting even though many were non-combatants. Violence and uncertainty was the norm as teak logs were seized in an opportunistic fashion.

Teak forests were also the focus of much more systematic and far-reaching military strategizing by well-organized insurgent armies intent on capturing an important source of revenue. The most notable case concerned the long-running insurgency by Karen and other ethnic minorities against the Burman-controlled state which began soon after independence was attained in 1948 (Smith 1999; Global Witness 2003). Indeed, the epic 50-year struggle by the Karen National Union (KNU) to establish the State of Kawthoolei – a sovereign state of the Karen people along the Thai-Burmese border – was partly reliant on teak revenue. This dependency was most notable after the 1960s as KNU forces were pushed back into the border region by the powerful *tatmadaw*. By the 1980s the KNU became ever more reliant on teak revenue to underpin the insurgency and a flourishing trade with Thai partners ensued (Falla 1991; Bryant 1997, 167–68).

Still, the greatest impetus to resource militarization was the effort by successive rulers of the Burmese State to assert direct central control over the teak forests. In pre-colonial times, the ability of the monarchical state to achieve such control was relatively limited. However, a series of organizational and technological innovations in colonial and post-colonial times meant that rulers of the Burmese State since the mid-nineteenth century have often achieved a greater degree of forest control than in the past. In particular, the combination of modern armaments and more systematic knowledge about the teak forests and their inhabitants (mainly courtesy of the forest department) were a boon for British and Burmese leaders keen to exploit the forests to the hilt (Adas 1982; Selth 1996; Bryant 1997).

Even then, teak exploitation was not easy for them. The assertion of official control as a basis for such exploitation was highly contingent – and indeed, involved much danger for forest officials caught in the line of fire. The level of violence has certainly fluctuated over time. However, chronic political upheaval and widespread social unrest – especially in the late colonial era/Second World War (e.g. 1920 to 1946) and post-colonial era (since 1947) – have been associated with extreme violence in the forests. Not surprisingly, state-led efforts to extract teak have often resembled a military campaign. Forest officials and/or private lessees would enter the forest only with heavily armed escorts – sometimes to the dismay of foresters afraid that these escorts were 'merely a succulent bait for the large bands of well armed rebels roaming the country' (BOF 1946). In the mid-1950s, the U Nu government mounted a large-scale military operation – code named 'Operation Teak' – in insurgent 'infested' southern Burma. In this campaign, units of the *tatmadaw* secured the banks of the Sittang River between Toungoo and Rangoon, even as they provided river escorts for the rafts themselves. Teak timber was rafted from the forests to Rangoon thereby earning the government precious foreign exchange (*The Nation* [Rangoon] 10 November 1955).

Such violence left a deep mark on local people. Livelihoods were disrupted while residents were sometimes forced to take up arms themselves, thereby inviting reprisals. The worst reprisals against forest dwellers have occurred since March 1962 as the *tatmadaw* has killed, tortured and raped countless thousands of ethnic minority villagers suspected of helping insurgents. Many thousands of villagers have also been forced to do highly dangerous work on behalf of the Burmese military (Smith 1999; Global Witness 2003).

A counter-insurgency campaign known as *Pya Ley Pya* ('four cuts') was at the core of this strategy. It targeted those who lived in or near to the forests and was designed to deprive insurgents of access to local food, funds, intelligence and recruits (Smith 1999). In military terms, this campaign was highly effective and enabled the *tatmadaw* to achieve a series of victories against insurgent armies beginning in the Irrawaddy Delta in the late 1960s. The Four Cuts campaign was subsequently extended to the teak-bearing Pegu Yoma with Operation Aung Soe Moe running from late 1973 to April 1975 when the last insurgent forces were cleared from these hills. During the 1980s and 1990s, the campaign moved to northern and eastern border areas where, again, military success was achieved

(Smith 1999; Global Witness 2003) and where teak was an important part of geopolitical strategizing by Burma's brutal military regime.

Anti-insurgency campaigns of this sort were not new – the British had mounted 'pacification' campaigns to sever insurgents from 'sympathetic' villages in the late nineteenth century (Aung-Thwin 1985). However, the sheer scale and brutality of the post-1962 campaign stands out with entire villages moved to 'secure' sites. In these sites, strict military surveillance was imposed while the displaced villagers faced a brutal forced labour regime (Doherty and Nyein Han 1994; Fink 2001). Extreme violence has been the norm (BCN and TNI 1999; Tucker 2001). In the teak forests, this campaign of terror has been accompanied by a strategy of large-scale and unsustainable extraction. The practice of 'cut and run' forestry over the years now threatens to eliminate the country's prime timber in only a matter of decades (Global Witness 2003). At the same time, of course, this devastation in the spaces of production amounts to a grand 'clearance sale' elsewhere, as spaces of consumption are flooded with timber. Indeed, this inter-linked process underpins the fine 'outdoor' living of Northern (and increasingly Southern) middle-classes.

If forced labour and relocation as well as arbitrary killing provide evidence of the brutal dimensions to this blood timber, there is also an entire set of forest management rules dating from colonial times that marks a systematic attempt to control forest and people. This process involved the imposition of forest access restrictions in aid of 'scientific' teak management that has been a serious blow to the subsistence and livelihood needs of residents (Bryant 1997; see also Guha 1989). The coercive nature of attempted state forest control in colonial and postcolonial times has been widely assessed (e.g. Boomgaard 1992; Peluso 1992; Sivaramakrishnan 2000). As such, only a brief and selective account is needed here in order to underscore the structural nature of the violence at play in the forests.

Much of what has transpired under the label 'scientific forest management' from the mid-nineteenth century was designed to introduce 'government' in the Foucauldian sense of the term – the introduction of disciplining and self-disciplining practices, the prevalence of widespread surveillance, and the elimination of antithetical social behavior (Dean 1999; Scott 1998). In Burma, as elsewhere, it entailed inter-linked processes of generating forest maps, resource inventories, and population censuses that provided a basis for draconian restrictions on popular access to timber and non-timber forest products. These restrictions prompted 'everyday' forms of popular resistance including arson, theft, illegal grazing and other local resource practices that contravened the law.

This political dynamic of control and resistance was often most intense in the teak forests. To reside in or near to teak-bearing forest was usually to invite systematic state intervention in one's life precisely because the regulation of teak was seen by many officials inevitably to involve the regulation of people – with the promotion of 'good conduct' by local people a key aim. The process of 'internal territorialization' that was part-and-parcel of the creation of a system of reserved forests (that were focused on teak-bearing areas) involved the creation of extensive borders that crisscrossed existing villages and land use patterns. These borders were designed to facilitate the disciplining of people – *where* they could go, *when* they could go there and *what* they could do whilst there. In one sense, these draconian measures reflected the fervent wish of forest officials in particular to design a comprehensive system of forest management that would permit long-term timber production within an intact forest estate. In another sense though, and that is a central concern in this chapter, the systemic disciplining of forest users in the spaces of production was an integral part of the dynamic associated with the 'safe' and 'reliable' extraction of one of the world's most valuable woods – and, from there, onwards to its appreciation and use in (often distant) spaces of luxury consumption.

Two examples taken from the colonial era provide some sense of how intensive teak management impacted on the lives of local people – an impact reflective of the structural violence implicated in producing a luxury timber. The first example relates to the fire prevention campaign that was mounted in the reserved forests in the late nineteenth century in the (mistaken) belief that fire inevitably harmed teak trees. In practice, this campaign sought to regulate local practices – such as game hunting, cattle grazing or honey gathering – involving the use of fire. Not surprisingly, it was bitterly resisted by residents. Such resistance joined with growing scientific doubts among foresters over the utility of the campaign leading ultimately to the demise of the policy (Slade 1896; Bryant 1997, 87–91).

The fire prevention campaign proved to be a serious additional burden on already hard-pressed villagers. On the one hand, the restrictions were a source of *individual* nuisance and concern. These restrictions were a nuisance inasmuch as villagers needed either to alter their practices in order to conform to the law or to ensure that those practices were safely hidden from view lest foresters catch and punish them for 'illegal' activities. The restrictions were a concern because of the omnipresent threat that they might be captured for violating the law – or indeed, that they might be pursuing practices unaware that their actions were in violation of the law. Ignorance of the law, though, was a weak defense given the publicity surrounding the fire prevention campaign – the more effective tactic was therefore to attribute fires to natural causes. Much depended here on the response of forest officials and local magistrates. The latter were often keen to moderate the punishment of villagers mindful of the need to avoid unrest. Nonetheless, many villagers were indeed convicted under the law and faced potentially heavy punishments including a 500-rupee fine, six months in prison, or both as well as court damages.

On the other hand, the fire prevention campaign imposed a potentially serious *collective* burden on villagers. Under the forest rules, villagers were required to assist forest officials in fighting local fires whatever the provenance of the fires. Not only did this legal requirement involve entire villages in unpaid dangerous work, it also meant that they needed to drop whatever they were doing at short notice – thereby disrupting a variety of livelihood activities. British complaints of peasant 'indolence' and 'negligence' in responding to fire-fighting duties were legion, and suggested an additional dynamic of collective imposition and resistance that negatively affected the lives of villagers in and around the teak forests.

The second example relates to the case of Karen shifting cultivators, many of who resided in valuable teak-bearing forests. For many British foresters, these forest dwellers were seen as *the* central obstacle to scientific teak management given that they cleared their *taungyas* (or hill clearings) with little or no thought to the protection of teak trees. As one observer remarked, the livelihoods of the Karen were 'altogether unconnected with an article which is the source of wealth and industry everywhere, but in the place where it is produced' (McClelland 1855, 13). To the sheer horror of the British, such indifference meant that teak trees were routinely fired, along with other tree species, in the clearance of new fields. An early aim of imperial forestry was thus to stop this practice. Official attention soon centered on the 5,000 or so Karen who lived in the Pegu Yoma (circa 1876).

Over the latter half of the nineteenth century, British foresters sought to resolve the 'problem' of the shifting cultivator through use of both 'sticks' and 'carrots' (Bryant 1994). The 'sticks' were associated with a series of punitive rules that forbade the destruction of teak trees and placed heavy restrictions on where and when the Karen could clear fields for agriculture (including the draconian fire-prevention rules noted earlier). An additional 'stick' was the ever more efficient collection of taxes – amounting to about 6 rupees per individual per year – from a populace among the poorest in the country. In effect, the quest for teak led the British into imposing a system of hill management that had not existed in anything near so comprehensive a form in pre-colonial times. The result was fierce resistance but also flight to less regulated areas, as British rule transformed Karen lifestyles.

The British also proffered 'carrots' as a way in which to elicit 'voluntary' cooperation from the Karen in teak areas. The system of *taungya* forestry was the classic example here. This system required cultivators to plant and tend teak seedlings alongside their own food crops for a modest payment with the aim being that, when cultivators moved on to clear a new patch of forest, teak plantations would grow up in their wake. Cultivators were aware right from the start that this system would undermine their way of life. As one forest official tellingly reported in 1864, Karen leaders that he had met:

> Openly admit that they look upon the sowing of teak in their [*taungyas*] as taking bread from the child's mouth. All they urge to prove this is true enough. Every one is aware of the fact of their returning to the same localities to cut [*taungyas*] after a lapse of from 10 to 15 years. (RFA 1864, 9)

With an estimated growing cycle of between 60 and 100 years until a teak tree was commercially ready for felling, both forest officials and Karen cultivators knew that the resulting plantations represented land irrevocably lost to shifting cultivation. These cultivators were also promised paid work as forest labourers – porters, fire wardens, etc. – working under the orders of forest officials. Here, too, there was a small supplementary income. Yet, for many, such payment could never adequately compensate for the loss of a locally valued way of life based on shifting cultivation.

In this way, a combination of carrots and sticks was used to transform a way of life that was not compatible with intensive teak management. In forestry accounts, the taungya forestry system has been held up as a classic example of international 'good practice' in reconciling forest dwellers to new management systems (e.g. Nisbet 1901; King 1968; Evans 1982). Seen from the perspective of many Karen cultivators, however, the system was all about punishing and disciplining them until they participated in teak planting arrangements that effectively 'planted out of existence' a way of life. Here too, living in close proximity to teak forest ended up being a recipe for trouble, as foresters promoted rules and practices that often blighted the lives of those affected by them.

Conclusion: Against the Grain

This chapter has examined selected themes in the production and consumption of one of the world's most valuable timbers. It has argued that the Burmese teak trail has been littered over time with policies and practices that involved patterns of brutal production, coercive conservation and seductive consumption. Such is the intertwined history of a commodity spanning spaces of consumption and production whose anatomy combines extinction and distinction in a way and to an extent that perhaps only violent luxury goods can do (blood diamonds being another example: see LeBillon 2001, 2006).

This account clearly goes against the grain of work that has sought to bring out the 'romance' of teak through an entire conservation and production history associated with its management, exploitation and consumption. Such work has – and still does – envelop teak in narratives of exotic place and 'progress' (Bryant 1996) that have profoundly shaped how the teak forests have been understood and consumed in colonial and post-colonial times. Here, then, is an example of a culture of consumption that is ripe with contradiction – one in which a dark underside of murder and mayhem is discreetly screened from view, all the better to appreciate the aesthetic and practical pleasures of a luxury wood par none.

Yet the unmasking of a violent luxury resource gathers momentum. By the early twenty-first century, the growing use of teak in garden furniture had prompted NGO campaigns against Northern retailers – for example, the Greenpeace campaign against Wyevale Garden Centre in the UK (Frith 2005) and EarthRights International's boycott campaign against the sale of Burmese teak in the US. The yachting industry is yet to receive the same in-depth treatment but probably will do so in the future. Soon this will mean that teak consumers will no longer be able to plead ignorance about the violent bases of their pleasurable lifestyles. Whether their consciences will be able to live with that fact or indeed whether the final elimination of teak in its natural habitat over the next few decades renders the entire debate 'historical' remains to be seen.

References

Adas, M. (1982), 'Bandits, Monks, and Pretender Kings: Patterns of Peasant Resistance and Protest in Colonial Burma, 1826–1941', in Weller, R. P. and Guggenheim, S. E. (eds), *Power and Protest in the Countryside* (pp. 75–105) (Durham, NC: Duke University Press).

Albion, R. G. (1926), *Forests and Sea Power: The Timber Problem of the Royal Navy 1652–1862* (Cambridge, MA: MIT Press).

Andrews, E. (1931), *The Bombay Burmah Trading Corporation Limited in Burmah, Siam and Java*, 3 volumes (London: n.p.).

Aung-Thwin, M. (1985), 'British "Pacification" of Burma: Order without Meaning', *Journal of Southeast Asian Studies* 16: 245–61.

Boomgaard, P. (1992), 'Forest Management and Exploitation in Colonial Java', *Forest and Conservation History* 36: 4–14.

Brandis, D. (1888), 'Teak' in *Encyclopaedia Britannica*, 9th Edition (pp. 103–105) (Edinburgh: Adam and Charles Black).

Bryant, R. L. (1994), 'Shifting the Cultivator: the Politics of Teak Regeneration in Colonial Burma', *Modern Asian Studies* 28, 225–50.

Bryant, R. L. (1996), 'Romancing Colonial Forestry: The Discourse of "Forestry as Progress" in British Burma', *Geographical Journal* 162, 169–78.

Bryant, R. L. (1997), *The Political Ecology of Forestry in Burma* (London: C. Hurst).

Burma Center Netherlands (BCN) and Transnational Institute (TNI). (1999), *Strengthening Civil Society in Burma: Possibilities and Dilemmas for International NGOs* (Chiang Mai: Silkworm Books).

Burma Office Files (BOF). (1946), 'Confidential Note on Illicit Extraction of Timber and General Lawlessness', Conservator of Forests, Sittang Circle, 8 May.

Cohen, D. (2006), *Household Gods: The British and their Possessions* (New Haven: Yale University Press).

Dean, M. (1999), *Governmentality: Power and Rule in Modern Society* (London: Sage).

Doherty, F. and Nyein Han (1994), *Burma: Human Lives for Natural Resources – Oil and Natural Gas* (n.p.: Southeast Asian Information Network).

EarthRights International. (2005), 'The Price of Luxury: The Global Teak Trade and Forced Labour in Burma', available at: http://www.earthrights.org/teak/indepth.shtml, accessed on 13 September 2007.

Evans, J. (1982), *Plantation Forestry in the Tropics* (Oxford: Blackwell).

Falla, J. (1991), *True Love and Bartholomew: Rebels on the Burmese Border* (Cambridge: Cambridge University Press).

Ferrars, M. and Ferrars, B. (1900), *Burma* (London: Sampson, Low, Marston and Co.).

Fink, C. (2001), *Living Silence: Burma under Military Rule* (London: Zed Books).

Foucar, E. C. V. (1956), *I Lived in Burma* (London: Denis Dobson).

Frith, M. (2005), 'Revealed: Sale of Garden Furniture in Britain is Propping up a Brutal Regime', *The Independent*, 24 March, pp. 16–17.

Gallant, M. N. (1957), *Report to the Government of Burma on the Teakwood Trade* (Rome: Food and Agriculture Organization).

Gardiner, J. R. (1942), 'The Teak Industry of Burma', *Australian Timber Journal* 8, 736–737, 749.

Geary, G. (1886), *Burma after the Conquest* (London: Sampson, Low, Marston, Searle and Revington).

Global Witness. (2003), *A Conflict of Interests: The Uncertain Future of Burma's Forests* (London: Global Witness).

Gloster. (n.d.), 'Gloster: Made for Life', Promotional Leaflet distributed at London Garden Centres, 2005–6).

Guha, R. (1989), *The Unquiet Woods: Ecological Change and Peasant Resistance in the Himalaya* (Delhi: Oxford University Press).

Hopwood, S. F. (1935), 'The Influence of the Growing Use of Substitutes for Timber upon Forest Policy with Special Reference to Burma', *Indian Forester* LXI, 558–72.

Htin Aung. (1968), *Epistles Written on the Eve of the Anglo-Burmese War* (The Hague: Martinus Nijhoff).

Kelly Talbot, R. (1912), *Burma: Painted and Described* (London: Adam and Charles Black).

King, K. F. S. (1968), *Agri-silviculture (The Taungya System)* (Ibadan).

Koenig, W. J. (1990), *The Burmese Polity, 1752–1819: Politics, Administration, and Social Organization in the Early Kon-baung Period* (Ann Arbor: University of Michigan Press).

Lambert, A. D. (1996), 'Empire and Seapower: Shipbuilding by the East India Company at Bombay for the Royal Navy, 1805–1850', in Haudrere, P. (ed.), *Les Flottes des Compagnies des Indes, 1600–1857* (pp. 149–71) (Vincennes).

LeBillon, P. (2001), 'The Political Ecology of War: Natural Resources and Armed Conflicts', *Political Geography* 20, 561–84.

LeBillion, P. (2006), 'Fatal Transactions: Conflict Diamonds and the (Anti) Terrorist Consumer', *Antipode* 38, 778–801.

Lieberman, V. B. (1984), *Burmese Administrative Cycles: Anarchy and Conquest c. 1580–1760* (Princeton: Princeton University Press).

Maung Maung. (1976), 'Nationalist Movements in Burma, 1920–1940: Changing Patterns of Leadership: From Sangha to Laity', MA thesis (Canberra: Australian National University).

McClelland, J. (1855), 'Report on the Southern Forests of Pegu', in *Selections from the Records of the Government of India (Foreign Department) IX* (Calcutta: Government of India).

Mills, J. A. (1979), 'Burmese Peasant Response to British Provincial Rule 1852–1885', in Miller, D. B. (ed.), *Peasants and Politics: Grass Roots Reaction to Change in Asia* (pp. 77–104) (London: Edward Arnold).

Morehead, F. T. (1944), *The Forests of Burma* (London: Longmans, Green and Co.).

Myat Tun, M. (1938), 'Some Asiatic Displays at the Glasgow Exhibition', *Asiatic Review* 34, 806–809.

The Nation [Rangoon]. (1955), 10 November.

Nisbet, J. (1901), *Burma under British Rule and Before* (London: Archibald Constable).

OriginAsia. (n.d.), 'Exotic Lifestyle Designs', Advertisement in *Expat Living* [Singapore] (January/February c. 2005).

Orwell, G. (1987 [1934]), *Burmese Days* (Harmondsworth: Penguin).

Pearn, B. R. (1939), *A History of Rangoon* (Rangoon: American Baptist Mission Press).

Peluso, N. L. (1992), *Rich Forests, Poor People: Resource Control and Resistance in Java* (Berkeley: University of California Press).

Peluso, N. L. and Watts, M. (eds) (2001), *Violent Environments* (Ithaca: Cornell University Press).

Report of Forest Administration (RFA). (1864), *Progress Report of Forest Administration in British Burma for 1863–64* (Rangoon: Government Printing).

Ribbentrop, B. (1900), *Forestry in British India* (Calcutta: Superintendent of Government Printing and Stationery).

Scott, C. W. (1945), 'Burma Teak Today', *Wood* 10: 4, 81–84.

Scott, J. C. (1998), *Seeing like a State* (New Haven: Yale University Press).

Selth, A. (1996), *Transforming the Tatmadaw: the Burmese Armed Forces since 1988* (Canberra: Australia National University).

Simmonds, P. L. (1885), 'The Teak Forests of India and the East, and our British Imports of Teak', *Journal of the Society of Arts* 33, 345–59.

Sivaramakrishnan, K. (2000), *Modern Forests: State-making and Environmental Change in Colonial Eastern India* (Cambridge: Cambridge University Press).

Slade, H. (1896), 'Too Much Fire Protection in Burma', *Indian Forester* 22, 172–76.

Smith, M. (1999), *Burma: Insurgency and the Politics of Ethnicity*, 2nd Edition (London: Zed Books).

Stebbing, E. P. (1922–1926), *The Forests of India* (3 volumes) (London: The Bodley Head).

Steinberg, D. I. (1981), *Burma's Road toward Development* (Boulder: Westview Press).

Taylor, R. H. (1987), *The State in Burma* (London: C. Hurst).

Teak Decking Limited UK. (2007), available at: http://www.teak-decking.co.uk, accessed 13 September 2007.

Tucker, S. (2001), *Burma: The Curse of Independence* (London: Pluto Press).

Williams, J. H. (1950), *Elephant Bill* (London: Rupert Hart-Davis).

Chapter 12
Space for Change or Changing Spaces: Exploiting Virtual Spaces of Consumption[1]

Angus Laing, Terry Newholm and Gill Hogg

Consumers, Information and Power

The internet driven information revolution is frequently cited as one of the key factors (re)shaping contemporary society. Within wider social changes (Laing et al. 2005a), the core of this revolution is a fundamental expansion of individual consumer access to information resources. In particular the internet, with its unprecedented breadth of interconnected information, offers consumers access to specialist product and market information which has conventionally been confined to organizations and professionals operating within that market. In the past decade the internet has variously been predicted to change working practices, lifestyles, personal relationships and even sense of community (Tambyah 1996; Jolink 2000). Yet many of these predictions have subsequently proved, along with some of the more high profile internet companies, to be at least premature. What is clear, however, is that the internet has a number of critical capabilities and recognizing these capabilities facilitates understanding of how, and where, the internet will ultimately impact on contemporary society. One of the critical areas where the internet has been perceived as disrupting established conventions is in respect of complex professional services where the core of the service product is information (Beck 2001).

Politically, the internet has been characterized as a resource through which citizens can become more self-reliant. Popularly, it has been viewed as a liberating medium, a mechanism by which consumers have been able to challenge the authority of professional and political establishments. The work of David (2001) in the context of health care exemplifies this notion that unprecedented access to information via the internet has empowered consumers and radically altered consumption relationships:

1 The research on which this chapter is based is funded by the Economic and Social Research Council and Arts and Humanities Research Council under the 'Cultures of Consumption' Programme. Project Title: Consuming Services in The Knowledge Society: The Internet and Consumer Culture. Award Reference: RES-143-25-0009.

> One of the main forces within the e-environment is *consumer empowerment*. With greater access to more readily available sources of information than their forefathers, consumers are assuming an increasingly active role. [...] Instead of being passive recipients of judgements and treatments handed out by the medical community, consumers will be actively involved in managing their own healthcare. They will demand a better quality of life, better care, personalized treatment, convenience, choice, and value for money. (2001, 6–7, emphasis added)

Yet there is extensive debate across professional services as to whether this is anything other than a metropolitan, middle class urban myth, and indeed whether the internet has not created a new monster which may in reality undermine consumer autonomy and reinforce dependence on professionals, albeit in a very different form (Newholm et al. 2006). The central thrust of this debate is around the quality of information available to consumers in an unregulated environment, and the capacity of consumers to handle the available information.

From a libertarian perspective the unregulated nature of the internet is precisely what empowers consumers. Conversely the professional perspective sees 'misleading' information as a bar to *properly* informed and genuinely empowered consumers (Impicciatore et al. 1997; Flanigan and Metzger 2000). This rests on a concept of professional knowledge skills as necessary to the 'correct' interpretation of information. From the competing consumer perspective, any restraint on content might be seen as a return to professional monopoly that would run counter to broader social changes. The description of the impact of the internet on medical services by Ham and Albetri (2002) as 'being akin to the translation of the Bible from Latin into English' is a powerful analogous image of the potential disruptive capabilities of the internet and in turn the implications for the professional priesthood. For services such as medicine and law, traditionally characterized by professional authority, what Foucault refers to as a 'regime of truth' (1980), in part, based on informational asymmetries and exclusive possession of specialist skills (Friedson 1986), this implies a fundamental challenging of established patterns of organizing and delivering services. Yet equally for consumers, this may be viewed as generating new uncertainties, anxieties and challenges in terms of negotiating or resisting a new settlement with professionals. Thus the plurality of the internet has the potential to destabilize traditional 'certainties'.

Fragmenting Behaviours and Virtual Spaces

The internet, as such, represents a new forum, a new environment within, and through which, consumers can contest the service domain with professionals. For some consumers this will offer valued opportunities to assert power over professionals and set the terms of engagement, for others it will pose questions as to the nature of their role, and for some it will generate doubts and uncertainties.

There is a fundamental danger in implicit assumptions that consumers will exhibit similar views towards the disruption of the 'established church' of professionalism arising from the internet, or at least will be facing in a common direction. Rather, evidence suggests increasing fragmentation of consumers and growing diversity of patterns of behaviour in engaging with professionals (Laing et al. 2005a). These diverging behaviours reflect differences in attitudes towards professionals among consumers (and towards consumers among professions), ranging from the sceptical to the convinced, and differences in behaviour in the consultation, spanning the active choice-maker to compliance with the professional service offering.

These differences are linked to divergent patterns of internet information usage, with differing usage of the internet in terms of type of forums and infomediaries being exploited (or not) by different groups of consumers in their reconfiguring of consumer-professional relationships (Laing et al. 2005b). For some consumers accessing the internet is about gathering a library of data to challenge professionals, for others the internet facilitates reflection on professional advice, or as their primary information source for professional reflection, while for others it is a communal discussion space to explore alternative service solutions. From the perspective of the consumer the internet is not a singular entity, rather it is a set of interconnected virtual environments, or spaces, within which consumers undertake different activities and perform in different ways reflecting how they interpret the nature of that space and the 'real' professional services. These spaces encompass a diversity of functions and formats with different consumers being comfortable in different types of space. Whilst some spaces are about protecting consumers from professional influence and pressure, entering and colonizing hitherto forbidden professional space, others are under the paternalistic guidance of professionals. Consumers gravitate towards, and utilize, those spaces that fit their style and values; they lurk or engage. Yet such identification is learnt through socialization to spatial norms and understanding of social cues (Kozinets 2002).

Across the virtual spaces of the internet, consumers face comparable choices and challenges. Challenges not only in assessing and assimilating information, but in understanding the nature and dynamic of particular online spaces (Kozinets 2002), and in balancing social pressures for empowerment (Newholm et al. 2006) with the need for reassurance in the face of individual vulnerability. The 'balancing paradigm' of consumer satisfaction posits that consumers constantly try to address a number of paradoxes in any consumption environment (Mick and Fournier 1998). Satisfaction derives from the degree to which they are successful in that ongoing process of balancing. Utilizing this perspective, professional service consumers need to resolve a number of paradoxes arising from the utilization of virtual spaces. These can be typified in healthcare for example as balances between the following characteristics in Figure 12.1:

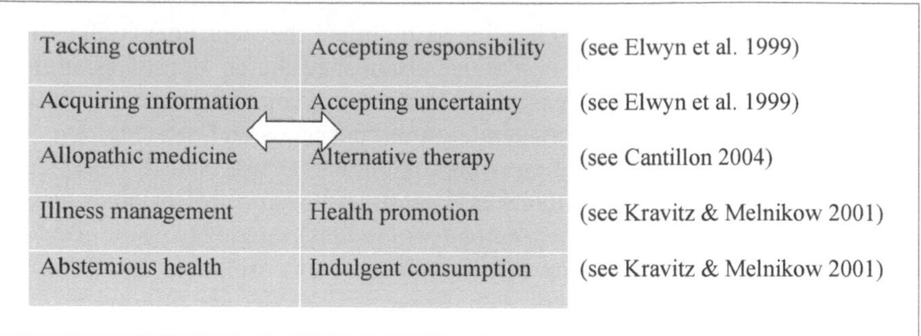

Figure 12.1 The 'balancing paradigm' in healthcare

Shankar et al. (2006, 1,021) note that, confronted by 'hyperchoice' in financial service markets, consumers have to balance between empowerment and paralysis. Similarly the greater use of information by consumers in legal services, especially for example in self-help programmes, is seen as empowering consumers but also 'harnesses the productive capacity of consumers' (Giddings and Robertson 2003, 102). Thus, as they take control they also paradoxically produce more of the service themselves.

Given this balancing act, these virtual spaces of consumption not only offer potential opportunities for seizing control of, and hence customizing, the professional service encounter, but equally confront the consumer with the challenges of accepting risk and personalized responsibility. The internet thus can be characterized as offering not only the spaces of opportunity which have been central to its contention as driving unprecedented consumer empowerment, but equally encompasses spaces of challenge which confront consumers with uncertainty and risk. It is in that the nature of such spaces that they may disable users' ability to form adequate judgments regarding the credibility of participants, the affiliation of sites and the veracity of the information. For consumers engaging with the internet, this ongoing tension between the internet being a space of support and being a space of confusion requires a personalized balancing of these countervailing forces, reflecting the circumstances and characteristics of the individual consumer.

Methodology

The research reported in this paper draws on data from interviews with professionals and consumer focus groups. Ten focus groups were conducted in six locations: two each in Aberdeen, Bristol, London, Manchester, and one each in Milton Keynes and Glasgow. Participants were recruited and compensated through a specialist fieldwork company with the requirement that each had, within the last 12 months consulted a qualified professional and made related use of the Internet in at least one of the three sectors forming the focus of this research. The groups comprised female (n=26) and male (n=27) participants. Their ages ranged from 18 to 61 and from unemployed through clerical to professional occupations. None was employed in legal, financial or primary healthcare services. Participants were asked to relate experiences they had had with relevant professionals and in particular their use of information in both addressing the underlying need which prompted the use of the professional and in their subsequent dealing with that professional. Other participants were encouraged to comment on and discuss these experiences (Morgan 1988) to increase the richness of the data and ensure relevant subjects were covered from the consumers' perspective. The geographically located, demographically mixed focus groups lasted between 70 and 90 minutes and were audio-recorded and later transcribed.

The datasets were analysed by two researchers using QSR N6 qualitative analysis software for recurring themes as well as contradictions across participants and sectors. General data categories derived from the conceptual literature, together with further categories and subcategories developed progressively from the data (Glaser and Strauss 1967) were used to facilitate analysis. The themes developed therefore arise from our categories. As such the data represents a 'two-level scheme' (Miles and Huberman 1994, 61) namely specific 'emic' consumer understandings nested in general 'etic' analytical themes.

In our interpretation it seems appropriate to draw on Foucault's work on power and knowledge (1980) to the extent that we view the coherence of professional and other discourses as, in part, a linguistic achievement. By eschewing an essentialist view we are able to account for the dynamic condition our data suggests. Shankar et al. (2006) present the broader case for this approach.

Spaces of Opportunity

At the core of the idea that the internet is a key agent of social change is its capability to enable consumers to perform tasks and undertake roles in which they were previously unable to engage. Popularly cited among these capabilities include the opportunity to form geographically unconstrained communities and undertake effective comparison of competing offerings. Within the context of professional services it is possible also to see the internet as creating an environment where consumers are able to generate the confidence to challenge professional judgement. Collectively these capabilities can be seen as creating an environment that offers the opportunity for the consumer to gain control of the service encounter.

Space for community

Consumer communities are not a new concept, what is new is the potential of the internet as a social space to facilitate communities that transcend traditional social or geographical boundaries, enabling consumers to communicate with individuals with whom they would not normally have contact (Muinz and O'Guinn 2001). Among professional service consumers this communal space is articulated as offering the opportunity to share lived service experiences as part of the coping with the challenges of understanding the service experience and the opportunity to create global communities of shared interest and experience (Hogg et al. 2004). The power of the individual story is a central feature of these communal spaces, offering consumers an experiential alternative to technical information.

> Well, one of the things, there is a guy who is actually 10 years older than me and he's gone through exactly the same thing as me, so we were just chatting, talking … he's 10 years further down the line than I was two years ago, so I was telling him what was happening and the pain I was getting, what tablets I was on, and you know what stage will I go through, and he was giving me his experiences as it happened to him 10 years ago … It's a website in America. (Manchester 1 Female – Health)

Alongside such experiential coping support is the role of this space in addressing gaps in the provision of technical information from professionals; an explicit recognition of the bounds of professional knowledge. There is a recurring sense of consumers attempting to exert control over the service professional based on a better understanding of the long-term implications of the situation faced by the individual consumer.

> … because when he was first born they (the doctors) didn't have a clue what his prospects would be. It's only sort of when you go onto these groups and talk you sort of discuss with people that have got similar conditions, you suddenly see what, you know, he's going to achieve. (Bristol 2 Male – Health)

The differing bases of knowledge across consumer and professional communities are exemplified in the discussions characteristic of these social spaces. Alongside the already highlighted importance of experiential information is the perceived depth of knowledge possessed by consumers in respect of a highly specific issue compared to professionals with a more generic knowledge base.

> People who are going through the same thing will trust somebody else who is going through the same thing, or has gone through the same thing, more than they will a GP, I certainly would. Most GPs are jacks-of-all-trades and masters of none. (Manchester 2 Male – Health)

The internet as a communal space, arguably characterized by a free flow of information, while providing consumers with unprecedented opportunities for information acquisition and interrogation, raises important questions in relation to the nature, veracity and orientation of such information. Not only is there the risk of consumer reliance on unverified information impacting on decision making and the service outcome, but if the information which consumers gather is judged to be biased by professionals, to lie outside the professional 'truth-regime' (Foucault 1980), the client-professional interaction will be affected, forcing the consumer into a changed relationship with the professional. The internet as a space for community cannot be seen as an unambiguously advantageous domain, rather the consumer is confronted with a set of challenges, with balancing a set of paradoxes.

Space for comparison

For consumers the breadth of informational resources available on the internet creates an unprecedented space for comparison. Although retailing has been characterized as the sector which has been revolutionized by the internet (Van der Poel and Leunis 1999) the capacity of the internet to enable consumers to compare offerings is, if anything, more significant in respect of professional services. Reflecting the core of service products being information and expertise (Mills and Moshavi 1999), the internet with its ability to present comparative information, along with facilitating the emergence of consumer communities of expertise, has significantly facilitated consumer choice in a market conventionally characterized by informational asymmetries. In this, the internet offers the consumer the opportunity for confirmation, or not, of service professionals advice at minimal financial or time cost. Critically this offers not only the opportunity to challenge the professional but also to seek confirmation of veracity of the advice provided.

> I would trust what I was told [by a professional] but I would maybe then go back and seek some sort of confirmation or more information from the Internet. I guess I've been quite fortunate, I haven't had any instances where me or someone in my family has been told A, and then we find out on the internet or from some other source that the information was incorrect. So generally the internet has confirmed what we've already been told. (Glasgow 1 Male – Legal)

The internet not only provides space for comparison following a consultation, but equally enables consumers to explore potential alternatives in advance of the service encounter. This, more than the post-hoc comparison, highlights the scope for consumers to challenge service providers in terms of the provision of alternatives, alternatives which may on occasions lie outwith the disciplinary bounds of the profession.

> I didn't want to have steroids and there was a bit of conflict going on there … I'm thinking, well hold on, its convenient for a quick fix but I tend to look at the long term effects … so a lot of the times I'm sort of like looking on the internet for other ways or alternative measures like homeopathic which, again, my GPs

> not to keen on. It's that battle. I tend sort of go to him knowing my back-up plan. (London 1 Female – Health)

Such behaviour cannot be seen as uniform (Newholm et al. 2006), with this pre-consultation comparison of options presupposing an ability to use available information effectively. In professional services it is evident the ability to engage with service providers on an equal footing, both socially and in the use of the language, is critical to consumers ability to use the internet's comparative capacity to shape service provision. In the absence of such cultural capital, the utilization of the comparative space of the internet has limited efficacy in enabling consumers to challenge professionals (Henry 2005). Where consumers possess the cultural capital they are able to use the comparative capacity of the internet to challenge service providers and negotiate enhanced service options.

> I've gone on the internet and I came up with a really good rate and I armed myself, got all the print offs, and I sat down with him [my financial adviser] and he was a bit flabbergasted that I'd done this because we normally just sit down and talk about what period of time we want to take the mortgage on and blah blah blah, and he actually went away and came back with an unbelievable rate. So I gave him competition, he came back with 3.5% over two years. I mean it was brilliant. (Manchester 2 Male – Financial)

This existence of such comparative space in itself is insufficient to rebalance the service encounter. Rather such rebalancing is dependent on consumers having the confidence to challenge professional advice and judgement. In part this is linked to cultural capital, but the internet itself is described by many consumers as creating the confidence to question service providers.

Space for confidence

Although current policy discourses praise the semi-self-sufficient responsible consumer, given the residual memory of paternalistic narratives of consumer-professional relationships, having the confidence to challenge a professional and the associated advice, in however modest a manner, is a key factor shaping consumers ability to wrest control of the service encounter from the professional. A recurring theme among service consumers is the scope the internet offers to develop the confidence to confront service providers. At the core of this confidence building space is the ability of consumers to gather information, ask questions anonymously, reflect on implications of advice, free from the interference of professionals. Reflecting the importance of cultural capital in shaping behaviour, a recurring theme is that of anonymity, with the internet offering anonymous space for 'vulnerable' consumers to ask the questions they would feel unable to explore in the face to face encounter.

> It's sometimes better to sort of have that anonymity, you know. I hate that word. You're not actually face to face with someone and you can look up all this stuff without any embarrassment or sort of feeling silly and sometimes its probably easier because you, when your face to face with someone, sometimes you get all choked up or you can't get out what you're trying to say, in the heat of the moment you sort of forget what half of what you're trying to say, so it gives you that time really. (Bristol 1 Female – Health)

It is not, however, simply the 'vulnerable' consumer for whom the internet offers space for confidence. There is awareness among a broad cross section of consumers of the pressures, both financial and temporal, in the service encounter. In publicly funded health care settings there is an evident sense of 'private constraint' (Walsh 1994) on the part of service users with such ideas inhibiting their behaviour in terms of the amount of time they have with the professional. Equally, there is a clear perception in private sector service settings that there are real financial costs associated with questioning professional advice.

> Marie: "You also get the feeling when you go home that you've got no pressure. You're not doing face to face. You can sit there at your own free will, taking your time, reading what you want to read and there's no pressure."
>
> Louise: "But at the doctor's, it's as if you've got one chance and one hit because you've got to make the five minutes count."
>
> Sean: "Same with financial because you know time is money with them. So at home on the internet you can spend as much time as you want and we all spend hours and hours." (Manchester 2 Male/Female – Health/Financial)

The internet offers consumers off-line space away from the service provider to assimilate to, and respond to, professional advice. In many circumstances this does not reflect vulnerability on the part of the consumer but rather a grounded, if occasionally cynical, understanding of the motivations and disciplinary perspective of professionals. The building of confidence is in part dependent on the confidentiality of such spaces and the opportunity for genuine autonomy.

> … [A]nd everybody [financial advisers] is going to try to sell you something and the thing I like about the net is obviously it's at my leisure, I can see all the [financial] details, I can try and consume it, I can even take a print-off and I can sit and digest it at my leisure without any pressure to buy or accept deals in front of me. I've always found that any time I went to see an adviser, they were just interested in getting me to sign, they would say this is the best deal you can get. (Aberdeen 2 Male – Financial)

By providing such confidential space where the consumer can operate independent of the professional, the internet has a significant role to play in engendering the consumer confidence that is critical to the ability of consumers to take effective control of the service encounter. Such consumer narratives, even allowing for variation in cultural capital and patterns of engagement with service professionals (see Laing et al. 2005b; Newholm et al. 2006) represent the dominant portrayal of the internet as a liberating and empowering space that offers the possibility of transforming the service consumption experience. Yet the essential paradox at the core of the internet is the fostering of fears and uncertainties alongside and in parallel with such empowerment.

Spaces of Threat

For many, if not all consumers the internet is a Janus-faced environment. On the one side it offers freedom and opportunity, on the other it stokes fear and uncertainty. The threatening dimension of the internet co-exists with and intermingles with opportunities in consumers utilization of the internet. These twin faces of the internet are not the preserve of distinct groups, the technophiles versus the technophobes, the cultural elite versus the marginalized. Rather they are reverse faces of the same spaces for the same consumers. The space for community is also the space for anxiety, the space for comparison is also the space for ambiguity, the space for confidence is also the space for alienation. For consumers the internet presents the ultimate forum for balancing paradoxes.

Space for anxiety

Set within a political and cultural milieu which questions the authority of professionals, professionals' advice to consumer not to consult the internet may all too readily be discounted as an attempt to preserve professional hegemony of the service encounter. Yet the internet has the capacity to sow doubts and anxiety in the minds of consumers as a result of the volume and complexity of information which can be accessed. Inextricably linked is uncertainty over the source and authenticity of information. While Pagliari and Gregor (2004) suggest that the risk of harm to patients as a result of the use of internet information maybe over-emphasized, the diversity of internet-based health information raises fundamental concerns over the quality of such information and the impact of such information on patient choice and behaviours. Where consumers lack the expertise to contextualize and to assess highly complex information, the risk is that the consumer will be unnerved, and far from being better placed to challenge the professional will be more dependent on the professional to draw them back from the brink.

> Started searching the net and then I saw all these scenarios about what she had wrong with her knee and that she was going to be flat on her back for six months,

> she'd have to have heel surgery, her leg opened up … I spent like three months waiting to see the consultant and I was panic stricken … I had all this information in my head that I found off the internet, and if I hadn't looked I wouldn't have been so worried. (Manchester 2 Female – Health)

Even where virtual communities have the capacity to generate communal expertise and facilitate understanding of technical information, the dynamic and nature of such communities can be highly unsettling for the consumer. For the novice participant, the language and practices of the community may effectively exclude and generate a sense of disconnection, further heightening anxieties. This is particularly striking in cross-cultural communal spaces where the cultural norms and service expectations may differ widely.

> My experience sort of two or three years ago, was not very good. I had an irregular heartbeat caused by stress and I went onto an American discussion group and they were so over the top about it and they were talking about all this stuff and they were getting bits of their hearts zapped away and I was absolutely petrified of this. So I did actually stop using that. (Bristol 2 Male – Health)

Such communities pose real challenges to consumers in assessing the quality of information, the credibility of advice generated, and the motivations of contributors. Specifically anonymity strips individuals of their 'status trappings' (Garrison 1994) and encourages frankness, allowing the development of what Tambyah (1996) calls the 'net self'. While this points to the democratic and relational nature of the internet, it also exposes the anarchic nature of the medium: freedom from control can also mean freedom from accuracy. Even for highly socialized consumers, engaging with virtual communities pose particular challenges in terms of identity, credibility, power and control given the absence of many of the cues available in 'real-life' contexts, leaving consumers adrift in an environment of uncertainty and ambiguity.

Space for ambiguity

Professional services are by their nature complex and characterized by esoteric language. The interpretation of such information requires judgement anchored in an understanding of the disciplinary field. A recurring theme among professionals is the lack of understanding on the part of consumers of the underpinning principles of the discipline, be that physiology or finance, which not merely constrains their ability to take control of decision making but can generate negative outcomes as a result of inappropriate decisions (Laing et al. 2005b). This is compounded by the frequent variance among professionals themselves, even when operating within a common disciplinary paradigm, as to the most appropriate options in addressing specific conditions or cases. When coupled to the sheer volume of information virtual spaces offer, the consumer is faced with sifting complex information, for which they lack the requisite knowledge base, while weighing up differing perspectives and advice. It is unsurprising that consumers report internet derived information as being highly ambiguous and generating significant uncertainty.

> The amount of information out there is staggering and it's very technical information, that generally speaking is in a language that we're not used to using because it's not designed to be used by us … Unless you go to a specific site like the Alzheimer's Society where they're actually interacting with the end user, if you like, but it's finding that information instead of finding some paper that's been written by some professor in America that's so specific you can't scratch the surface of it. (Glasgow 1 Female – Health)

This proliferation of alternatives is exacerbated by discussions within community spaces. In such spaces the range of perspectives presented both extends beyond conventional disciplinary boundaries to embrace a greater diversity of (alternative) opinion, and lacks clear clues as to the underlying standpoint of contributors.

> If you submit a question to a forum, you're then getting so many different people's opinions back and you never know quite, quite which one to go by and which one is necessarily the right one to go with … once you start questioning it's very hard to then work out whether you should believe any of them and ask anything more than someone else's opinion, and there are so many opinions that you may as well go with your own. (London 1 Male – Legal)

Not only does this diversity engender further ambiguity and uncertainty but it also pushes consumers towards relativism, that one viewpoint is not inherently more valid than another. Such 'de-prioritiziation' of perspectives manifests an assumption that all contributors validate their perspective, with the conventional justification of professional status not automatically being sufficient to convince consumers (Lyotard 1984). The challenge facing consumers operating in such information rich spaces is to select appropriate information and triangulate this with a range of other information sources.

> You have to be really careful to pick out bits, and it's almost like piecing it together from the different pages. You can't take just one as that's what I'm going to go on, because another site may say something slightly different and another site may say some people say this and some people feel that. You need to pick out certain pieces of information and facts and whatever from several different sites I think. (Bristol 2 Female – Health)

This ambiguity can be seen as a contributory factor in undermining professional authority. By highlighting the existence of divergent opinions among professionals, the internet contributes to the awareness among consumers that professionals within a disciplinary field do not automatically hold uniform views but rather that professional advice may have varying degrees of heterogeneity and perspectives that may once have been seen as certain are in fact contested (see Elam and Bertilsson 2003). Such uncovering of professional disagreement alongside recognition of the bounded nature of advice and the impact of budgetary or commercial pressures contributes to a diminution of trust in professionals, and a sense of alienation.

Space for alienation

The utilization of the multiple virtual spaces offered to consumers by the internet has seen a diversity of consumer narratives evolving semi-independently of established professional discourses. Conventionally consumer narratives drew primarily on the professional discourse for its construction and consequently were closely aligned with, and indeed mirrored, that discourse. However, drawing increasingly on independent sources of information, and critically independent consumer mediated means of knowledge construction, within this virtual space there is evidence that these consumer narratives are increasingly diverse and distinct from the professional discourse. The consequence is a clash of cultures and an emergent sense of alienation. In the context of healthcare, Thomson (2003) has expressed this eloquently: 'The patient as consumer desires to produce his/her own medico-administrative identity through interaction with physicians, nurses and technologies. … Yet these post-modern currents inevitably collide with the more intractable, modernist features of the medico-administrative system' (103). Although professionals have reacted to these changes, with notable exceptions, these responses have often been slow, limited and only reluctantly conceded, reflecting the innate conservatism of professions (Aldridge and Evetts 2003). Such grudging response to the interests of consumers further undermines consumer confidence in professional values and the validity of the professional discourse (Abernethy and Stoelwinder 1995).

This emergent sense of alienation is anchored in increasing awareness and questioning of the constraints on professionals and the limits to the professional discourse as a result of consumer engagement with the virtual spaces of the internet. A number of constraining factors are consistently articulated by consumers as contributing to a sense of alienation. In certain professional settings the role of professionals as gate-keepers places clear bounds on the extent to which consumers can use information to gain control over the service encounter, generating dissatisfaction and erosion of belief in the professional discourse.

> The doctor could say well, I think that's wrong, then you're up against the wall, the internet is saying this, but you're saying this. You've still got the doctor's decision, at the end of the day your still up against the barrier of they've got to make the last decision. You cannot get something without them saying so. (Bristol 2 Female – Health)

In both public and private sector service settings this gate-keeping function is associated with diminished professional autonomy, from either the state or corporate interests, with this generating increasing scepticism of professional advice and a reinforced perception of the need to acquire independent information to hold professionals to account.

> I would say that, you know, some of the information that they're telling me is perhaps not true, and they're not being quite so open as they should be, because it all comes back to money. (Bristol 2 Male – Legal)

> They've got huge lists. They don't want to discuss anything with you. You know what? They just give you a prescription and go. And that's why you then start looking, people will then look, you know, all right, where do I get this information because the doctor hasn't told me, didn't explain to me, didn't take enough time to listen. (London 2 Male – Health)

The notion of consumer choice inevitably sits uneasily with the maintenance of coherent professional discourses. Consumers' desire to extend the range of service options beyond the established disciplinary boundaries is a recurring theme in emerging alienation of consumers. The unrestricted nature of the internet spaces and the colonization of such spaces by quasi-professions has dramatically increased consumer exposure to alternative conceptualizations of service. Professional responses being constrained by the parameters of the professional discourse of necessity consistently draw consumers back into a narrower service range, proscribing options which the consumer sees as valid.

> I'll leave him things to read and he'll give me back his comments. I find that useful because he's aware that I'm not accepting blindly what he says … look, you see, he's confined to BMA medicine. You know, he makes it very clear that he's not going to go beyond, he's going to say, look, forget it … if you do it too regularly you'll have to go off my list. Not in a threatening way. Meaning if you're questioning his clinical judgement. (London 2 Male – Health)

The differing perceptions of what constituted valid options appears to be a key factor in this sense of alienation with many consumers embracing broader definitions of what is valid and acceptable than professionals. What is striking from accounts of consumer-professional interaction is that the rejection by professionals of options advanced by consumers did not in itself lead to a sense of alienation. Rather it was the manner in which the professional participated in the encounter which is critical. The idea of respect, of being listened to, appears central to whether consumers accept the professional discourse and submits to the authority of the professional.

> I think we don't tend to investigate if we find we are respected. You know, if you're speaking to a medical professional, you can tell in some respects, they either think it's rubbish, but they respect the right to have it. You back off because you think this person is listening to me. (Glasgow Male – Legal)

This emphasis on the dialogue of the encounter as critical to consumer alienation draws out the importance of consumers having a perception of control within the professional consultation. The willingness of professionals to step away from established paternalistic modes of operation, when appropriate, and negotiate a more egalitarian encounter in the face of consumer exposure to diverse and contradictory information represents a sustainable mode of engagement which potentially integrates the internet as a space of enhancement rather than as a space of alienation and disruption for those consumers seeking active participation in decision making.

Conclusion

Prior to the emergence of the internet, spaces of consumption might stereotypically be presented as exhibiting stable and coherent discourses of professional services. Even in the political arena there was consensus around the discourse of the professional as a paternalistic authority figure that transcended competing ideologies and political parties (Kavanagh and Morris 1994). Figure 12.2 represents the nature of such relatively settled spaces of consumption.

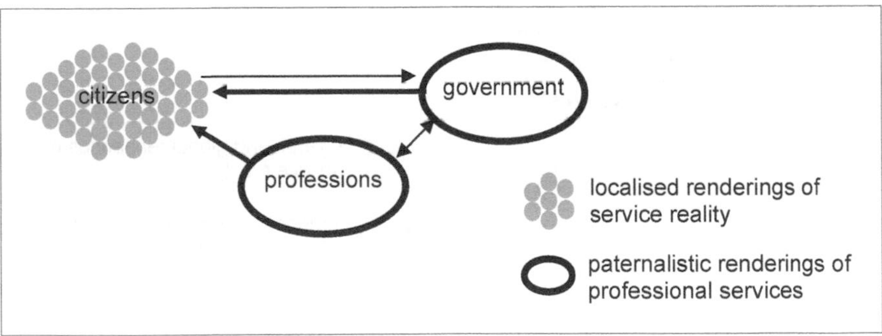

Figure 12.2 Settled spaces

In the UK this consensus has been seen as breaking down since the early 1980s as a result of both the political and broader socio-economic changes occurring during that decade. The spread of internet from the mid-1990s requires to be seen as enabling that process, providing consumers with the means, the spaces, to challenge this consensus.

Among service consumers the internet is variously characterized as space for community, for comparison and for confidence building, enabling them to gain greater control of encounters with professionals. Alternatively and sometimes concurrently, the internet is characterized as an ambiguous space threatening alienation and inducing anxiety. However, in contrast to this heterogeneity of spaces, the internet also offers the possibility of coherent, if temporary, renderings of reality. For example, the government is able to develop very credible policy discourses around the 'good citizen' as responsible for their own finances and health. This 'self-governing' subject (Shankar et al. 2006, 1019) is in part possible because information on which to make 'rational' judgments is understood to be readily available on the internet. For the professions while better informed clients make their job easier, the internet can be dangerously misleading to the ill advised. Such discourses are always contested (Foucault 1980) and so are inherently unstable. The internet has swiftly become a prime space for contestation, 'a space for change', uncovering the paradoxes previously mystified by professionalism. Figure 12.3 represents such post-internet contested spaces.

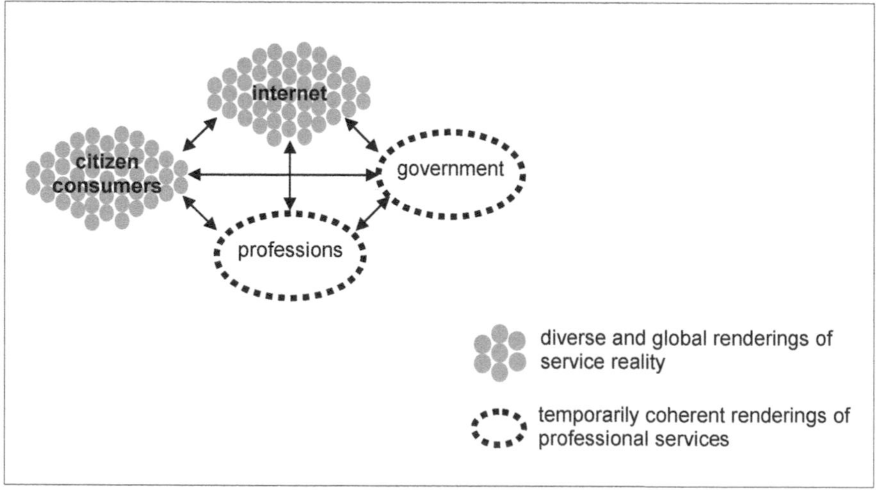

Figure 12.3 Contested spaces

The fragmented and multifaceted nature of internet spaces has played a central role in destabilizing professional discourses and the patterns of consumer-professional interaction. Innovative consumers by confronting service professionals with empirically anchored challenges are forcing a fundamental renegotiation of the terms of service consumption, both in respect of the specific encounter and in the practice of the encounter generically (Laing 2005b; Newholm 2006). Consumer exploration of the spaces created by the internet reshapes consumer-professional relationships whether or not the parties to a specific service encounter are themselves internet users. It is a 'changing space' both in terms of being itself dynamic, and in that it changes 'real' spaces by affecting the conduct of the substantive service encounter. Opportunities are always presented by a disjuncture in civil society such as that engendered by the internet-driven information revolution. Some will relish the pervasive plurality of the internet spaces, some will be profoundly uncomfortable with the greater ambiguity that inevitably accompanies this, most, we would argue, will experience the internet as profoundly paradoxical space. This seems to be our trajectory whether or not we plan or choose that course.

References

Abernethy, M. and Stoelwinder, J. (1995), 'The Role of Professional Control in the Management of Complex Organizations', *Accounting, Organizations and Society* 20: 1, 1–17.

Aldridge, M. and Evetts, J. (2003), 'Rethinking the Concept of Professionalism: The Case of Journalism', *British Journal of Sociology* 54: 4, 547–564.

Beck, H. (2001), 'Banking is Essential, Banks are Not: The Future of Financial Intermediation in the Age of the Internet', *Netnomics* 3, 7–22.

Cantillon, E. (2004), 'Is Evidence Based Patient Choice Feasible?', *British Medical Journal* 329, 39.

David, C. (2001), 'Marketing to the Consumer: Perspectives from the Pharmaceutical Industry', *Marketing Health Services* 21: 1, 24–31.

Elam, M. and Bertilsson, M. (2003), 'Consuming, Engaging and Confronting Science: Emerging Dimensions of Scientific Citizenship', *European Journal of Social Theory* 6: 2, 233–252.

Elwyn, G., Edwards, A., Gwyn, R. and Grol, R. (1999), 'Towards a Feasible Model for Shared Decision Making: Focus Group Study with General Practice Registrars', *British Medical Journal* 319, 753–756.

Flanigan, A. and Metzger, M. (2000), 'Perceptions of Internet Information Credibility', *Journalism and Mass Communication Quarterly* 77: 3, 515–540.

Foucault, M. (1980), *Power/Knowledge* (Brighton: Harvester).

Friedson, E. (1986), *Professional Powers: A Study of the Institutionalization of Formal Knowledge* (Chicago: University of Chicago Press).

Garrison, P. (1994), 'Liberty, Equality and Fraternity', *Netguide* December, 50–53.

Giddings, J. and Robertson, M. (2003), 'Large-scale Map or the A-Z? The Place of Self-help Services in Legal Aid', *Journal of Law and Society* 30: 1, 102–119.

Glaser, B. and Strauss, A. (1967), The Discovery Of Grounded Theory: Strategies for Qualitative Research (Chicago: Aldine).

Ham, C. and Alberti, K. (2002), 'The Medical Profession, the Public, and the Government', *British Medical Journal* 324, 838–842.

Henry, P. (2005), 'Social Class, Market Situation, and Consumers' Metaphors of (Dis)empowerment', *Journal of Consumer Research* 31: 4, 766–778.

Hogg, G., Laing, A. W. and Newholm, T. J. (2004), 'Talking Together: Consumer Communities in Healthcare', *Advances in Consumer Research* 31, 67–73.

Impicciatore, P., Pandolfini, C., Casella N. and Bonati M. (1997), 'Reliability of Health Information for the Public on the World Wide Web: Systematic Survey of Advice on Managing Fever in Children at Home', *British Medical Journal* 314, 1875–1881.

Jolink, D. (2000), *Virtual Communities* (Groningen: Gopher Publishers).

Kavanagh, D. and Morris, P. (1994), *Consensus Politics from Attlee to Major* (London: Blackwell).

Kozinets, R.V. (2002), 'The Field Behind the Screen: Using Netnography for Marketing Research in Online Communities', *Journal of Marketing Research* 39, 1: 61–72.

Kravitz, R. and Melnikow, J. (2001), 'Engaging Patients in Medical Decision Making', *British Medical Journal* 323, 584–585.

Laing, A. W., Newholm, T. J. and Hogg, G. (2005a), 'Regulating in the Information Society', *Consumer Policy Review* 16: 6, 122–128.

Laing, A. W., Newholm, T. J. and Hogg, G. (2005b), 'Crisis of Confidence: Re-narrating the Consumer-Professional Discourse', *Advances in Consumer Research* 32: 1, 514–521.

Lyotard, J. F. (1984), *The Post-Modern Condition: A Report on Knowledge* (Manchester: Manchester University Press).

Mick, D. G. and Fournier, S. (1998), 'Paradoxes of Technology: Consumer Cognizance, Emotions, and Coping Strategies', *Journal of Consumer Research* 25, 123–143.

Mills, P. and Moshavi, D. (1999), 'Professional Concern: Managing Knowledge-based Service Relationships', *International Journal of Service Industry Management* 10: 1, 48–67.

Miles, M. B. and Huberman, M. A. (1994), *Qualitative Data Analysis*, 2nd Edition (Thousand Oaks: Sage).

Morgan, D. L. (1988), *Focus Groups as Qualitative Research* (Newbury Park: Sage).

Muinz, A. and O'Guinn, T. (2001), 'Brand Community', *Journal of Consumer Research* 27: 4, 412–432.

Newholm, T. J., Laing, A. W. and Hogg, G. (2006), 'Assumed Empowerment: Consuming Professional Services in the Knowledge Economy', *European Journal of Marketing* 40: 9/10, 994–1012.

Pagliari, C. and Gregor, P. (2004), 'Review of the Traditional Research Literature', Report on Scoping Exercise EH1, NHS SDO, London.

Poel, D., van der and Leunis, J. (1999), 'Customer Acceptance of the Internet as a Channel of Distribution', *Journal of Business Research* 45, 249–256.

Shankar, A., Chyerrier, H. and Canniford, R. (2006), 'Consumer Empowerment: A Foucauldian Interpretation', *European Journal of Marketing* 40: 9/10, 1013–1030.

Tambyah, S. (1996), 'Life on the Net: The Reconstruction of Self and Community', *Advances in Consumer Research* 23, 172–177.

Thomson, C. J. (2003), 'Natural Health Discourses and the Therapeutic Production of Consumer Resistance', *Sociological Quarterly* 44: 1, 81–107.

Walsh, K. (1994), 'Marketing and Public Sector Management', *European Journal of Marketing* 28: 3, 63–71.

Index

'abandoned spaces' 83, 86–88, 90, 93
Acona Limited 130–132
aesthetic critique 13, 98, 101, 114, 115, 200–202
  geographical indications 105–108, 109
AFNs *see* alternative food networks
agrarian reform 100, 101, 102, 103, 190–191
agrofood system 97, 98, 101, 102–105, 106, 110, 115
alienation, space for 269–271, 272
alternative food networks (AFNs) 12–13, 17–18, 22, 97–98, 99, 112, 113, 189–191, 205–206
  cultural identity 193–194
  economic rent 195–197
  elusive consumer 199–202
  localized production 192–193
  public sector procurement 197–198
  resistance 193
  territorial valorization 194–195
alternative hedonism 28
ambiguity, space for 267–269, 272
anti-consumption 26
anxiety, space for 266–267, 272
appéllation d'origine products 20, 100, 105–108
Arcades Project 14, 43–44
artisan food production 101, 105–107, 112, 115, 204

Barnett, C. and Land, D. 22–23
Baudrillard, J. 58–59, 68, 72, 73
Bauman, Z. 57–58, 65, 66
Bell, D. and Valentine, G. 13
Benjamin, W. 14, 15, 43–44
biofuels 99, 105, 115
biotechnologies 97, 98, 102–103
Blowfield, M. 129–130, 140
Boltanski, L. and Chiapello, E. 98, 102, 110
boutique hotels 18, 147–151, 156–158
  Schrager, I. 147, 149, 150, 151–154, 155–156
  Starwood W Hotels 147, 154–155, 157
brands 19–20, 21–22, 27, 107–108, 196–197, 201–202
  boutique hotels 147, 149, 154–155
  ETI 128
  GVC 125
  own-label 107, 183, 195, 196–197, 221–222, 223–224, 225, 232–233
  palate geographies 218–219
Brazil 12–13, 99, 104, 105, 109n2
Brunori, G. 98, 113
Bryant, R. L. 29–30
Bryant, R. L. and Goodman, M. 202, 205
Burma (Myanmar)
  teak 243–247
  teak industry 29–30, 239–243, 247–253
buying practices 18, 123–125, 127, 138–139
  ethical trade 129–139
  ETI 124, 134, 135–139
Buying Your way into Trouble (Acona 2004) 130–132

Café Direct 112
Callon, M. 98, 99, 113, 150, 157
campaigns 16–17, 26, 48–49, 112–113
  ethical 123, 124–125, 127, 128–129, 133–135, 137, 140
Cancun 81, 84, 89–92
CAP (Common Agricultural Policy) 190–191
carbon offsets 26
Carmenère grape 232
Carrefour 103, 107

Caste War 83, 85, 86, 89
castration anxiety 70, 71
Catherwood, F. 83–84
Chacchoben 92–93
chewing gum 12, 83, 84–86, 87–88, 91
chicken supply chain 5, 24, 164–168, 184–185
  'manufacturing meaning' 169–183
chicle 12, 83, 84–86, 87–88, 89
chicleros 85, 86
Chilean wine 215, *221*, 223–226, 233–235
  producers 29, 216, 217, 218, 226–228, 231–233
  regions 228–231, *229*
Clarke, D. B. 4–5, 12, 25, 29
Cohen, D. 245
Cohen, L. 45
Colchagua Valley 217, 219, 227, 228, 230–231, 232
commercial spaces 44–45
  chicken supply chain 164, 165, 169, 174–178
commodities 3, 10–11, 19–20, 59–61
commodity biographies 11, 19–20, 41, 52
commodity chains 18, 123–124, 125–127, 140, 141, 142, 164, 192, 216
  wine value chain 220–221, 226–228, 234
commodity fetishism 12, 22, 57, 59–64
commodity production 24, 59, 99
commodity spaces 164, 169, 170–174
Common Agricultural Policy *see* CAP
communal space 261–263, 267
comparison, space for 263–264, 266, 272
confidence, space for 264–266, 272
connection, consumption as 19–22, 184, 194, 201, 219, 267
consumer anxieties 24, 30, 169, 182, 183, 184, 258, 266–267
consumer communities 261–263, 267
consumer culture 29, 41, 43, 44, 45, 52
consumerism 22, 26, 28, 45, 57, 65–67, 68–69, 70, 72, 74
consumer society 43, 52, 58, 64–65, 66
consumer, the 4, 5, 12, 142
  mental space 41, 45–47
  moral spaces 47–49
consumer trust 104, 163, 178, 184, 185, 190, 201
consumption 4–6, 25–29, 64–65
consumption, spaces of 4, 8, 12, 57, 62, 66, 72, 178–183, 199–202
  mental 45–47
  moral 47–49
  virtual 260, 271–273
Cook, I. 216–217
Cook, I. and Crang, P. 11, 19, 22
Cozumel 84, 86–87, 88–89, 90, 93
Crang, P. 13, 14–15, 22
cruise ships 81, 93
culinary cultures 11
cultural identity 193–194
Curry Commission (report) 163, 194

Day Chocolate Company 112
de Certeau, M. 12, 50, 51
de Grazia, V. 43
delayed innovation 103, 107
department stores 14, 43, 44, 45
Derrida, J. 61–62
Dolan, C. and Humphrey, J. 126
dreams 66–68, 69
DuPuis, E. M. and Goodman, D. 23–24, 199, 204

economic rent 13, 191, 195–197, 198, 205–206
eco-tourists 92, 93–94
elusive consumer 199–202
ethical campaigns 123, 124–125, 127, 128–129, 133–135, 137, 140
ethical consumers 22, 24, 27, 46, 47, 123, 127, 130, 140
ethical consumption 4, 22, 26, 28, 48–49, 123, 124, 127–129, 139–140, 142, 204–205
ethical trade 18, 110, 112, 123–125, 127–129, 140–142, 253
  UK retailers 129–139, 141–142
Ethical Trade Initiative (ETI) 110, 114, 128, 129, 130, 133–135, 140
  Principles of Implementation 128n1, 134
  Purchasing Practices Project 124, 125, 134, 135–139

fair-trade 11, 16, 20, 24, 105, 112–113, 115, 198, 202
  alternative food networks 17, 18, 97, 98, 204
  commodity fetishism 22
FairTrade Movement 13, 48, 49, 101, 102, 108–110, 112, 113, 114, 202
fantasy 12, 57–58, 68, 69–70, 71–72
FDI *see* foreign direct investment
fetishism 12, 57, 59, 70, 72–74, 201–202
  brands 19, 20, 22
  commodity 12, 22, 57, 59–64
Food and Agriculture Organization (FAO) 49
foodie gentrification 202–204
food knowledge 10–11, 22–23, 24, 178–183, 199–200, 202
food safety 98, 103, 104, 163, 177, 184
food scares 104, 163, 190, 201
food security 191
foreign direct investment (FDI) 102, 105
'framing' 98, 99, 113, 114
France, wine consumers 215–216
Freud, S.
  dreams 66–68
  fantasy 69
  fetishism 57, 70, 71, 74
  pleasure principle 65–66
From Farming to Biotechnology (1987) 97

Gade, D. W. 218–219
GCC (global commodity chains) 18, 125, 216
geographical indications (GI) 98, 100, 105–108, 109, 115
  wine 220, 227, 228, 230–231
Gereffi, G. 18, 125, 220
globalization 9, 82, 98–99, 102–105, 114, 115
global value chains *see* GVC
governmentality 46, 50–51
Guthey, G. 219
Guthman, J. 21–22, 200, 202
GVC (global value chains) 29, 125, 216, 226, 233
Gwynne, R. N. 29

Habermas, J. 44
Halstead, R. 222
Hobson, J. A. 47–48
Hudson Hotel 152–154
Hughes, A., Wrigley, N. and Buttle, M. 18
Humphrey, J. 220–221
hunger 49

Impactt Limited 130–131
imperial consumerism 9, 29, 41, 48–49, 245
International Fair Trade Association (IFAT) 112, 113
internet 16, 30, 98, 112, 113, 257–260, 271–273
  spaces of opportunity 261–266
  spaces of threat 266–271

Jackson, P., Ward, N. and Russell, P. 24

Karen people 240, 248, 252–253
knowledge transfer 217, 219, 220–221, 226–228, 233–234

Lacan, J. 57, 63, 70–72
  fantasy 69–70, 71–72
  objet petit a 69–70, 71
  vel of alienation 63–64, *64*, 69
Laing, A., Newholm, T. and Hogg, G. 30
Lefebvre, H. 3, 7–8, 13, 41–42, 44, 50, 51, 52
lobbies 18, 150, 153, 156, 156–157
  Hudson Hotel 153–154
  Royalton Hotel 152, 157
  Starwood W Hotels 155
localized production 191, 192–193, 194–195, 196–198
  cultural identity 193–194
  resistance 193
  territorial valorization 194–195

mainstreaming 11, 24, 98, 99, 101, 108, 113–114, 196–197, 198, 205
malls 14–15, 16, 45
Mansvelt, J. 26–27, 30
'manufacturing meaning' 164, 165, 169–183, 184
  commercial spaces 174–178
  commodity spaces 170–174

spaces of consumption 178–183
Marks & Spencer 127, 134
chicken supply chain 168, 172–174, 177–178, 181–183
Marsden, T. 195–196, 198
Marx, K. 8, 19, 67
commodity fetishism 57, 59–60, 61, 62–63, 64
fetishism 72–73, 199
Mayan Riviera *see* Mexican Caribbean
Mayan World 83–84, 85, 86, 89, 93
McNeill, D. and McNamara, K. 18
mental spaces 12, 41, 45–47, 52
Mexican Caribbean 12, 81, 82–83, 87, 89, 93–94
Cancun 81, 84, 89–92
Cozumel 84, 86–87, 88–89, 90, 93
Playa del Carmen 87–88, 92, 93
Mexico 81–83, 86–89, 93–94, 109n2
Cancun 81, 84, 89–92
Chacchoben 92–93
chicle 12, 83, 84–86, 87–88, 89
Cozumel 84, 86–87, 88–89, 90, 93
Playa del Carmen 87–88, 92, 93
tourist 'pioneers' 83–84
Micheletti, M. 27–28, 201
Miller, D. 63, 72, 140, 150
Mintz, S. 43
Montgras 230, 232–233
moral geographies 29, 41, 47, 48, 191, 202–205, 241
moral spaces 47–49, 52
municipalization 46, 52
Murdoch, J. and Miele, M. 200–201

New World wines 29, 216, 217, 218, 219, 220, 223, 228, 233
NGOs (non-governmental organizations) 100, 130, 133 *see also* Oxfam
ethical trade 112, 124, 125, 140, 141, 253
purchasing practices 18, 127–128, 134, 136, 138
non-consumption 26

objet petit a 69–70, 71
opportunity, spaces of 30, 260, 261–266
organic production 11, 20, 100–101, 112, 196–198, 202, 204
'overflowing' 98, 99, 113, 114
own-label brands 107, 183, 195, 196–197, 221–222, 223–224, 225, 232–233
Oxfam 102, 112, 127, 134, 135, 140, 141
Trading Away Our Rights 129, 130, 132–133

palate geographies 29, 215, 216, 217–223, 226, 230, 231–232, 233, 234
palates 29, 215, 220, 226, 232, 233, 234
Paterson, M. 13–14, 16, 27
place 6–13, 42–45, 42–45
place confirmation 8–9, 10
place construction 9–10
Playa del Carmen 87–88, 92, 93
pleasure principle 57–58, 65–66, 68, 69, 72
Ponte, S. and Gibbon, P. 220
Principles of Implementation (ETI) 128n1, 134
private spaces 12, 41, 44, 46, 50–52, 200
production 25–26
professional services 30, 258, 259–260, 261, 271–273
alienation 269–271
ambiguity 267–269
anxiety 266–267
communal space 262–263
space for comparison 263–264
space for confidence 264–266
public sector procurement 103, 197–198, 205
public spaces 12, 41, 44–45, 46, 50–52, 200
purchase effort 221–222, 223, 224
purchasing practices 18, 123–125, 139–142
ethical trade 127–129, 133–135, 140–142
ETI Purchasing Practices Project 124, 135–139
UK retailers 129–135
Purchasing Practices Project (ETI) 124, 125, 134, 135–139

Rancière, J. 60–61
Rappaport, E. 44
reality principle 57, 58, 65, 66, 69, 72

Redclift, M. R. 12, 25
reflexive consumer 199, 201
relative risk 221–222, 223, 224
resistance 193
retail chains, specialist 217, 224–225, 226, 233–234
retail places 13–19, 42–45
retail spaces 13–19, 42–45
retail supply chains 123–124, 126–127, 137
  ethical trade 127–128, 129–135, 140–142
Reverend Billy, Church of Stop Shopping 16–17
Rose, N. 46
Royalton Hotel 152, 153, 156, 157
Rubell, S. 151–152
rural development 97, 191, 195
  cultural identity 193–194
  localized production 192–193
  resistance 193
  territorial valorization 194–195
rural rents 97, 98
rural spaces 12–13, 97, 98, 100

Sainsbury's 127, 168, 183, 195, 196, 224, 232–233
Schrager, I. 147, 149, 150, 151–154, 155–156
seduction 12, 58–59, 74
SFSCs (Short Food Supply Chains) 191, 192–193, 194–195, 196, 197
Slow Food Movement (SFM) 20, 98, 101, 108, 113, 193, 200
Smith, M. 15–16
social space 59, 82, 261–263
Soper, K. 28–29, 200
space 6–13, 82–83
Spawton, T. 221, 222, 223
Starbucks 15–16, 19, 27, 110
Starck, P. 150, 152, 153
Starwood W Hotels 147, 154–155, 157
Stephens, J. 83–84
supermarkets 14, 16, 18, 103, 107, 126, 196, 197, 198, 205
  chicken supply chain 168, 183
  ethical trade 127, 137
  mainstreaming 18, 24, 113
  own-label 107, 183, 195, 196–197, 221–222, 223–224, 225, 232–233
  UK consumers 223–226
  wine 29, 216, 217, 220–223, 225, 226–227, 231, 234
sustainable consumption 7, 21–22, 25n4, 26, 27, 29
sustainable production 100, 189, 197–198, 199–200

Taking the Temperature (Impactt 2004) 130–131
teak industry 29–30, 239–243, 247–253
  uses for 243–247
territorial valorization 107–108, 194–195
terroir 10, 29, 108, 114, 218–219, 222, 227
  Chile 219, 224, 228, 230–231, 233, 234
Tesco 14, 16, 103, 107, 198
  chicken supply chain 168, 183
  ethical trade 127, 129, 132
  wine 216, 222, 224, 227, 232
The Production of Space (1974) 41–42, 50
threat, spaces of 266–271
tourism 7–8, 81, 86, 87–89
  Cancun 89–92
  Chacchoben 92–93
tourist 'pioneers' 83, 87–88, 89, 92, 93
trade unions 124, 125, 128, 130, 134, 140, 141
Trading Away Our Rights (Oxfam 2004) 129, 130, 132–133, 135
transnationalization 102–103
Trentmann, F. 12

UK consumers 215, 221–222
  Chilean wine 216, 217, 223–226, 232–233, 234–235
UK retailers 129–139, 141–142
Unilever 107
urban economy 148
urban networks 44, 50, 51–52

van der Ploeg, J. 194
vel of alienation 63–64, *64*, 69
virtual consumer 140
virtual spaces 7, 30, 112, 257–260, 271–273

spaces of opportunity 261–266
spaces of threat 266–271

Waitrose 198, 225, 232, 233
Wal-Mart 16, 103, 168, 196
Wilkinson, J. 12–13
wine 29, 215–217, 226–228, 233–235
Chilean wine 215, 216, 218, 219, 223–226, 228–231, *229*, 233–235
palate geographies 217–223, 231–233
UK consumers 215, 216, 221–222, 223–226
wine value chain 220–221, 226–228, 234
WTO (World Trade Organization) 102, 190–191

Žižek, S. 62